D0935766

COMPARATIVE HEMOSTASIS IN VERTEBRATES

COMPARATIVE HEMOSTASIS IN VERTEBRATES

Jessica H. Lewis

University of Pittsburgh School of Medicine
and Central Blood Bank
Pittsburgh, Pennsylvania

Plenum Press • New York and London

Library of Congress Cataloging-in-Publication Data

Lewis, Jessica H.
Comparative hemostasis in vertebrates / Jessica H. Lewis.
p. cm.
Includes bibliographical references and index.
ISBN 0-306-44841-6
1. Hemostasis. 2. Blood--Coagulation. 3. Physiology, Comparative. 4. Vertebrates--Physiology. I. Title.
QP90.4.L48 1996
596'.0113--dc20 95-43745
CIP

ISBN 0-306-44841-6

A Division of Plenum Publishing Corporation
233 Spring Street, New York, N. Y. 10013

10 9 8 7 6 5 4 3 2 1

Printed in the United States of America

PREFACE

This volume results from a curious type of off-and-on research that was interspersed with long periods of teaching or clinical responsibilities and research concerning human disease. Early on, because of my interest in the fibrinolytic enzyme system, I was fortunate to work with John H. Ferguson, M.D., Chairman of the Department of Physiology at the University of North Carolina. From him I learned a high respect for knowledge derived from animal research using basic physiological techniques.

After we moved to the University of Pittsburgh, this research became more focused on the phylogenetic development of hemostasis. An early article surveyed coagulation in many animals (Didisheim *et al.*, 1959). Deciding to limit comparative studies to vertebrate animals and to begin at the beginning, my laboratory staff, two hemophiliac friends (who provided fresh substrate for factors VIII and IX), and I trekked to the Marineland Research Laboratory near St. Augustine, Florida. There, the Marineland staff helpfully provided us with blood from sharks, many fishes, toads, and large marine turtles. This was a good start, but the most important step—publishing the results—was easily postponed. Studies continued sporadically, and results were published at a slow rate of one per year. This research would have ended long ago without the encouragement of colleagues and friends, in particular Franklin A. Bontempo, M.D., the Director of the Clinical Coagulation Laboratory at Central Blood Bank in Pittsburgh. His enthusiasm stimulated plans to write this volume.

Hemostasis, the combined mechanisms that prevent excessive bleeding following injury, has been widely studied in humans, but has received far less attention in "lower" animals. The 30-year series of opportunistic studies presented in this volume has involved animals from all seven major vertebrate classes and has sought insight into the phylogenetic development of the various phases or components of hemostasis. There are an estimated 70,000 species of vertebrates. In choosing those to study, availability, convenience, and size have been important factors. Ideally, one should study a large cohort of each species. This being impossible, an attempt was made to study six or more representative animals, three female and three male, that appeared to be normal adults in good health. In order to update the original research, an attempt was made to examine or reexamine an example from each vertebrate class and some additional mammals. These 1990–1993 investigations included the following animals: dogfish, sturgeons, sculpins, bullfrogs, alligators, baboons, chimpanzees, ferrets, hamsters, manatees, pigs, and rats.

The results have been organized in three parts. In Part I, introductory Chapter 1 presents a phylogenetic classification and an alphabetical list of the vertebrates examined and an evolutionary time scale. Chapter 2 discusses laboratory methods and presents model tables for the tests. Chapter 3 discusses human hemostasis.

Part II comprises Chapters 4–26, which discuss animals of each class, order, or species. These chapters contain results of general coagulation tests, of coagulation factor assays, of fibrinolytic tests, and of platelet/thrombocyte counts and functions, as well as scanning electron micrographs (SEMs) and transmission electron micrographs (TEMs) showing find structures. They also include additional available material such as serum proteins, biochemical tests, and cell counts.

Part III presents the results of comparative studies. Chapter 27 compares hemostatic results, including bleeding times, clotting times, prothrombin times with ordinary reagents and with animal tissues, thrombin times, thromboplastin generation tests, coagulation factors, fibrinolytic activities, and platelet counts, as well as sections on species characteristics vs. disease and hereditary hemorrhagic/coagulation diseases in mammals. Chapter 28 covers hematological data, including erythrocyte sizes and shapes and osmotic fragility. Chapter 29 presents serological results, including serum proteins and effects of animal plasmas on human platelets and on staphylococcal clumping.

The Appendix lists the manufacturers and suppliers, reagents and tests, and instruments used in our studies.

References cited in the text and suggested readings for the reader interested in further pursuing the subject matter of the chapter are listed at the end of each chapter. The Bibliography comprises general references for further suggested reading, as well as articles that describe investigations on animals from two or more different classes or orders.

This monograph is intended to serve as a reference for students, researchers, and teachers of biology, zoology, veterinary science, and human medicine. In particular, it is hoped that the volume will be a guide to the choice of vertebrate animal models for studies of the treatment and correction of human disease. Through these models, human life may be bettered.

REFERENCE

Didisheim, P., Hattori, K., and Lewis, J. H., 1959, Hematologic and coagulation studies in various animal species, *J. Lab. Clin. Med.* **53:**866.

ACKNOWLEDGMENTS

The coagulation, hematological, electron-microscopic, and special protein studies were done in the Clinical and Research Coagulation Laboratory, which was first located within the University of Pittsburgh Department of Medicine and later moved to the Central Blood Bank (Pittsburgh). Most of the biochemical and paper electrophoretic studies were done at the Presbyterian University Hospital Laboratories.

The title *Comparative Hemostasis in Vertebrates* implies the study of blood from a variety of animals with the purpose of understanding differences from humans and from each other. Because many of the studies required freshly drawn blood, animals were brought to the laboratory; if it was impossible to do so, the laboratory was brought to the animals.

Research such as this requires financial support, availability of animal subjects, and technical and intellectual support. The author wishes to acknowledge with grateful thanks the enormous support that has made this volume possible.

Financial support was received from and indebtedness is acknowledged to: U.S. Public Health Services–National Institute of Health Grants HL-002254, and HE-003036; Health Research and Services Foundation, Pittsburgh, Pennsylvania, Grants A-35 and A-48; Hemophilia Foundation of Western Pennsylvania; The Cassandra Mellon Henderson Foundation, Pittsburgh, Pennsylvania; and The Blood Research Foundation, Pittsburgh, Pennsylvania.

Blood for these animal studies was usually obtained from animals that were being bled for another purpose, for example, routine examinations, other research, or preparation for market. Some animals (e.g., frogs) were brought directly to our coagulation laboratory. Others were housed in local facilities where blood could be obtained and rapidly brought back to the laboratory. With still others, particularly marine animals, it was necessary to go to the sites, obtain the blood, and do some of the studies or prepare the samples for transportation. Local and distant sites at which blood was obtained and some studies done are listed below, and their assistance is gratefully acknowledged.

Local sites included: Magee Womens Hospital Animal Facility, Pittsburgh; Pennsylvania State University, State College; Pittsburgh Zoo; poultry farms; Resuscitation Laboratory, Pittsburgh; riding academies; Ringling Brothers and Barnum & Bailey Circus (on location in Pittsburgh); Transplantation Surgery Laboratory, Pittsburgh; University of Pittsburgh; and Weiman Mink Farm, Wexford, Pennsylvania.

Distant sites and research laboratories include: Aquarama Theatre of the Sea, Philadelphia; Biological Laboratories, Woods Hole, Massachusetts; Charles River Breeding Laboratory, Windham, Maine; Gatorland, Orlando, Florida; Institute of Medical and Veterinary Science, Adelaide, Australia; Marineland Research Laboratory, St. Augustine, Florida; Montreal Aquarium, Montreal, Canada; Mt. Desert Island Biological Laboratory, Salisbury Cove, Maine; National Institutes of Health, Bethesda, Maryland; Northern Diagnostic Laboratories, Portland, Maine; Ringling Brothers and Barnum & Bailey Circus, Sarasota, Florida; Sea World, Orlando, Florida; U.S. Fish and Wildlife Service, Orlando, Florida; and Wometco Seaquarium, Miami, Florida.

Co-authors of publications concerning animals include: William L. Bayer, M.D.; Franklin A. Bontempo, M.D.; John M. Brandon, M.D.; Dario Curiel, M.D.; Paul Didisheim, M.D.; Alfred P. Doyle, M.D.; Elizabeth E. Ferguson, B.A.; John H. Ferguson, M.D.; Judith S. Gavaler, Ph.D.; C. Hann, Ph.D.; Ute Hasiba, M.D.; Kenichi Hattori, M.D.; John L. Myers, M.D.; Louise L. Phillips, Ph.D.; Mitzura Shirakawa, M.D.; Joel A. Spero, M.D.; Isabel L. F. Szeto, M.D.; David H. Van Thiel, M.D.; Evan Vosburgh, M.D.; John Wilson, B.S.; and Zbigniew A. Zawadzski, M.D.

Co-workers—skilled technicians, assistants, or word processors—to whom I am deeply grateful include: Tom Ankowiak, George Arvay, Rosemary Bartoszewicz, Brenda Briscoe, Beverly Caputo, Connie Chew, Carol Church, Cathy Clay, Willie Cooks, Jean Coram, Jim Donahue, Jean Draude, Libby Ferguson, Kay Fitz, Sue Gonzowiski, Toni Gorenc, Ginny Jackson, Carol Jeffreys, Susan Johnson, Trudy Maxon, Gerri Morici, Arlene Neve, Chris Petrowski, Talisa Reed, Erin Rice, Rich Robinson, Fran Shagas, Miles Simon, Mike Snyder, Cathy Thomas, Mary Vogel, Anne Wallace, John Wilson, and Judy Wuchina.

Co-workers—individuals who have helped in the preparation of this volume—to whom the author expresses very special thanks include: Bonnie Chojnacki, Chris Deible, Lisa Garbin, Mike Grimes, Mary Keane, and Bruce Nelson.

Collaborators or colleagues who provided greatly valued aid and assistance include: Joseph H. Amenta, M.D.; Michael J. Bechet, M.D., Ph.D.; Paul H.

Bramson, D.V.M.; James Clugston, Ph.D.; William M. Cooper, M.D.; Robert H. Fennel, M.D.; Kendra Folker, Ph.D.; J. Y. Henderson, D.V.M.; Richard Iammarino, M.D.; Peter C. Johnson, M.D.; Craig Kitchens, M.D.; Bernard L. Klionsky, M.D.; John R. Krause, M.D.; Albert D. Lansing, Ph.D.; Robert E. Lee, M.D.; Thomas McManus, M.D.; William Medway, D.V.M.; Daniel K. Odell, Ph.D.; Samuel Ridgeway, D.V.M.; Eugene Robin, M.D.; Peter J. Safar, M.D.; J. Swart, D.V.M.; Jace Trickey; Mohamed A. Virji, M.D., Ph.D.; Michael T. Walsh, D.V.M.; Jesse R. White, D.V.M.; and Warren Zeiller.

CONTENTS

PART I
INTRODUCTION

CHAPTER 1
INTRODUCTION

CHAPTER 2
LABORATORY METHODS AND TEST PARAMETERS

CHAPTER 3
HUMAN HEMOSTATIC/COAGULATION MECHANISMS

PART II
STUDIES IN REPRESENTATIVE VERTEBRATES

CHAPTER 4
THE SEA LAMPREY

CHAPTER 5
THE CARTILAGINOUS FISH

CHAPTER 6
THE BONY FISH

CHAPTER 11
THE OPOSSUM

CHAPTER 12
THE SPINY HEDGEHOG

CHAPTER 13
THE FRUIT BAT

CHAPTER 14
THE PRIMATES

CHAPTER 15

THE NINE-BANDED ARMADILLO

CHAPTER 16

RABBITS

CHAPTER 17

RODENTS

CHAPTER 18

THE BOTTLE-NOSED PORPOISE

CHAPTER 19

THE CARNIVORES

CHAPTER 20

SEALS

CHAPTER 21

THE INDIAN ELEPHANT

CHAPTER 22

THE MANATEE

CHAPTER 23

THE THOROUGHBRED RIDING HORSE

CHAPTER 24

PIGS

CHAPTER 25

THE CAMEL FAMILY

CHAPTER 26

THE HOLLOW-HORNED DOMESTIC RUMINANTS

PART III

COMPARATIVE STUDIES

CHAPTER 27

COMPARATIVE HEMOSTASIS

CHAPTER 28

COMPARATIVE HEMATOLOGY

CHAPTER 29

COMPARATIVE SEROLOGY

APPENDIX

PART I

INTRODUCTION

CHAPTER 1

INTRODUCTION

1. CLASSIFICATION OF VERTEBRATES

The first comprehensive attempt to study and classify animals was made by Aristotle (384–322 BC). He grouped animals according to structure, habitat, and customs. For the most part, his observations were accurate, but one startling error was his classification of porpoises and whales with fish and sharks. During the Roman period and through the ensuing Dark Ages, there was little or no progress in the biological sciences. At the beginning of the Renaissance (1200 AD), the oppressive religious influence lessened and universities commenced and grew, spreading scientific knowledge. The great Swedish botanist Carolus Linnaeus (1707–1778) established a definitive nomenclature for animals and plants, much of which is still in use today.

The classification of vertebrates presented here considers extant animals and stresses those that are included in this study. Most of this classification was taken from Storer and Usinger (1957). The major exception is that Pinnipedia is considered a separate order (order 13), rather than a suborder of Carnivora.

1.1. Cyclostomata

This class includes lampreys, hagfishes, and slime eels. These most primitive of living vertebrates are poikilothermic (cold-blooded), exchange oxygen with the surrounding water through gills, and live in fresh water, salt water, or both. They are jawless and have long snakelike bodies with median fins, cartilaginous bones, soft, slimy skin, and suctorial mouths. Lampreys may pass from sea water to lakes and streams, where they lay large numbers of small eggs that develop into larvae that grow during a long period (3–5 years) and then metamorphose. As adults, most of them are parasitic, attaching to and feeding on fish.

Hagfish and slime eels separated from lampreys early in their evolutionary path. They live in salt water, are hermaphroditic, and lay rather large eggs.

1.2. Chondrichthyes

This class includes Elasmobranchii (sharks, skates, and rays) and Chimaeras, almost all of which live in sea water. They are poikilothermic and have cartilaginous skeletons with paired movable jaws. They exchange oxygen through paired gills. Their skins are thick and tough, with small sandpaperlike placoid scales composed of grainy cartilaginous material. Fertilization is internal, and the animals may be viviparous (bearing living young) or oviparous (producing eggs).

1.3. Osteichthyes

Osteichthyes is the first class to have bony skeletons. They, too, are poikilothermic and live in fresh water or sea water. Their spindle-shaped or flattish bodies are covered with scales. They have paired or single (tail) fins, and most can swim at great speed. Fertilization is external, and they usually spawn large numbers of small eggs. In a few species, the females are ovoviviparous (retaining the eggs and producing living young).

1.4. Amphibia

This class is the first to emerge from a life spent entirely in the water, and some are adapted to spend part of their lives on land. The animals are poikilothermic; have thin, moist skins, two pairs of limbs, and bony skeletons; and exchange oxygen through gills, lungs, or skin. The Urodelas, or tailed amphibians, include amphiuma, cryptobranchus, necturus, and salamanders. The first three retain gills throughout life. The Anura, or tailless amphibia, include frogs and toads. They hatch in water and have gills and tails until metamorphosis, when they lose their tails, develop lungs, and emerge on land. Most amphibians lay small eggs enclosed in gelatinous coverings in freshwater streams or ponds.

1.5. Reptilia

This class is the first group totally adapted for life on land. It includes turtles (Chelonia), snakes and lizards (Squamata), and alligators (Crocodilia). All orders have dry, horny skin, two pairs of limbs (sometimes vestigial), an ossified bony skeleton, and lungs. They are poikilothermic, and most lay large eggs with soft, leathery shells. In a few, the young are retained in the mother and born alive. Neither eggs nor live young are nurtured by the parents.

1.6. Aves

The animals in this class are the first to be homoiothermic (warm-blooded) and to fly. Their skin is covered with feathers, and they have two pairs of limbs. The front pair are wings; the hind pair have four toes adapted for perching, walking, or swimming. Their bony skeletons are strong, but the bones are hollow to decrease their weight, thus allowing them to fly. Fertilization is internal, and almost all lay a limited number of eggs with big yolks and hard shells. Most birds build nests for the eggs, keep them warm, and nurture the young until they can fly.

1.7. Mammalia

This class is characterized by heavily boned skeletons, two pairs of appendages, lungs, and hair. They are homoiothermic and except for Monotremata produce living offspring that are nurtured, often for many years. In all except the most primitive, the females produce milk in mammary glands to feed their infants. There are 19 living (not fossil) orders. This study included no examples from the following four orders: Dermoptera (flying lemurs), Pholidota (pangolins), Tubulidentata (aardvarks), and Hyracoidea (coneys).

1.8. Taxonomy of Animals Studied

Table 1.1 shows the common and scientific names of the vertebrate classes, subclasses, orders, and species studied.

2. ANIMALS STUDIED

The animals studied are presented in alphabetical order by common name for the convenience of interested readers. Table 1.2 lists all the animals except the fishes, which are adequately listed in Table 1.1 under Osteichthyes. Table 1.2 lists the numbers of each species studied, the average weight, the animal's ancestry and geographic origin, the site at which the species was studied, and the phlebotomy done, including the anesthesia used, if any, the site of blood collection, and the approximate volume of blood collected.

As noted, Table 1.2 lists the animal's ancestry. Wild animals have a minimal chance of perpetuating an unfavorable gene because their gene pool is large and restricted only by the numbers of animals present in the limited living area. The

TABLE 1.1
Classification of Vertebrates Studied

Class Subclass/Order Common name	*Species*
Cyclostomata (Chapter 4)	
Petromyzontia	
Sea lamprey	*Petromyzon marinus*
Chondrichthyes (Chapter 5)	
Elasmobranchii	
Dogfish, spiny	*Squalus acanthias*
Guitarfish	*Rhinobatus lentiginosus*
Shark, black-tipped reef	*Carcharhinus maculipinnis*
Shark, great blue	*Prionace glauca*
Shark, mako	*Isurus oxyrinchus*
Skate	*Raja eglanteria*
Osteichthyes (Chapter 6)	
Actinopterygii	
Sturgeon	*Acipenser oxyrinchus*
Teleostei	
Amberjack	*Seriola dumerili*
Arctic grayling	*Thymallus arcticus*
Arctic trout	*Salvelinus namaycush*
Cobia	*Rachycentron canadus*
Flounder	*Paralycthys lethostigmus*
Goosefish	*Lophius americanus/piscatorius*
Grouper	*Mycteroperca bonaci*
Mullet, striped	*Mugil cephalus*
Orange filefish	*Alutera schoepfii*
Porgy	*Stenotomus aculeatus*
Sculpin	*Myoxocephalus octodecimspinosus*
Sea bass	*Centropristes striatus*
Sea robin	*Prionotus carolinus*
Sheepshead	*Archosargus probatocephalus*
Snapper, red	*Lutianus aya*
Spadefish	*Chaetodipterus faber*
Tautog	*Tautoga onitus*
Tilefish	*Lopholatilus chamaeleonticeps*
Toadfish	*Opsanus tau*
Triggerfish	*Balistes carolinensis*

TABLE 1.1 (*Continued*)

Class Subclass/Order Common name	*Species*
Amphibia (Chapter 7)	
Urodela	
Congo eel	*Amphiuma means*
Hellbender	*Cryptobranchus alleganiensis*
Mud puppy	*Necturus maculosus*
Salamander	*Ambystoma tigrinum*
Anura	
Bullfrog	*Rana catesbeiana*
Grass frog	*Rana pipiens*
Giant toad	*Bufo marinus*
Toad	*Bufo terrestris*
Reptilia (Chapter 8)	
Chelonia	
Freshwater/land	
Box turtle	*Terrapene carolina triunguis*
Musk turtle	*Sternotherus odoratus*
Painted turtle	*Chrysemys picta marginata*
Pond (PSE) turtle	*Pseudemys scripta elegans*
Soft-shelled turtle	*Amyda ferox spinifera*
Saltwater/marine	
Green turtle	*Chelonia mydas*
Loggerhead (LH) turtle	*Caretta caretta caretta*
Squamata	
Sauria	
Iguana	*Basilicus americanus*
Colubridae	
Garter snake	*Aglypha thamnophis*
Crocodilia	
Alligator	*Alligator mississippiensis*
Aves (Chapter 9)	
Domesticated fowl	
Chicken	*Gallus gallus (domesticus)*[a]
Duck	*Anas platyrhynchos*[a]
Goose	*Anser albifrons*[a]
Pigeon	*Columbia livia*[a]
Turkey	*Meleagris gallopovo*[a]

(*continued*)

TABLE 1.1 (*Continued*)

Class Subclass/Order Common name	*Species*
Mammalia (Chapters 10–26)	
Monotremata (Chapter 10)	
Echidna or spiny anteater	*Tachyglossus aculeatus*
Marsupialia	
Quokka (Chapter 10)	*Setonix brachyurus*
Wallaby (Chapter 10)	*Thylogale eugenii*
Opossum (Chapter 11)	*Didelphis marsupialis virginiana*
Insectivora (Chapter 12)	
Hedgehog, European	*Erinaceus europaeus*
Chiroptera (Chapter 13)	
Bat, fruit (Indian)	*Pteropus giganteus*
Primates (Chapter 14)	
Baboon	*Papio cynocephalus/anubis*
Chimpanzee	*Pan troglodytes*
Man	*Homo sapiens*
Rhesus monkey	*Macaca mulatta rhesus* (*Catarrhini macacus*)
Edentata (Chapter 15)	
Armadillo	*Dasypus novemcinctus*
Lagomorpha (Chapter 16)	
Rabbit	*Oryctolagus cuniculus*[a]
Rodentia (Chapter 17)	
Guinea pig	*Cavia porcellus*[a]
Hamster	*Cricetus auratus*[a]
Rat	*Rattus norvegicus* (*rattus*)[a]
Cetacea (Chapter 18)	
Porpoise, bottle-nosed	*Tursiops truncatus*
Carnivora (Chapter 19)	
Cat	*Felis catus*[a]
Dog	*Canis familiaris*[a]
Ferret	*Mustela putorius furo*[a]
Red fox	*Vulpes vulpes*
Grey fox	*Vulpes urocyon*
Mink	*Mustela vison*
Raccoon	*Procyon lotor*

TABLE 1.1 (*Continued*)

Class Subclass/Order Common name	*Species*
Pinnipedia (Chapter 20)	
Seal, harbor	*Phoca vitulina*
Seal, harp	*Pagophilus groenlandica*
Proboscidea (Chapter 21)	
Elephant, Indian	*Elephas maximus*
Sirenia (Chapter 22)	
Manatee, West Indian	*Trichechus manatus*
Manatee, Florida	*Trichechus manatus latirostris*
Perissodactyla (Chapter 23)	
Horse	*Equus caballus*[a]
Artiodactyla	
Pig (Chapter 24)	*Sus scrofa*[a]
Camel (Chapter 25)	*Camelus dromedarius*
Guanaco (Chapter 25)	*Lama guanicoe*
Llama (Chapter 25)	*Lama guanicoe gama*
Cow (Chapter 26)	*Bos taurus*[a]
Goat (Chapter 26)	*Capra musimon*[a]
Sheep (Chapter 26)	*Ovis aegagrus*[a]

[a]This is the wild species from which the domesticated animal descended.

gene pool of farm-bred animals is usually more confined because only a few sires are used for many females. Artificially bred animals may have very restricted gene pools because it is desirable to reproduce certain characteristics (e.g., fur color in mink). A few laboratory animals (e.g., mice, rats) are totally inbred so that the animals are genetically identical (i.e., they are clones).

3. A BRIEF WORD ON EVOLUTION

Current geologic evidence suggests that our planet originated some 4–4½ billion years ago. The Earth was first a fiery ball of flaming gases, and almost 2 billion years passed as the surface cooled and solidified into various plates and masses. Water formed, and an atmosphere containing nitrogen and a little oxygen gradually surrounded the earth. Roughly 2 billion years ago, life started with a

TABLE 1.2
Animals Studied

Animal	N	Weight (kg)	Ancestry[a]	Origin[b]	Study site[c]		Phlebotomy[d]		
					Obt.	Proc.	Anesth.	Site	Vol. (ml)
Alligator	6	28	W	FL	1	16	0	Tail	45
Armadillo	6	6	W	TX	C	C	N	H	20
Baboon	9	30	B	M	2	C	N	Arm	10
Bat	5	0.6	W	India	C	C	0	H	5
Bullfrog	14	<1	W	WI	C	C	E	H	3
Cat	8	7.5	F	M	3	C	N	J	10
Camel	3	600	F	India	4, 5	C	0	J	45
Chicken	9	4.0	F	M	6	C	0	Wing	5
Chimpanzee	5	45	W	Africa	7	C	K	Arm	40
Congo eel	3	2	W	NC	C	C	0	Tail	5
Cow	7	325	F	M	8	C	0	J	45
Dog	6	14	F	M	9	C	K	J	20
Dogfish	7	2	W	Atl.	10	10	0	Tail	5
Duck	4	3	F	M	6	C	N	Wing	4
Echidna	3	6	W	Austr.	11	11	0	H	10
Elephant	9	4500	W	India	4, 5	C	0	Ear	45
Ferret	6	6	B	M	12	C	0	H	5
Fish	See Table 1.1								
Fox	2	14	W	PA	P	C	0	J	15
Frog, grass	5	0.3	W	WI	C	C	E	H	1
Goat	6	52	F	M	P	C	0	J	45
Goose	6	6	F	M	6	C	0	Wing	4

Green turtle	1	50	W	Atl.	13	13	0	H	45
Guanaco	1	60	F	S.Am.	5	C	0	J	30
Guinea pig	11	0.4	B	M	C	C	N	H	5
Guitarfish	1	3.5	W	Atl.	13	13	0	H	2
Hamster	21	0.15	B	M	2	C	N	H	4
Hedgehog	2	0.6	W	Europe	C	C	N	H	5
Hellbender	4	1.5	W	PA	C	C	0	Tail	3
Horse	8	425	F	M	14	C	0	J	45
Lamprey	3	3	W	L. Erie	C	C	0	Tail	20
Llama	1	75	F	S.Am.	5	C	0	J	45
Loggerhead turtle	2	42	W	Atl.	13	13	0	H	45
Man	100+	60	·	Africa	C	C	0	V	45
Manatee	7	200	W	FL	15, 16	C, 16	0	Flipper	10
Mink	3	6	B	M	17	C	0	H	10
Monkey	14	17	B	India	P	C	N	Arm	4
Mud puppy	3	1	W	NC	C	C	0	Tail	10
Opossum	11	8	W	PA	C	C	N	H	15
Pig									
Standard	6	50	F	M	P	C	♀ 0, ♂K	VC	45
Mini	4	30	B	M	18	5	♀ 0, ♂K	VC	45
Micro	4	20	B	M	18	5	♀ 0, ♂K	VC	45
Pigeon	2	2	W	M	P	C	0	Wing	4
Pond (PSE) turtle	11	2	W	M	C	C	N	H	15
Porpoise	12	170	W	Atl.	19, 20	20	0	Flipper	45
Quokka	1	8	W	Austr.	11	11	N	H	15
Rabbit	6	3	F	M	P	C	N	H	15
Raccoon	3	15	W	PA	P	C	N	A	20
Rat									
Sprague-Dawley	6	<1	B	M	C	C	N	A	2
Wister	6	<1	B	M	2	C	N	H	2

(*continued*)

TABLE 1.2 (*Continued*)

Animal	N	Weight (kg)	Ancestry[a]	Origin[b]	Study site[c]		Phlebotomy[d]		
					Obt.	Proc.	Anesth.	Site	Vol. (ml)
Salamander	1	<1	W	SC	C	C	0	H	2
Seal									
Harbor	3	60	W	Atl.	10	10	0	Flipper	40
Harp	2	85	W	Atl.	21	C	0	Flipper	40
Shark									
Black-tipped	1	20	W	Atl.	13	13	0	Tail	45
Blue	2	60	W	Atl.	22	22	0	H	45
Mako	1	43	W	Atl.	22	22	0	H	45
Sheep	10	70	F	M	C	C	0 or K	J	45
Skate	3	3	W	Atl.	C	C	0	H	5
Snake, garter	6	3	W	WI	C	C	0	Tail	5
Toad	3	<1	W	WI	C	C	E	H	1
Toad, giant	5	<1	W	WI	C	C	E	H	3
Turkey	8	9	F	M	6	C	0	Wing	6
Wallaby	1	14	W	Austr.	11	11	N	H	20

[a] Ancestry: (B) bred; (F) farm; (W) wild.

[b] Origin: States are indicated by conventional two-letter abbreviations; (Atl.) Atlantic Ocean; (Austr.) Australia; (M) mixed, often worldwide; (S.Am.) South America.

[c] Laboratory at which blood was obtained (Obt.) and processed (Proc.): (C) Coagulation Laboratory; (P) University of Pittsburgh; (1) Gatorland (Orlando, FL); (2) Transplantation Surgery Laboratory (Pittsburgh); (3) University of Pittsburgh; (4) Ringling Brothers and Barnum & Bailey Circus (on location in Pittsburgh); (5) Northern Diagnostic Laboratories (Portland, ME); (6) poultry farms; (7) National Institutes of Health (Bethesda, MD); (8) Pennsylvania State University (State College); (9) Resuscitation Laboratory (Pittsburgh); (10) Mt. Desert Island Biological Laboratory (Salsbury Cove, ME); (11) Institute of Medical and Veterinary Science (Adelaide, Australia); (12) Magee Women's Hospital Animal Facility (Pittsburgh); (13) Marineland Research Laboratory (St. Augustine, FL); (14) riding academies; (15) Pittsburgh Zoo; (16) Ringling Brothers and Barnum & Bailey Circus (Sarasota, FL); (17) Weiman Mink Farm (Wexford, PA); (18) Charles River Breeding Laboratory (Windham, ME); (19) Aquarama Theatre of the Sea (Philadelphia); (20) U.S. Fish and Wildlife Service (Orlando, FL); (21) Montreal Aquarium; (22) Biological Laboratories (Woods Hole, MA).

[d] Anesthetic (Anesth.) administered: (0) none; (E) ether; (K) ketamine; (N) Nembutal. Site of blood collection: (A) abdominal aorta; (H) heart; (J) jugular vein; (VC) vena cava; (V) superficial vein.

carboniferous molecule that had the astonishing, *new* ability to self-copy. This novel, extraordinary capacity to propagate or reproduce is the basis of all life. It can never be known whether the first replication occurred at a single site or at multiple sites. To me, the single-site theory is more acceptable, although it almost eliminates the possibility of finding other life forms in what the poet John Milton called "the vast of boundless space."

The first molecules multiplied hugely, changed in structure and size through a multitude of mutations, and were spread widely by the winds and waters. After many millions of years, recognizable life forms evolved, each having the capacity to reproduce, grow, metabolize, and react to its environment.

Charles Robert Darwin (1809–1882), together with other naturalists of that time, developed a theory that helps to explain the enormously complex processes that have resulted in the bewildering abundance and diversity of life forms. Darwin's theory is expounded in his book, *On the Origin of Species by Means of Natural Selection, or the Preservation of Favoured Races in the Struggle for Life*.

In essence, the theory of natural selection states: (1) Variations occur among members of all species; (2) some variations are detrimental and some are advantageous; (3) when a species becomes crowded, those with advantageous variations survive. The conclusion is that natural selection results in "survival of the fittest or preservation of favoured races."

There must have been a time in the late Silurian–early Devonian Age, about 300 million years ago (MYA), when the land and sea were essentially empty and a great expansion of vegetable and animal life occurred. Vegetable life—for example, grasses, bushes, and trees—produced oxygen, consumed carbon dioxide, and provided foodstuff for some animals. Animal life—for example, bugs, fishes, and frogs—produced carbon dioxide, consumed oxygen, and when they decomposed provided food (fertilizer) for plants. Thus, parallel development of the animal and vegetable kingdoms proceeded rapidly for a million years or so. Incredibly intricate processes have enabled life to progress from a single self-copying molecule to a multibillion-celled creature, allowing many favored species to survive.

Table 1.3 presents a simplified geologic–biological time scale derived from many sources including *Webster's Dictionary.*

A species is defined as a group within which natural cross-fertilization may occur. Table 1.4 gives estimated numbers of vertebrate species extant today. New species are being discovered or are evolving; old species are becoming extinct. Of the estimated 70,000 species of vertebrates, over half are teleosts.

TABLE 1.3
A Simplified Geologic Time Scale

Era/period	Million years ago (MYA)	Life
Archeozoic	**4500–1400**	None
Proterozoic	**1400–620**	Bacteria, algae
Paleozoic	**620–200**	Trilobites, sponges
Cambrian	600–420	Shellfish, coral
Ordovician	420–360	Fishes
Silurian	360–330	Sharks, teleosts
Devonian	330–290	Insects, plants, amphibians
Carboniferous	290–230	Reptiles
Permian	230–200	Expansion, vegetable world
Mesozoic	**200**	Early dinosaurs
Triassic	200–160	Birds
Jurassic	160–128	Early mammals
Cretaceous	128–63	
Cenozoic	**63**	Extinction of dinosaurs
Tertiary	*63–2.5*	Monotremes, marsupials
Paleocene	63–58	Insectivores, rodents
Eocene	58–36	Carnivores, ungulates
Oligocene	36–25	Edentates
Miocene	25–13	Many mammals
Pliocene	13–2.5	
Quarternary	**2.5–present**	
Pleistocene	2.5–0.01	Early man
Holocene (Recent)	0.01–present	Modern man and animals

TABLE 1.4
Estimated Numbers of Vertebrate Species

Class	Species
Cyclostomata	50
Chondrichthyes	800
Osteichthyes	40,000
Amphibia	3,300
Reptilia	10,000
Aves	8,850
Mammalia	7,000
TOTAL:	70,000

REFERENCES

Darwin, C., 1859, *On the Origin of Species by Means of Natural Selection, or the Preservation of Favoured Races in the Struggle for Life,* John Murray, London.

SUGGESTED READINGS

Doolittle, R. F., 1976, The evolution of vertebrate fibrinogen, *Fed. Proc. Fed. Am. Soc. Exp. Biol.* **35:**2145.

Doolittle, R. F., 1983, The structure and evolution of vertebrate fibrinogen, *Ann. N. Y. Acad. Sci.* **408:**15.

Erdos, E. G., Miwa, I., and Graham, W. J., 1967, Studies on the evolution of plasma kinins: Reptilian and avian blood, *Life Sci.* **6:**2433.

Heilbrunn, L. V., 1961, The evolution of the haemostatic mechanism, in: *Functions of the Blood* (R. G. Macfarlane and A. H. T. Robb-Smith, eds.). Academic Press, New York.

Hewitt-Emmett, D., Czelusniak, J., and Goodman, M., 1981, The evolutionary relationship of the enzymes involved in blood coagulation and hemostasis, in: *Contributions to Hemostasis* (D. A. Walz and L. E. McCoy, eds.), *Ann. N. Y. Acad. Sci.* **370:**511.

Kimura, E., 1969, Phylogenetic studies on blood platelets: Comparative physiology of spindle cells, *Acta Haematol. Jpn.* **32:**1.

Moody, P. A., 1970, *Introduction to Evolution,* Harper and Row, New York.

Ratnoff, O. D., 1987, The evolution of hemostatic mechanisms, *Perspect. Biol. Med.* **31:**4.

Rebuck, J. W., 1971, Historical perspectives and the blood platelets, in: *The Circulating Platelet* (S. A. Johnson, ed.). Academic Press, New York.

Romer, A. S., 1960, *Man and the Vertebrates,* Penguin Books, Harmondsworth, Middlesex, England.

Sonderqvist, T., and Blomback, B., 1971, Fibrinogen structure and evolution, *Naturwissenschaften* **58:**16.

CHAPTER 2

LABORATORY METHODS AND TEST PARAMETERS

1. INTRODUCTION

Over the 30 years during which these studies were done, the laboratory methods have changed greatly. At first, all were manual–eyeball techniques. Currently, most clotting–counting tests are automated. Despite these changes in technique, the reagents used and the results obtained have changed very little, and comparisons between old and new test values are quite valid. On the other hand, comparisons between species are fraught with technical difficulties. It has long been recognized that many protein lipid interactions are species-specific.

In general, the methods used on animals are the same as those used on humans. In mammals, such comparative tests may have some significance, but in premammalian vertebrates, they may tell us little or nothing. For example, that some of these animal plasmas do not decrease the clotting time of human hemophilic plasma does not necessarily mean that these animals have no analogue of factor VIII.

Animal coagulation is best assessed with tests that do not involve admixture with materials from other species, such as the clotting time (Clot T), recalcification time (Recal T) with homologous tissue, and the thromboplastin generation test (TGT).

Ideally, many examples of both sexes in a single species should be studied. Because it was impossible to do so, an attempt was made to examine six healthy adult animals, three of each sex. For small animals, the number tested was greater, but not all tests were done on each animal. For some of the premammalian vertebrates, only a single example of the species was available.

2. METHODS

2.1. Anesthesia and Blood Collection

In order to prevent anxiety and possible pain and to temporarily immobilize the animals, some form of anesthesia was often used. Intraperitoneal Pentothal sodium or intravenous ketamine was used in most of the mammals. Rodents were usually anesthetized and the abdominal skin opened surgically. Birds were easily quieted by covering their heads and gently patting or rubbing their bellies. Ether was used for amphibians and some reptiles. Fishes, elasmobranchs, and lampreys were not anesthetized, but were quieted, almost hypnotized, by turning them ventral side up in a confining water box and gently rubbing the ventral surface. Table 1.2 in Chapter 1 lists the number and average weight of each species studied and the site of blood sampling, the anesthetic administered, and the volume of blood obtained.

The accuracy of coagulation tests is highly influenced by the technique. Free-flowing blood without tissue fluid is necessary for clotting times and Recal Ts. Obviously, the technique must vary with the size of the animal and the accessibility of veins and arteries.

From large animals (see Table 1.2), approximately 45 ml of blood was collected in the same order and number of syringes as is customary for human coagulation studies. If Vacutainer tubes were used, the vacuum was released. A No. 18, 19, or 20 needle or butterfly is inserted into the vein, and blood flow is started with slight suction on the new polystyrene syringe. Blood samples are collected from large animals as follows:

Syringe				Tube	
No.	Size (ml)		Blood (ml)	No.	Purpose[a]
1	5		3–4		Serum tube
2	20				Clot tubes
		2		1	New glass
		2		2	13 × 100 mm
		2		3	Siliconized glass
		2		4	13 × 100 mm
		2.7		5	SPT tube—new glass
		4.5		6	Cit tube
		4.5		7	Cit tube
			19.7		
3	20	3		8	EDTA tube
		4.5		9	Cit tube
		4.5		10	Cit tube
		4.5		11	Cit tube
			16.5		
			TOTAL VOLUME:	39.2	

[a](SPT) serum prothrombin time; (Cit) 1 part trisodium citrate + 9 parts blood.

With small animals, it was sometimes necessary to use one animal for each test or preparation, for example, four for clotting time, one for factor assays, one for prothrombin times (PTs), three for TGTs, one for counts, and two or three for platelet-rich plasma (PRP).

2.2. Blood Processing

Tubes 1–4	37°C. Start stopwatch and note time on clock. Observe for clotting every 2 min for 10 min, then every 5 min for 50 min. Observe at 4 hr for clot retraction and at 24 hr for lysis.
Tube 5	37°C. At 2 hr, add 0.3 ml 0.15 M sodium citrate, mix, centrifuge, and chill. This is the serum for serum prothrombin time (SPT).
Tubes 6 and 7	Ice Bath. Centrfuge 10 min at 2000 *g*. Remove plasma and store in an ice bath or freeze promptly and store at −70°C.
Tube 8	Room temp. EDTA tube for cell counts. Prepare smears and proceed to counts.
Tube 9	Room temp. Process for electron micrographs (EMs) (see "EM Prep" in Section 2 of the Appendix) as soon as possible.
Tubes 10 and 11	Room temp. Centrifuge 8 min at 800 RPM. Remove PRP to plastic tube. Use for platelet aggregations as soon as possible. Put the remainder in a tube and recentrifuge 10 min at 2000 *g*. Remove and store plasma.

3. TEST PARAMETERS

Model Tables 2.1 through 2.7 present the abbreviations, units, and names of the tests and the range of normal human results for the animal test results presented in the corresponding tables in Chapters 4–27 and 29. Details of the tests are to be found in the Appendix.

TABLE 2.1
General Coagulation Tests

Abbreviation	Unit	Test	Normal human range
Clot T		Clotting time	
Glass	min	Glass (new borosilicate, 13 × 100 mm)	6–12
Silicone	min	Siliconized	20–59
Clot Retr[a]	0–4+	Clot retraction	3–4+
Clot Lys[b]	0/+	Clot lysis	0
SPT		Serum prothrombin time	
Human	sec	Human substrate	20+
Animal	sec	Animal substrate	—
PT		Prothrombin time	
Simpl	sec	Simplastin	10–13
Hu Br	sec	Human brain	11–15
Homolog Br	sec	Animal brain	—
RVV	sec	Russell's viper venom	13–18
APTT	sec	Activated partial thromboplastin time	24–34[c]
Recal T	sec	Recalcification time	90–180
Lys[b]	0/+	Lysis	0
Lys MCA	0/+	Lysis in 1% monochloroacetic acid	0
Thromb T		Thrombin time	
Bov Thr	sec	Bovine thrombin	11–18
Hu Thr	sec	Human thrombin	11–18
CHH	sec	*Crotalus horridus horridus* venom	11–18
Atroxin	sec	*Bothrops atrox* venom	11–18

[a]Some results are stated, not as degree of clot retraction (0–4+), but as number of clots retracted/number observed.
[b]Some results are stated, not as no lysis (0) or lysis (complete dissolution) (+), but as number of clots lysed/number observed.
[c]Before 1978, the APTT reagent gave a range of 45–55 sec with normal human plasma.

TABLE 2.2
Thromboplastin Generation Test

Generating mixture		Substrate clotting time (sec)[a]	
Human	Animal	Human	Animal
Human	—	**H on H**[b]	H on A
—	Animal	A on H	**A on A**

[a]Homologous systems are in **boldface**.
[b]Normal human range: 9.8–11.2 sec.

TABLE 2.3
Coagulation Factors

Abbreviation	Factor	Normal human range (units)
I	Fibrinogen	150–450 mg/dl
II	Prothrombin	0.70–1.30 U/ml
V	Proaccelerin, accelerator globulin (AcG)	0.65–1.45 U/ml
VII	Proconvertin, serum prothrombin conversion accelerator (SPCA)	0.50–1.30 U/ml
X	Stuart factor	0.75–1.25 U/ml
VIII	Antihemophilic factor (AHF)	0.75–1.40 U/ml
IX	Plasma thromboplästin component (PTC)	0.65–1.70 U/ml
XI	Plasma thromboplastin antecedent (PTA)	0.75–1.40 U/ml
XII	Hageman factor	0.50–1.45 U/ml
Flet	Fletcher [prekallikrein (PK)]	0.50–1.50 U/ml
Fitz	Fitzgerald [high-molecular-weight (HMW) kininogen]	0.50–1.50 U/ml
ATIII (A[a])	Antithrombin III (A)	0.80–1.20 U/ml
ATIII (I[a])	Antithrombin III (I)	0.80–1.20 U/ml
Prot C (A[a])	Protein C (A)	0.60–1.40 U/ml
Prot C (I[a])	Protein C (I)	0.60–1.40 U/ml
Prot S (I[a])	Protein S (I)	0.60–1.40 U/ml
Plgn	Plasminogen	0.80–1.20 U/ml
Antipl	Antiplasmin	0.80–1.20 U/ml
RCF	Ristocetin cofactor	0.50–1.50 U/ml
XIII	Fibrin-stabilizing factor (FSF)	4–16 (R)[a]

[a](A) activity; (I) immunoassay; (R) reciprocal of the dilution that prevents dissolution of XIII-free clot in 1% MCA.

TABLE 2.4
Serum Proteins by Paper Electrophoresis

Abbreviation	Protein	Normal human range
TP	Total protein	6.3–7.9 g/dl
Alb	Albumin	3.2–4.4 g/dl
α_1-Globulin	Alpha$_1$-globulin	0.2–0.4 g/dl
α_2-Globulin	Alpha$_2$-globulin	0.4–1.0 g/dl
β-Globulin	Beta-globulin	0.5–1.0 g/dl
γ-Globulin	Gamma-globulin	0.6–1.8 g/dl

TABLE 2.5
Biochemical Tests

Abbreviation	Substance	Normal human range
TP	Total protein	6.0–8.0 g/dl
Alb	Albumin	3.5–5.0 g/dl
Ca	Calcium	8.5–10.5 mg/dl
P	Phosphorus	2.5–4.5 mg/dl
—	Cholesterol	150–200 mg/dl
—	Triglyceride	45–200 mg/dl
—	Glucose	65–110 mg/dl
—	Uric acid	2.5–8.0 mg/dl
—	Creatinine	0.7–1.4 mg/dl
T bilirubin	Total bilirubin	0.1–1.4 mg/dl
Fe	Iron	80–160 mg/dl
TIBC	Total iron-binding capacity	250–420 mg/dl
BUN	Blood urea nitrogen	10–20 mg/dl
Cl	Chloride	90–105 mg/dl
CO_2	Carbon dioxide	24–32 meq/dl
K	Potassium	3.5–5.0 meq/dl
Na	Sodium	135–145 meq/dl
Alk Phos	Alkaline phosphatase	30–85 IU/liter
HBD	Hydroxybutyric dehydrogenase	14–185 IU/liter
CPK	Creatine phosphokinase	0–110 IU/liter
LDH	Lactic dehydrogenase	60–200 IU/liter
GGT	γ-Glutamyl transferase	<44 IU/liter
SGOT	Serum glutamic-oxalacetic transaminase, aspartine aminotransferase (AST)	0–41 IU/liter
SGPT	Serum glutamic-pyruvic transaminase, alanine aminotransferase (ALT)	10–50 IU/liter

TABLE 2.6
Cellular Elements of the Peripheral Blood

Abbreviation	Element	Normal human range (units)
HCT	Hematocrit (or packed cell volume)	37–52%
HGB	Hemoglobin	12–18 g/dl
RBC	Red blood cell (count)	$4.2–6.2 \times 10^6/mm^3$
MCV	Mean corpuscular volume	79–97 fl
MCH	Mean corpuscular hemoglobin	27–31 pg
MCHC	Mean corpuscular hemoglobin concentration	32–36 g/dl
WBC	White blood cell (count)	$5–10 \times 10^3/mm^3$
Neutr	Neutrophil(s)	55–75%
Lymph	Lymphocyte(s)	20–40%
Mono	Monocyte(s)	2–10%
Eos	Eosinophil(s)	0–3%
Bas	Basophil(s)	0–3%
Plat C	Platelet count	$150–450 \times 10^3/mm^3$
MPV	Mean platelet volume	5.6–10.4 μm^3

TABLE 2.7
Platelet Activities

Abbreviation		Activity	Normal human range
Plat Glass Ret Ind		Platelet glass retention index	≥20%
Aggregation[a]		Aggregation in:	
ADP:	20 μM	Adenosine 5-diphosphate of different strengths	70–100%
	10 μM		70–100%
	5 μM		50–100%
	2.5 μM		40–100%
Collagen:	0.19 mg/ml	Bovine soluble collagen	80–100%
Risto:	0.9 mg/ml	Ristocetin	80–100%
Risto ½:	0.45 mg/ml	Ristocetin half-strength	<18%
Arach A:	0.5 mg/ml	Arachidonic acid	60–100%
Bov Thr:	0.4 U/ml	Bovine thrombin	60–100%
Bov Fib:	0.1%	Bovine fibrinogen	60–100%
CHH:	0.05 mg/ml	*Crotalus horridus horridus* venom	60–100%
Pig Pl:	1 : 10	Pig plasma	60–100%
Sh Pl:	1 : 10	Sheep plasma	60–100%
Hu Pl:	1 : 10	Human plasma	<10%

[a]The strengths indicated for the reagents are final concentrations.

SUGGESTED READINGS

Amiral, J., Adalbert, B., and Adam, M., 1984, Application of enzyme immunoassays to coagulation testing, *Clin. Chem.* **30:**1512.

Bang, N. V., Beller, F. K., Deutsch, E., and Mammen, E. F., eds., 1971, *Thrombosis and Bleeding Disorders: Theory and Methods*. Academic Press, New York.

Born, G. V. R., and Hardisty, R. M., 1976, Platelets, in: *Human Blood Coagulation, Haemostasis and Thrombosis,* 2nd ed. (R. Biggs, ed.), pp. 168–201, Blackwell, London.

Boyer, C., Rothschild, C., Wolf, M., Amiral, J., Meyer, D., and Larrieu, M. J., 1984, A new method for the estimation of protein C by ELISA, *Thromb. Res.* **36:**579.

Brandt, J. T., and Triplett, D. A., 1981, *Am. J. Clin. Pathol.* **76**(Suppl.)**:**530.

Coleman, R. W., Hirsh, J., Marder, V. J., and Salzman, E. W., 1987, *Hemostasis and Thrombosis,* 2nd ed., J. B. Lippincott, Philadelphia.

Comp, P. C., Nixon, R., and Esmon, C. T., 1984, Determinations of functional levels of protein C, an antithrombotic protein, using thrombin/thrombomodulin complex, *Blood* **63:**16.

Day, H. J., and Holmsen, H., 1972, Laboratory tests of platelet function, *Ann. Clin. Lab. Sci.* **2:**63.

Esmon, C. T., 1983, Protein C biochemistry, physiology and clinical implications, *Blood* **62:**1155.

Esmon, C. T., and Esmon, N. L., 1984, Protein C activation, *Semin. Thromb. Hemost.* **10:**122.

Gallimore, M. J., Amundsen, E., Assen, A. O., Laarsbraten, M., Lyngass, K., and Svendsen, L., 1979, Studies on plasma antiplasmin activity using a new specific chromogenic tripeptide substrate, *Thromb. Res.* **14:**51.

Gaspar, H., 1978, Collagen–platelet interaction, *Thromb. Haemost.*, Supplement No. 63.

Kasper, C. K., *et al.*, 1975, A more uniform measurement of factor VIII inhibitors, *Thromb. Haemat.* **34:**869.

Lewis, J. H., 1961, Coagulation defects, *J. Am. Med. Assoc.* **178:**1014.

Lewis, J. H., Iammarino, R. M., Spero, J. A., and Hasiba, U., 1978, Antithrombin Pittsburgh: An α_1-anti trypsin variant causing hemorrhagic disease, *Blood* **51:**129.

Miale, J. B., 1977, *Laboratory Medicine: Hematology,* 5th ed., C. V. Mosby, St. Louis.

Miale, J. B., and LaFond, D., 1969, Prothrombin time standardization, *Am. J. Clin. Pathol.* **52:**154.

Murano, G., and Bick, R. L., 1980, Basic concepts of hemostasis and thrombosis, in: *Platelet Disorders* (D. Triplett, ed.), pp. 101, CRC Press, Boca Raton.

Naito, K., and Aoki, N., 1978, Assay of alpha-2 plasmin inhibitor activity by means of a plasmin specific tripeptide substrate, *Thromb. Res.* **12:**1147.

Odegard, O. R., Lie, M., and Abilgaard, U., 1975, Heparin cofactor activity measured with an amidolytic method, *Thromb. Res.* **6:**287.

Olson, J. D., Brockway, W. J., Fass, D. N., Magnuson, M. A., and Bowie, E. J. W., 1975, Evaluation of ristocetin–Willebrand factor assay and ristocetin-induced platelet aggregation, *Am. J. Clin. Pathol.* **63:**210.

Owen, W. G., and Esmon, C. T., 1981, Functional properties of an endothelial cell cofactor for thrombin catalyzed activation of protein C, *J. Biol. Chem.* **256:**5532.

Penner, J. A., 1974, Experience with a thrombin clotting time assay for measuring heparin activity, *Am. J. Clin. Pathol.* **61:**645.

Pesce, A. J., Ford, D. J., and Gaizutis, M. A., 1978, Qualitative and quantitative aspects of immunoassays. *Scand. J. Immunol.* **8**(Suppl.)**:**1.

Ramsey, R., and Evatt, B. L., 1979, Rapid assay for von Willebrand factor activity using formalin-fixed platelets and microtitration technic, *Am. J. Clin. Pathol.* **72:**996.

Reddy, K. N. N., and Markus, G. J., 1972, Mechanism of activation of human plasminogen by streptokinase: Presence of active center in streptokinase–plasminogen complex, *J. Biol. Chem.* **247:**1683.

Rosenberg, R. D., 1975, Actions and interactions of antithrombin and heparin, *N. Engl. J. Med.* **292:**146.

Salem, A., and Cieslica, R., 1976, Comparison of the Coag-a-mate dual channel with the Fibrometer, *Am. J. Med. Technol.* **42:**39.

Samama, M., Schleguel, N., Cazenave, B., Horellou, M. M., Conrad, J., Castel, M., and Douenias, R., 1980, Alpha-2-antiplasmin assay: Amidolytic and immunological method. Critical evaluation results in a clinical material, in: *Synthetic Substrates in Clinical Blood Coagulation Assays* (H. R. Lijnen, D. Collen, and M. Verstraete, eds.), pp. 93–99, Martinus Nijhoff, The Hague.

Stenflo, J., 1984, Structure and function of protein C, *Semin. Thromb. Hemost.* **10:**109.

Stenflo, J., and Jonsson, M., 1979, Protein S, a new vitamin K–dependent protein from bovine plasma, *FEBS Lett.* **101:**377.

Stocker, K., and Meier, J., 1985, Thrombin-like snake venom enzymes, paper presented at the Symposium on Animal Venoms and Haemostasis, San Diego, July 20–21.

Suzuki, K., Nishioka, J., and Hashimoto, S., 1983, Regulation of activated protein C by thrombin-modified protein S, *J. Biochem.* **94:**699.

Svendsen, L., 1981, Synthetic peptide substrate: A critical perspective, in: *Perspective in Hemostasis* (J. Fareed, H. L. Messmore, J. W. Fenton II, and K. M. Brinkhous, eds.), pp. 343–353, Pergamon Press, New York.

Teger-Nilson, A. C, Friberger, P., and Gyzander, E., 1977, Determination of a new rapid plasmin inhibitor in human blood by means of a plasmin specific tripeptide substrate, *Scand. J. Clin. Lab. Invest.* **37:**403.

Thomson, J. M. (ed.), 1980, *Blood and Coagulation and Haemostasis,* Churchill Livingstone, London.

Triplett, D. A., Harms, C. S., Newhouse, P., and Clark, C., 1978, *Platelet Function: Laboratory Evaluation and Clinical Application,* American Society of Clinical Pathologists, Chicago.
Walker, F. J., 1980, Regulation of activated protein C by a new protein, *J. Biol. Chem.* **255:**1260.
Walker, F. J., 1984, Protein S and the regulation of activated protein C, *Semin. Thromb. Hemost.* **10:**131.
Weiss, P., Soff, G. A., Halkin, H., and Seligsohn, U., 1987, Decline of proteins C and S and factors II, VII, IX and X during the initiation of warfarin therapy, *Thromb. Res.* **45:**783.
Wohl, R. C., Sinio, L., and Robbins, K. C., 1982, Methods for studying fibrinolytic pathway components in human plasma, *Thromb. Res.* **27:**520.

CHAPTER 3

HUMAN HEMOSTATIC/ COAGULATION MECHANISMS

1. INTRODUCTION

There are hundreds if not thousands of factors that have been shown to play some role in human hemostasis. This short discussion deals only with those that can be measured or easily detected and that appear to be analogous to hemostatic mechanisms in other mammals or lower vertebrates.

Following injury to a vessel wall, there are three major closely related mechanisms that cause prompt stoppage of blood flow:

1. **Vasoconstriction**
2. **Platelet plug formation**
3. **Hemostatic clot stabilization**

2. VASOCONSTRICTION AND DECREASED BLOOD FLOW

Vasoconstriction has two phases: The first is mechanical and results from reflex contraction of muscle fibers in the injured vessel wall following the direct stimulus of the injury. The second is humoral and results from the release of serotonin from the dense bodies of the degranulating platelets as they aggregate and agglutinate to form the platelet plug. Serotonin diffuses into the surrounding tissue and constricts not only the injured vessel but also neighboring vessels, thus slowing blood flow to the entire injured area.

3. PLATELET PLUG FORMATION

Damage to vascular walls releases or exposes many components of hemostasis. Of importance to the simplified version described here are collagen fibers

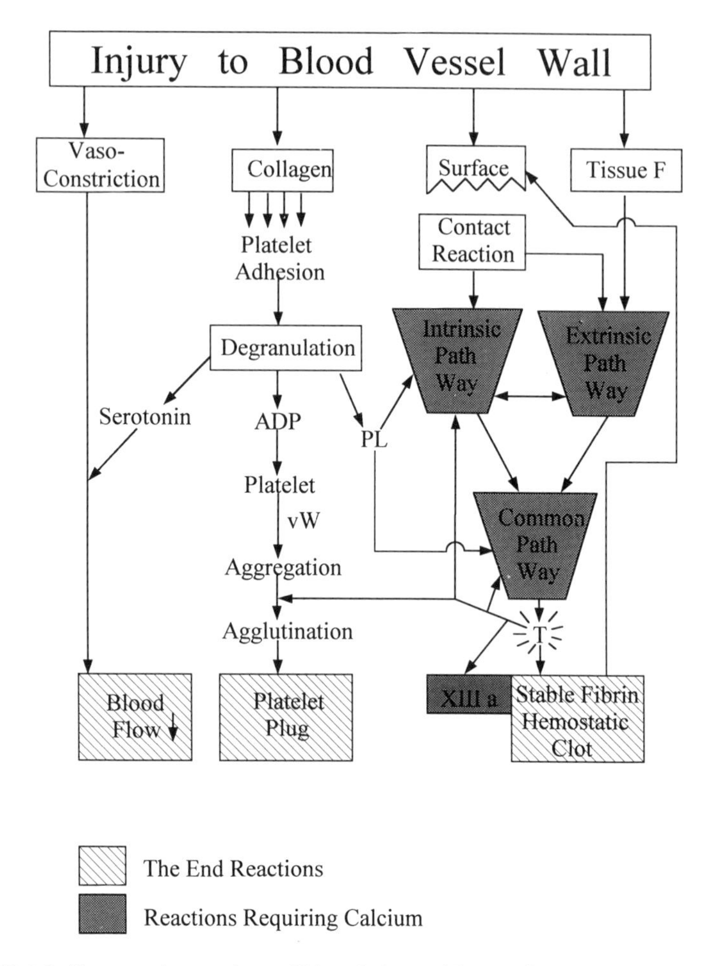

FIGURE 3.1. Hemostatic reactions. Abbreviations: (Tissue F) tissue factor, (PL) phospholipid complex; (T) activated thrombin; (vW) von Willebrand factor.

and tissue factor. Platelets are highly structured cytoplasmic buds that are derived from megakaryocytes and enter the circulation from the bone marrow. In the injured area, they adhere to the exposed collagen fibers. They then degranulate, releasing their contents, including serotonin, ADP, and exposed platelet phospholipid. ADP causes the platelets to adhere or stick to each other. Von Willebrand factor (vWF), from the plasma or vessel wall, is essential to the firm aggregation of platelets and thrombin to the irreversible agglutination that forms the platelet plug. This plug or white thrombus may be sufficient to permanently stop blood leakage from a small wound or may form the nidus of the stable fibrin clot.

4. STABLE FIBRIN CLOT

Among vertebrates, plasma fibrinogen is always present, but shed blood may not clot unless it is modified by the addition of tissue factor or calcium. In man, early investigators (MacFarlane, 1964; Davie and Ratnoff, 1964) pictured a cascade or waterfall in which one factor was activated and then activated the next, which in turn activated the next, until fibrin, the insoluble structural substance of the clot, was formed. Each step proceeds with more force—thus amplifying the reaction. Three pathways—*intrinsic, extrinsic,* and *common*—were described. The same pathways are included in Fig. 3.1, although we now realize that there are cross-activations and most of the reactions involve phospholipid complexes.

5. SURFACE REACTIONS

Figure 3.2 shows some of the recognized reactions that occur on the injured endothelial surface. On the left are the platelet reactions. Prostaglandin inhibitor 2 (PGI_2) released from endothelium is a modifier or regulator of the platelet reactions. Collagen and vWF are necessary to platelet aggregation.

The center section of Fig. 3.2 shows the early coagulation pathways. In the contact reaction, factor XII is activated on the endothelial surface by the effects of Fitzgerald (Fitz) (HMW kininogen) and Fletcher (Flet) (prekallikrein) factors and then activates factor XI to XIa, which initiates the intrinsic pathway of coagulation. After insoluble fibrin begins to be formed it may act as a foreign surface for XII activation, thus promoting the extension of the clot. It should be noted that this group of contact factors is not at all or is only partially involved in hemostasis. Patients with deficiencies or dysfunctions (D/Ds) of factors XII, Fitz, or Flet do not have excessive bleeding tendencies. Only about half of patients with factor XI D/D have excessive bleeding, usually mild.

Tissue factor released from endothelium starts the extrinsic pathway. On the right side of Fig. 3.2 are some of the inhibiting or modulating reactions. Fibrinolysin or plasmin is activated from its precursor, profibrinolysin or plasminogen, by tissue plasminogen activator (t-PA). Also, it can be at least partially activated by either factor XIIa or activated protein C. Plasma protein C is activated by protein S and thrombin bound to endothelial thrombomodulin. Both active plasmin and protein C can modulate coagulation by inhibiting or modifying factors V and VIII.

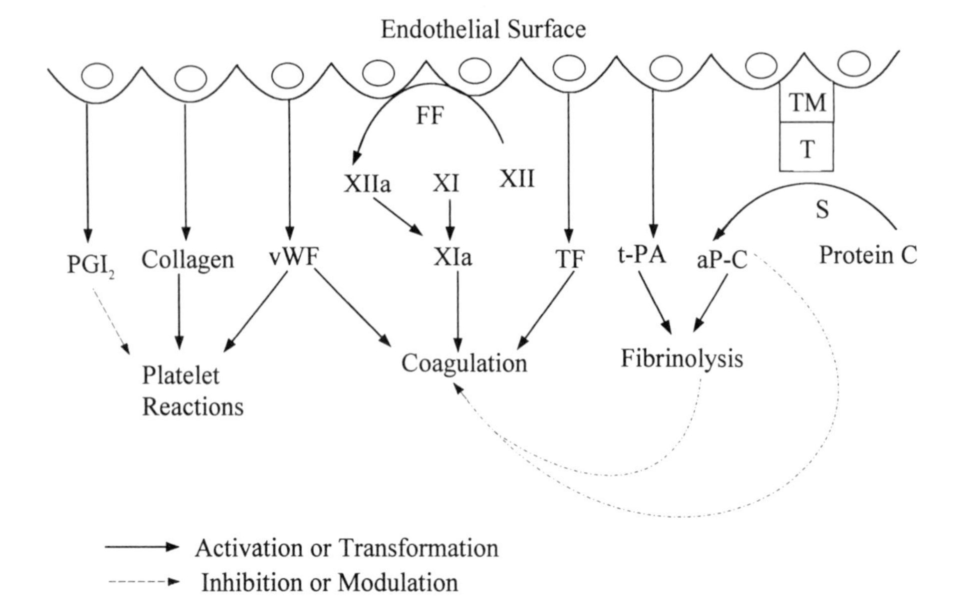

FIGURE 3.2. Surface reactions. Abbreviations: (FF) Fletcher (prekallikrein) and Fitzgerald (HMW kininogen) factors; (TM) thrombomodulin; (T) thrombin; (PGI_2) prostaglandin inhibitor 2; (vWF) von Willebrand factor; (TF) tissue factor; (t-PA) tissue plasminogen activator; (aP-C) activated protein C; (S) protein S.

6. COAGULATION PATHWAYS

The exact mechanisms of coagulation and hemostasis are poorly understood despite a good deal of research. Two major differences from the original cascade–waterfall theory have become apparent. These are the major "feedback" role of thrombin in activating factors VIII and V and the importance of platelet-derived phospholipid (PL) as a surface on which factors may complex and react.

My own interpretation of the bare bones of the coagulation pathways is presented in Fig. 3.3. Coagulation is initiated by the release of tissue factor—perhaps only a minute amount—from injured endothelium. The first reactions follow the so-called extrinsic pathway and are shown on the right side of Fig. 3.3 by outlined arrows (⇒). Tissue factor (TF), in the presence of calcium, complexes with and activates factor VII to VIIa. This complex slowly activates prothrombin (factor II) to thrombin (T or IIa). Thrombin proceeds to its many functions. Of particular import to the coagulation system is the thrombin activation of factors VIII and V to VIIIa and Va, which complex with factors X and II, respectively, on the PL surfaces.

The second series of reactions (indicated by solid arrows ⟶) involves

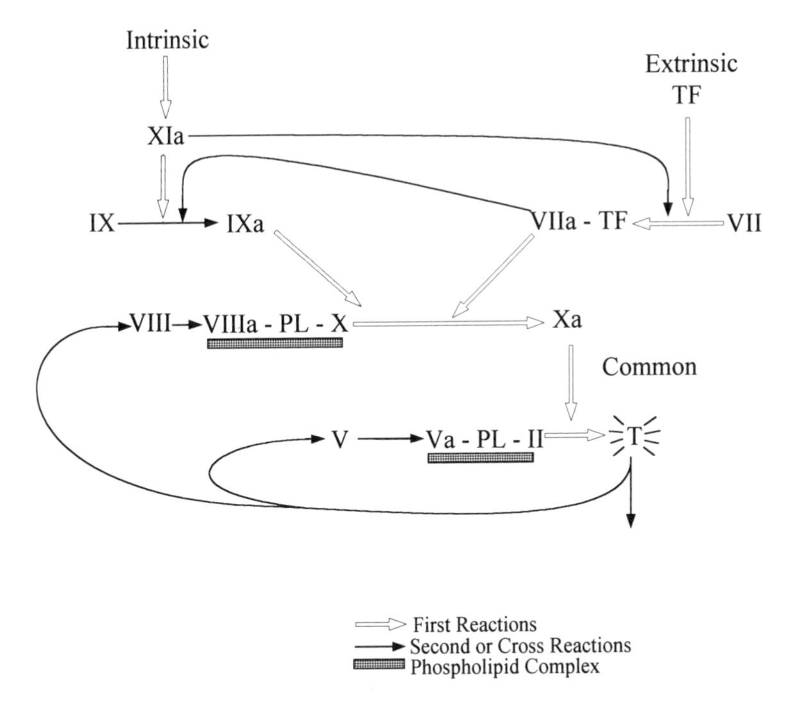

FIGURE 3.3. Coagulation pathways. Abbreviations: (TF) tissue factor; (PL) phospholipid complex; (T) activated thrombin.

the old so-called "intrinsic" system, wherein factor XIa converts factor IX to IXa. Calcium is necessary, but it is unclear whether or not PLs are active here. The conversion of factor IX to IXa is speeded by the VIIa–TF complex, and in a cross-reaction, factor XIa plays a role in the formation of this VIIa–TF complex. The next two steps are similar. Thrombin-activated VIIIa and Va complex with X and II on PL surfaces. These complexes greatly speed the conversion rates of factor X to Xa and factor II to activated thrombin.

Some of the hemostatic functions of activated thrombin (T) are listed here:

1. The conversion of fibrinogen to insoluble fibrin

T splits two fibrinopeptides (A and B) from the fibrinogen molecule, leaving fibrin monomer that polymerizes rapidly to insoluble fibrin.

2. The activation of XIII to XIIIa

T, in the presence of calcium, activates XIII to XIIIa, which cross-links the polymerized fibrin to form stable fibrin.

3. The agglutination of platelets

T causes the firm agglutination of aggregated platelets into an irreversible clump called the platelet plug or white thrombus.

4. The activation of VIII and V
T causes proteolysis of these large molecules, which in both cases leaves a major fragment with greatly enhanced activity.

5. The binding of thrombin to thrombomodulin (TM) exposed after injury to the endothelium
T binds to TM and in this conformation activates Protein C to aPC. The T–TM loses most of its other characteristics (e.g., T–TM can no longer clot fibrinogen, agglutinate platelets or activate VIII or V). Protein S acts as a rate-regulator for Protein C actitvation.

7. INHIBITORS AND MODULATORS

Specific inhibitors exist in the plasma milieu in active or precursor form. Damaged endothelial structures are involved in the *in situ* activations: Thrombomodulin (TM) for the Protein C - Protein S system and tissue plasminogen activator (tPA) for the fibrinolytic system (See Chapter 27 Section 1.7 Page 342–344). Activated Protein C (aPC) apparently does not directly destroy Factor V or VIII but rather destroys or inhibits the inhibitor (? s) of Protein C activation. The active enzymes formed in theses two distinct systems can, by different mechanisms, destroy the coagulation functions of Factors V and VIII. Of particular interests are some of the additional functions of fibrinolysin: it can destroy fibrin or, if potent enough, fibrinogen. The latter reaction is limited by the antithrombic (anti-T) ability of the large fibrinogen degradation products.

The other major types of inhibitors are those which block the enzymes (serine proteases) formed by the coagulation reaction. Antithrombin III (ATIII) binds T to form inactive thrombin-antithrombin complex (TAT). ATIII also inactivates IXa, Xa and XIa. Heparin greatly enhances these ATIII activities. There appear to be other large plasma proteins which have antithrombin-like properties.

Major inhibitor systems are summarized below:

Factor	Inhibitors
1. V and VIII	1. aPC + Prot S
	2. Plasmin (also called fibrinolysin)
2. T, VIIa, IXa, Xa, XIa	1. ATIII + heparin
	2. Fibrinogen degradation products
3. Plasmin	1. ATIII
	2. antifibrinolysin ($\alpha 2$ antiplasmin)
	3. Fibrinogen degradation products

8. FINAL STEPS OF HEMOSTASIS

Clot retraction is the next step in completion of hemostasis. The clot pulls itself together, squeezing out about 50% of its volume in serum. Because the clot is attached to the vessel wall, this process narrows the vessel, making permanent closure simpler. Neutrophils and macrophages migrate into the clot and start the healing process by ingestion and digestion of fibrin. Fibroblasts develop, and in 5–10 days the wound is converted to a scar.

9. SUMMARY

Normal hemostasis is an exquisitely balanced reaction that is limited by inhibitors and modulators so that it effectively prevents excessive blood loss and is confined to the injured area and does not extend to other areas to cause dangerous and life-threatening thrombi or emboli.

REFERENCES

Davie, E. W., and Ratnoff, O. D., 1964, Waterfall sequence for intrinsic blood clotting, *Science* **145:**1310.

MacFarlane, R. G., 1964, An enzyme cascade in the blood clotting mechanism and its function as a biochemical amplifier, *Nature (London)* **202:**498.

SUGGESTED READINGS

Colman, R. W., Bagdasarian, A., Talamo, R. C., Scott, C. F., Seavey, M., Guinaraes, J. A., Pierce, J. F., and Kaplan, A. P., 1975, Williams trait: Human kininogen deficiency with diminished levels of plasminogen proactivator and prekallikrein associated with abnormalities of the Hageman factor–dependent pathways, *J. Clin. Invest.* **56:**1650.

Comp, P. C., and Esmon, C. T., 1984, Recurrent venous thromboembolism in patients with a partial deficiency of protein S, *N. Engl. J. Med.* **311:**1525.

Doolittle, R. F., 1984, Fibrinogen and fibrin, *Annu. Rev. Biochem.* **53:**195.

Esmon, C. T., and Owen, W. G., 1981, Identification of an endothelial cell cofactor for thrombin-catalyzed activation of protein C, *Proc. Natl. Acad. Sci. U.S.A.* **78:**2249.

Fulcher, C. A., Gardiner, J. E., Griffin, J. H., and Zimmerman, T. S., 1984, Proteolytic inactivation of human factor VIII procoagulant protein by activated human protein C and its analogy with factor V, *Blood* **63:**486.

Furie, B., and Furie, B. C., 1988, The molecular basis of blood coagulation, *Cell* **53:**505.

Gilbert, G. E., Furie, B. C., and Furie, B., 1990, Binding of human factor VIII to phospholipid vesicles, *J. Biol. Chem.* **265:**815.

Griffin, J. H., Evatt, B., Zimmerman, T. S., Kleiss, A. J., and Wideman, C., 1981, Deficiency of protein C in congenital thrombotic disease, *J. Clin. Invest.* **68:**1370.

Hathaway, W. E., Belhasen, L. P., and Hathaway, H. S., 1965, Evidence for a new plasma thromboplastin factor. I. Case report, coagulation studies and physiochemical properties, *Blood* **26:**521.

Mann, K. G., Jenney, R. J., and Krishnaswamy, S., 1988, Cofactor proteins in the assembly and expression of blood clotting enzyme complexes, *Annu. Rev. Biochem.* **57:**915.

Naito, K., and Fujikawa, K., 1991, Activation of human blood coagulation factor XI independent of factor XII: Factor XI is activated by thrombin and factor XIa in the presence of negatively charged surfaces, *J. Biol. Chem.* **266:**7353.

Osterud, B., and Rapaport, S. I., 1977, Activation of factor IX by the reaction product of tissue factor and factor VII: Additional pathway for initiating blood coagulation, *Proc. Natl. Acad. Sci. U.S.A.* **74:**5260.

Rao, L. V., and Rapaport, S. I., 1988, Activation of factor VII bound to tissue factor: A key early step in the tissue factor pathway of blood coagulation, *Proc. Natl. Acad. Sci. U.S.A.* **85:**6687.

Walker, F. J., Sexton, P. W., and Esmon, C. T., 1979, The inhibition of blood coagulation by activated protein C through the selective inactivation of activated factor V, *Biochim. Biophys. Acta* **571:**333.

PART II

STUDIES IN REPRESENTATIVE VERTEBRATES

CHAPTER 4

THE SEA LAMPREY

Class: Cyclostomata

1. INTRODUCTION

The earliest vertebrate class, Cyclostomata, consists of two orders: Petromyzontia and Myxinoidia. Lampreys belong to the first and slime eels and hagfish to the second. These may be called the most primitive fish—if the term "fish" is used to denote any poikilothermic aquatic craniate vertebrate. Almost all are parasitic. Lampreys are jawless and have mouths that consist of a large suctorial funnel with many horney teeth capable of rasping through the skin and scales of their hosts. Their bodies are long snakelike cylinders with smooth, slimy skin and both median and tail fins. They have seven pairs of gill slits well posterior to the mouth. The skeleton is completely cartilaginous. Adult lampreys seek clean fresh water in which to reproduce. The female clears a shallow nest from the stream bottom and attaches herself upstream. The male then attaches to the female. Fertilization is external, and the numerous small eggs sink to the bottom and are covered with silt and sand. The parents die after spawning. The eggs hatch a few weeks later, and the young pass through a hermaphroditic larval stage that may last for years. Characteristically, lampreys are euryhaline and thus able to migrate between salt water and fresh water. Recently, many have become landlocked, particularly in Lake Erie, and have destroyed many prized eatable lake fish by their parasitic attacks.

Coagulation studies on Cyclostomata are limited to those of Doolittle and associates (Doolittle, 1965a,b; Doolittle and Wooding, 1974; Doolittle *et al.*, 1962, 1963), which have dealt with the clotting of fibrinogen and with fibrin–fibrinogen split products. A few studies of blood cells and serum proteins of various Cyclostomata have also been published.

2. SOURCE

Six sea lampreys (*Petromyzon marinus*) weighing 2–4 kg were obtained from a reliable dealer and kept in a plastic tank (laundry cart) in about 1 foot of aerated artificial salt water. They were not fed. Satisfactory blood samples were obtained from three by venipuncture of the caudal vein. These unattractive, slimy animals were poorly tolerated by the laboratory staff.

3. TESTS CONDUCTED

3.1. Bleeding Times

After the lower ventral skin surface of each of the six lampreys was cleaned with an alcohol sponge, shallow cuts about 1 cm long were made with a Bard-Parker No. 9 blade. Times to cessation of bleeding in the six animals were: 0, 1½, 1¼, 2, 0, and 1½ min (mean: 1.04 min).

3.2. General Coagulation Times and Factors

Table 4.1 shows the very long coagulation times of the blood of the three lampreys. Two glass and one siliconized tube were used for each animal. Only one glass tube showed any clotting by 1 hr. The others showed soft clots at 2 hr. One of the siliconized tubes was clotted at 2 hr; the other two had soft clots at 6 hr. Clot retraction (Clot Retr) was decreased (1–2+), and the clots did not lyse.

Prothrombin times were done at room temperature (23°C) for lamprey and at 37°C for the human control. Simplastin and human brain did not produce clotting in less than 120 sec in lamprey plasma, but were within their usual range

TABLE 4.1
General Coagulation Tests in the Sea Lamprey

	Lamprey No.[a]						Human	
Test	1	2	3	*N*	Mean	SD	Control	Normal range
Clot T (min)								
Glass	120	120	120	3	120	0	—	6–12
Silicone	360	120	360	3	280	139	—	20–59
Clot Retr[b]	2–2	1–2	1–1	3	1–2	—	—	3–4+
Clot Lys[b]	0–0	0–0	0–0	3	0–0	0	—	0
PT (sec)								
Simplastin	c	c	c	3	c	—	12.7	10–13
Human brain	c	c	c	3	c	—	13.4	11–15
Lamprey skin	30.4	28.8	31.4	3	30.2	1.3	96.0	—
RVV	23.8	21.2	20.0	3	21.7	1.9	14.2	13–18
APTT (sec)	c	c	c	3	c	—	51.4	45–55
Recal T (min)	6½	8	8	3	7½	0.9	2¼	1½–3
Lys	0	0	0	3	0	—	0	0
Lys MCA	0	0	0	3	0	—	0	0
Thromb T (sec)								
Bov Thr	11.9	12.4	14.2	3	12.8	1.2	15.8	11–18

[a]At 23°C. [b]At 24 hr. [c]More than 120 sec.

TABLE 4.2
Coagulation Factors in the Sea Lamprey

Factor	Lamprey No. 1	2	3	N	Mean	D	Normal range
I (mg/dl)	125	214	187	3	175	46	150–450
II (U/ml)	[a]	[a]	[a]	3	[a]	—	0.70–1.30
V (U/ml)	[a]	[a]	[a]	3	[a]	—	0.65–1.45
VII (U/ml)	[a]	[a]	[a]	3	[a]	—	0.50–1.30
X (U/ml)	[a]	[a]	[a]	3	[a]	—	0.75–1.25
VIII (U/ml)	[a]	0.10	0.05	3	0.05	0.04	0.75–1.40
IX (U/ml)	0.05	0.10	0.05	3	0.07	0.03	0.65–1.70
XI (U/ml)	0.05	—	—	1	0.05	0	0.75–1.40
XII (U/ml)	[a]	[a]	[a]	3	[a]	—	0.50–1.45
XIII (R)	2	4	4	3	3	1.2	4–16

[a]Less than 0.01 U/ml counted as 0 in calculations.

in human plasma. Lamprey skin suspension clotted lamprey plasma in about 30 sec, but was much less active on human plasma. Russell's viper venom (RVV) was active in clotting both lamprey and human plasmas. Activated partial thromboplastin time (APTT) on lamprey plasma was longer than 120 sec. Recal Ts were slow and are recorded in minutes. The clots were solid and did not lyse alone or in 1% monochloroacetic acid (MCA), indicating that factor XIII (fibrin-stabilizing factor) was present in lamprey plasma. Rather amazingly, the clotting time with bovine thrombin was faster than human.

Table 4.2 lists the coagulation factor assay results in tests using regular human assay systems at 37°C. Fibrinogen (factor I) was within the human range. Most coagulation factors (II, V, VII, X, and XII) were not detectable. Factors VIII, IX, XI, and XIII were present with less than 10% of human activity (<0.10 U/ml).

3.3. Serum Proteins

Total serum protein in one lamprey was 2.7 g/dl. The paper electrophoretic distribution showed no albumin, a high α_1 peak (1.62 g/dl), no α_2, β at 0.38 g/dl, and γ at 0.72 g/dl.

3.4. Cellular Elements

Table 4.3 shows counts of peripheral blood cells in two lamprey. The red cells were large, roundish, nucleated cells with high concentrations of hemo-

TABLE 4.3
Cellular Elements of the Peripheral Blood in the Sea Lamprey

Element	Lamprey No. 1	2	N	Mean	Normal human range
HCT (%)	39	46	2	43	37–52
HGB (g/dl)	8.0	8.2	2	8.1	12–18
RBC ($\times 10^6/mm^3$)	0.77	0.82	2	0.82	4.2–6.2
MCV (fl)	506	561	2	534	79–97
MCH (pg)	103	100	2	101	27–31
MCHC (g/dl)	21	18	2	20	32–36
WBC ($\times 10^3/mm^3$)	48	71	2	59	5–10
Hetero (%)	12	14	2	13	55–75
Lymph and Mono (%)	33	51	2	62	22–50
Bas (%)	24	28	2	26	0–3
Thromb ($\times 10^3/mm^3$)	35	60	2	48	150–450

globin. WBC counts were high, and it was difficult to differentiate lymphocytes and monocytes on stained smears.

4. SUMMARY

Lampreys are primitive parasitic fish with no known tendency to easy bleeding, although their shed blood clots very poorly in glass or siliconized tubes. Bleeding times are short, indicating effective hemostasis. The RBCs are large, roundish, nucleated cells with high hemoglobin content.

REFERENCES

Doolittle, R. F., 1965a, Differences in the clotting of lamprey fibrinogen by lamprey and bovine thrombins, *Biochem. J.* **94:**735.

Doolittle, R. F., 1965b, Characterization of lamprey fibrinopeptides, *Biochem. J.* **94:**742.

Doolittle, R. F., and Wooding, G. L., 1974, The subunit structure of lamprey fibrinogen and fibrin, *Biochim. Biophys. Acta* **271:**277.

Doolittle, R. F., Oncley, J. L., and Surgenor, D. M., 1962, Species differences in the interaction of thrombin and fibrinogen, *J. Biol. Chem.* **237:**3123.

Doolittle, R. F., Lorand, L., and Jacobsen, A., 1963, Some comparative aspects of the fibrinogen–fibrin conversion, *Biochim. Biophys. Acta* **69:**161.

SUGGESTED READINGS

Cottrell, B. A., and Doolittle, R. F., 1976, Amino acid sequences of lamprey fibrinopeptides A and B and characterization of the junctions split by lamprey and mammalian thrombins, *Biochim. Biophys. Acta* **453:**426.

Fine, J. M., Boffa, G. A., and Drilhon, A., 1964, Electrophoretic and immunological investigation of the serum proteins of the sea lamprey (*Petromyzon marinus*), *C. R. Soc. Biol.* **158:**2021.

Hardisty, M. W., 1979, *Biology of the Cyclostomes,* Chapman and Hall, London.

Hardisty, M. W., 1982, Lampreys and hagfishes: Analysis of cyclostome relationships, in: *The Biology of Lampreys,* Vol. 4B (M. W. Hardisty and I. C. Potter, eds.), Academic Press, London.

Jordan, H. E., and Speidel, C. C., 1930, Blood formation in cyclostomes, *Am. J. Anat.* **46:**355.

Linthicum, D. S., 1975, Ultrastructure of hagfish blood leucocytes, in: *Immunologic Phylogeny* (W. Hildemann and A. A. Benedict, eds.), p. 241, Plenum Press, New York.

Lorand, L., 1961, The lamprey eel blood clot, *Biol. Bull.* **121:**396 (abstract).

Mattisson, A. G., and Fange, R., 1977, Light and electron microscopic observations on the blood cells of the Atlantic hagfish, *Myxine glutinosa, Acta Zool.* **58:**205.

Page, M., and Rowley, A. F., 1983, A cytochemical, light and electron microscopical study of the leucocytes of the adult river lamprey, *Lampetra fluviatilis, J. Fish Biol.* **22:**503.

Paleus, S., and Liljeqvist, G., 1972, The hemoglobins of *Myxine glutinosa.* II. Amino acid analyses, end group determinations and further investigations, *Comp. Biochem. Physiol. B* **42:**611.

Percy, R., and Potter, I. C., 1976, Blood cell formation in the river lamprey, *Lampetra fluviatilis, J. Zool.* **178:**319.

Percy, R., and Potter, I. C., 1977, Changes in haemopoietic sites during the metamorphosis of the lampreys, *Lampetra fluviatilis* and *Lampetra planeri, J. Zool.* **183:**111.

Percy, R. P., and Potter, I. C., 1981, Further observations on the development and destruction of lamprey blood cells, *J. Zool.* **193:**239.

Potter, I. C., Robinson, E. S., and Brown, I. D., 1974, Studies on the erythrocytes of larval and adult lampreys (*Lampetra fluviatilis*), *Acta Zool.* **55:**173.

Potter, I. C., Percy, R., Barber, D. L., and Macey, D. J., 1982, The morphology, development and physiology of blood cells, in: *The Biology of Lampreys,* Vol. 4A (M. W. Hardisty and I. C. Potter, eds.), p. 233, Academic Press, London.

Quast, R., and Vesterberg, O., 1968, Isoelectric focusing and separation of hemoglobins of *Myxine glutinosa* L. in a natural pH gradient, *Acta Chem. Scand.* **22:**1499.

Rall, D. P., Schwab, P., and Zubrod, C. G., 1961, Alteration of plasma proteins at metamorphosis in the lamprey (*Petromyzon marinus dosatus*), *Science* **133:**279.

Sekhon, S. E., and Maxwell, D. S., 1970, Fine structure of developing hagfish erythrocytes with particular reference to the cytoplasmic organelles, *J. Morphol.* **131:**211.

Tomonaga, S., Shinohara, H., and Awaya, K., 1973, Fine structure of the peripheral blood cells of the hagfish, *Zool. Mag.* **82:**211.

CHAPTER 5

THE CARTILAGINOUS FISH
Class: Chondrichthyes

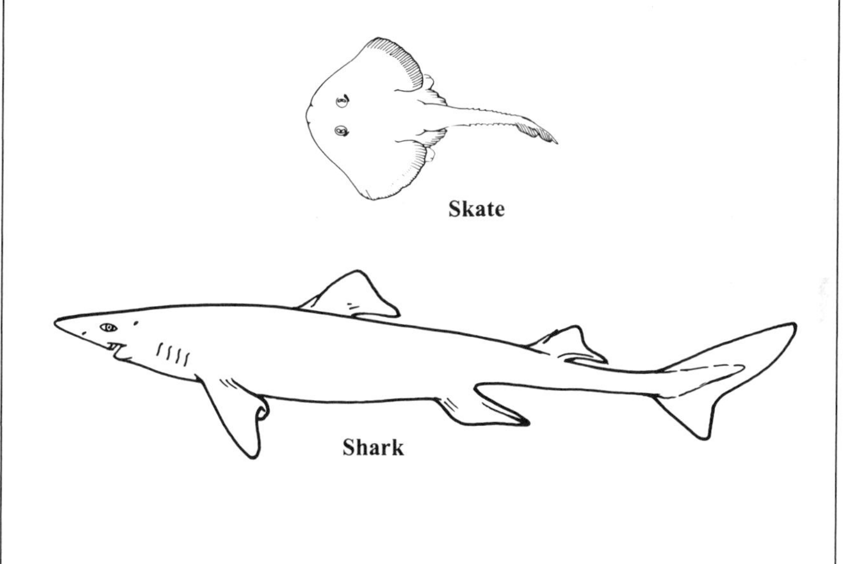

1. INTRODUCTION

The cartilaginous fish probably evolved about the same time as or just before the bony fish some 300 MYA during the great growth burst of animal and vegetable life on this planet. Chondrichthyes have skeletons composed entirely of cartilage and movable lower jaws usually with teeth. They are divided into two subclasses:

Elasmobranchii
 Order: Selachi (sharks and dogfish)
 Rajiformes (guitarfish and skates)
Chimaeras (ratfish)

Elasmobranchs are found worldwide and are almost entirely marine. They have two pairs of fins and additional unpaired ones. Sharks swim by undulating movements of their torpedo-shaped bodies in which the caudal and anal fins supply the main propulsive power and the dorsal fins the directional force. Skates have flattened bodies and swim by vertical beating of their dorsal fins. Chondrichthyes have no air sacs or swim bladders and therefore must swim continuously in order to avoid sinking to the ocean floor. They are voracious eaters, consuming whatever is available: fish, crustacea, squids, carrion, diving birds, mammals—even man. Reproduction is most frequently oviparous, and fertilization is always internal. A few species are viviparous or ovoviviparous.

Chondrichthyes are of moderate economic importance. Many species are edible, although the high urea levels make them unpalatable.

Coagulation and hematological studies of Chondrichthyes are rare. Doolittle and Surgenor (1962) and Doolittle (1963) showed that blood from these cartilaginous fish clots very slowly and that added calcium will cause more rapid clotting and, at some concentrations, will also cause clot lysis. Various tissue extracts also hasten coagulation. Belamarich *et al.* (1966, 1968) reported that thrombocytes from dogfish as well as turtles, alligators, and chickens did not aggregate when ADP, serotonin, or epinephrine was added. RBC counts, size measurements, hemoglobins, and osmotic fragilities have been described by a number of investigators. Previous studies from this laboratory (Lewis, 1972) have been published, and some data are included here with the kind permission of the editor of *Comparative Biochemistry and Physiology*. In the 1990s, seven additional spiny dogfish (*Squalus acanthias*) were studied, and the results are included as "new" dogfish.

2. SOURCE

Blood of a mako shark (*Isurus oxyrinchus*) (44 kg) and two blue sharks (*Prionace glauca*) (50–70 kg) was obtained by cardiac puncture from animals caught off Woods Hole, Massachusetts, with Albert Lansing, M.D. (University of Pittsburgh). A black-tipped reef shark (*Carcharhinus maculipinnis*) (20 kg), three skates (*Raja eglanteria*) (2–4 kg), and a guitarfish (*Rhinobatus lentiginosus*) (3.5 kg) were caught by the collecting vessel for the Marineland Research Laboratory near St. Augustine, Florida. On two occasions, blood of dogfish (1–3 kg) was collected at the Mt. Desert Island Biological Laboratory in Salsbury Cove, Maine. The dogfish were kept in holding tanks with circulating fresh sea water for 2–3 days before blood was obtained from caudal vein punctures.

3. TESTS CONDUCTED

3.1. Bleeding Times

Bleeding times were done on two skates, the black-tipped shark, and six dogfish by making small cuts with a Bard-Parker No. 11 knife blade. The two cuts on the skates, the cut on the black-tipped shark, and three of the cuts on the dogfish bled for less than 1 min or not at all. On the other three dogfish, the bleeding times were 2, 2¾, and 2¾ min.

3.2. General Coagulation Tests and Factors

Table 5.1 shows the very long whole blood clotting times found with elasmobranch blood. Some samples were still liquid at 48 hr and could be clotted by adding strong thrombin solution or calcium. Among the dogfish, there were a few with relatively short clotting times (19–30 min in glass tubes), but these were usually those in which difficulty with the venipuncture had been encountered and are not included in the table. Clot retraction could not be assessed because the blood in the tubes did not clot. PTs with Simplastin and RVV were much longer than on human plasma. When homologous tissue extracts were used in the PTs, fish but not human PTs were accelerated. The fish APTTs were more than 120 sec. For the fish Recal Ts 0.1M $CaCl_2$ was used. These were not clotted at 120 sec, but did clot by 10 min. They did not lyse in MCA after 1 hr. Recal clots from dogfish and shark lysed by 24 hr; skate and human clots did not. Thrombin times (Thromb Ts) were moderately long in the fish plasmas. ATroxin

TABLE 5.1
General Coagulation Tests in Chondrichthyes[a]

	Sharks						Human	
	Dogfish (7)	Black (1)	Blue (2)	Mako (1)	Skates (2)	Guitarfish (1)	Control (4) (mean)	Normal range
Clot T (min)								
Glass	60–120	1440+	240+	240+	240+	60	—	6–12
Silicone	600+	1440+	240+	240+	240+	120	—	20–59
Clot Retr	3[b]	[b]	[b]	[b]	[b]	1	—	3–4+
Clot Lys	+[b]	[b]	[b]	[b]	[b]	+	—	0
PT (sec)								
Simplastin	57.7	[c]	[c]	[c]	[c]	65.0	12.1	10–13
Homologous								
Brain	8.6	38.4	27.8	—	44.4	—	[c]	—
Gill	10.3	32.4	25.4	—	68.2	—	[c]	—
Skin	31.4	51.7	57.2	—	39.0	—	[c]	—
RVV	[c]	115	118	—	118	58.7	15.0	13–18
APTT (sec)	[c]	[c]	[c]	—	[c]	[c]	36.6	24–34
Recal T[d] (sec)	[c]	[c]	[c]	—	[c]	[c]	140	90–180
Lys	7/7	1/1	2/2	—	0	1/1	0	0
Lys MCA	0	0	0	—	0	?[e]	0	0
Thromb T (sec)								
Bov Thr	57.4	33.8	29.8	—	44.4	29.4	14.5	11–18
Atroxin	[c]	—	97.4	—	—	—	15.5	11–18

[a]Fish blood was tested at an ambient temperature of 23–26°C, human blood at 37°C.
[b]The blood was not clotted, but did clot when strong thrombin was added.
[c]More than 120 sec.
[d]Fish blood was recalcified with 0.1 M $CalCl_2$, human blood with 0.025 M $CaCl_2$.
[e]The clot lysed before the MCA test.

times were performed only on dogfish and blue shark and were much longer than on human plasma.

Table 5.2 shows that the fibrinogen and sometimes factors XI and XII were measurable in the chondrichthyes. Other coagulation factors II–X were undetectable.

Table 5.3 shows fibrinogen and some of the “newer” factors assayed on dogfish. Fibrinogen varies considerably from fish to fish. ATIII, ProtC, Prot S, and antiplasmin were all low but measurable.

Table 5.4 shows the results of the TGT. Human mixture clotted only the human plasma. Fish mixtures did not clot human or fish substrate plasmas.

TABLE 5.2
Coagulation Factors in Chondrichthyes

Factor	Dogfish New (7) means	Dogfish Old (4) means	Sharks BT (1)	Sharks Blue (2) (mean)	Sharks Mako (1)	Skates (2) (mean)	Guitarfish (1)	Normal human range
I (mg/dl)	216	110	150	260	190	160	212	150–450
II (U/ml)	[a]	[a]	[a]	[a]	[a]	[a]	[a]	0.70–1.30
V (U/ml)	[a]	[a]	[a]	[a]	[a]	[a]	[a]	0.65–1.45
VII (U/ml)	[a]	[a]	[b]	[a]	[a]	[a]	[b]	0.50–1.30
X (U/ml)	[a]	[a]	[b]	[a]	[a]	[a]	[b]	0.75–1.25
VIII (U/ml)	[a]	[a]	[a]	[a]	[a]	[a]	[a]	0.75–1.40
IX (U/ml)	[a]	[a]	[a]	[a]	[a]	[a]	[a]	0.65–1.70
XI (U/ml)	[a]	[a]	[a]	0.15	—	0.05	[a]	0.75–1.40
XII (U/ml)	[a]	[a]	[a]	0.02	0.05	0.05	0.40	0.50–1.45

[a]Less than 0.01 U/ml.
[b]Factors VII and X assayed in the proconvertin test: <0.01 U/ml.

3.3. Effects of Calcium on Chondrichthye Clotting

Table 5.5 shows the Recal Ts of dogfish plasma using varying strengths of calcium. The optimum is 0.1M, but if dogfish brain is added, the fastest clots are found with 0.05–0.025M $CaCl_2$. Lysis of clots is seen with 0.2, 0.1, and 0.075M without brain and 0.3, 0.2, and 0.075M with brain.

Figure 5.1 shows the effects of adding $CaCl_2$ at various molarities to shark whole blood. The shortest clotting times were after addition of 0.1 and 0.2M $CaCl_2$. These tubes clotted and the clots lysed completely (dissolved) in a few hours. At lower or higher calcium concentrations, the clots retracted but did not lyse. A simular calcium-related shortening of clotting times was found in three different sharks, two skates, and seven dogfish.

3.4. Effects of Buffy Coat

Table 5.6 shows the effects of decreasing amounts of buffy coat in cell-poor dogfish plasma. Clotting times became longer and clot retraction lessened as the amount of buffy coat decreased. These results show that retraction was dependent on WBC or thrombocyte concentration.

TABLE 5.3
Other Coagulation Factors of Dogfish

Factor	Dogfish No.										Normal human range
	1	2	3	4	5	7[a]	8	*N*	Mean	SD	
I (mg/dl)	270	145	270	230	230	130	240	7	216	57	150–450
ATIII (U/ml)	0.16	0.19	0.22	0.24	0.25	0.03	0.03	7	0.16	0.10	0.80–1.20
Prot C (U/ml)	0.05	0.07	0.67	0.36	0.05	0.05	0.05	7	0.19	0.24	0.60–1.40
Prot S (U/ml)	0.07	0.05	0.07	0.04	0.06	0.10	0.07	7	0.07	0.02	0.60–1.40
Antipl (U/ml)	0.37	0.52	0.62	0.23	0.20	0.27	0.05	7	0.32	0.20	0.80–1.20

[a]Number 6 was lost to the experiment.

TABLE 5.4
Thromboplastin Generation Test in Chondrichthyes

Generating mixture				Substrate Clot T (sec)[a]			
Human (mean)	Dogfish	Shark	Skate	H	D	Sh	Sk
H (2)	—	—	—	**9.4**	40+	40+	40+
—	D	—	—	40+	**40+**	—	—
—	—	Sh	—	40+	—	**40+**	—
—	—	—	Sk	40+	—	—	**40+**

[a]Homologous systems are in **boldface**.

3.5. Effects of Dilutions of Tissue Factor

Dogfish brain in serial dilutions (Table 5.7) was added to tubes to which fresh whole blood was subsequently added. The tubes with brain tissue factor all clotted promptly, and the clots retracted well. They were not saved for lysis. These results show that even highly diluted tissue factor was an active hemostatic agent in Chondrichthyes and that whole blood, when clotted rapidly, retracted well.

TABLE 5.5
Effect of Calcium Molarity on Recalcification Time and Clot Lysis of Dogfish Plasma With or Without Tissue Factor[a]

	Saline		Dogfish brain		
$CaCl_2$ molarity	Recal T (min)	Clot Lys	Recal T (sec)		Clot Lys
0.5	60	0	180+		No clot
0.3	48	0	105		+
0.2	13	+	68		+
0.1	12	+	40		+
0.075	13	+	40	Soft gels	+
0.05	14	0	35		0
0.035	18	0	36		0
0.025	48	0	34		0
0.01	60+	0	39	Flecks	0
0.005	60+	0	42	Flecks	0

[a]Test system: 0.1 ml citrated dogfish plasma + 0.1 ml 0.85% saline or dogfish brain homogenate + 0.1 ml $CaCl_2$ at 20°C. Lysis was read at 24 hr.

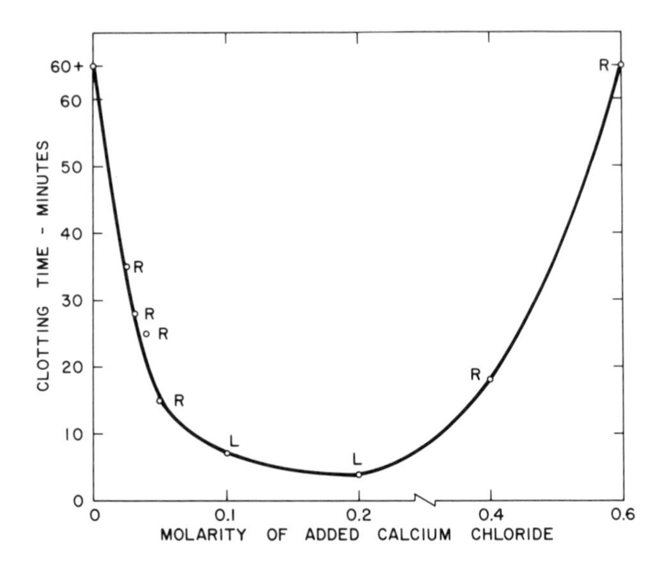

FIGURE 5.1. Effects of calcium on black-tipped shark blood clotting (0.1 ml $CaCl_2$ + 0.2 ml blood). Labels: (R) clot retraction; (L) clot lysis.

3.6. Thrombin Inhibitor

Table 5.8 shows that dogfish plasma mixed with human plasma caused a prolongation of the thrombin and prothrombin times. If the dogfish plasma was first heated to 56°C for 5 min, the inhibitory effect was destroyed. Barium

TABLE 5.6
Effects of Buffy Coat on Recalcification Time and Clot Retraction in Dogfish Plasma/Buffy Coat Mixtures (Temp = 24°C)[a]

Buffy coat in citrated plasma (ml)	Cell-poor in citrated plasma (ml)	Clotting time (min)	Clot retraction at 1 hr.
0.1	0.1	$4^1/_4$	4+
0.05	0.15	5	3+
0.025	0.175	$6^3/_4$	2+
0.01	0.19	$9^1/_4$	1+
0	0.2	11	0

[a]Technique: 2ml of fresh dogfish blood was centrifuged at 400 × g for 5 min. The slightly opaque plasma was pipetted to a new test tube and centrifuged again at 2000 × g for 10 min. Cell-poor plasma was removed with a pipette and the buffy coat was resuspended in 0.2 ml plasma. The mixtures were prepared in new glass tubes as detailed above, and an equal volume of 0.1M $CaCl_2$ added, mixed, and observed for 1 hr. Clotting time was recorded and clot retraction observed at 1 hr. These tubes were not retained because lysis was anticipated.

TABLE 5.7
Effects of Dilution of Dogfish Brain Homogenate on Clotting of Whole Blood[a]

Dogfish brain dilution	Clot T (min)	Appearance at 24 hr
Full strength		
1 : 2	All clotted in 2 min	All good clot with 6+ retraction
1 : 4		
1 : 8		
1 : 16		
1 : 32		
1 : 64		
Saline	No clot	Liquid[b]

[a]Test system: 0.1 ml brain homogenate or dilution + 1.0 ml fresh whole dogfish blood at 25°C. Brain homogenate was prepared by snipping and mashing 1½ dogfish brains, adding them to 5 ml I/S, and allowing to settle 10 min. Blood was obtained by venipuncture with a No. 19 needle; 2 ml blood was obtained in a small syringe and discarded, then 10 ml blood was obtained in a fresh syringe and 1 ml was distributed to each of eight 10 × 75 mm new glass tubes containing dogfish brain or dilution. The tube contents were mixed and observed at 2, 10, 240, and 1440 min.
[b]No clots, but clotted when strong thrombin was added.

treatment of the plasma did not destroy the inhibitory effect. This inhibitor may be fibrinogen itself or a large fibrinogen breakdown product.

3.7. Effects of EACA

Table 5.9 shows the effects of an amino acid—EACA—known to have inhibitory action against mammalian plasmin and other serine proteases. The

TABLE 5.8
A Heat-Labile Inhibitor in Dogfish Plasma

Plasma (ml)					
	Dogfish				
Human (whole)	Whole	56°C	Ba	Thromb T (sec)	PT (sec)
0.2	—	—	—	19.8	12.4
—	0.2	—	—	57.4	120+
0.1	0.1	—	—	48.8	120+
0.1	—	0.1	—	19.4	11.4
0.1	—	—	0.1	43.4	120+

TABLE 5.9
Inhibition of Clot Lysis in Blue Shark Blood by ε-Aminocaproic Acid[a]

	CaCl$_2$ in NS			CaCl$_2$ in 2% EACA		
CaCl$_2$ (M)	Clot T (min)	Clot Retr at 24 hr	Clot Lys (hr)	Clot T (min)	Clot Retr at 24 hr	Clot Lys (hr)
0.5	5	—	8	20	++++	0
0.4	20	—	8	20	++++	0
0.4	20	—	8	20	++++	0
0.3	20	—	8	30	++++	0
0.2	60	—	24	60	++++	0
0.1	120	—	24	120	++++	0
0.075	120	—	24	120	++++	0
0.025	120	++++	None	120	++++	0
0.01	240+	++++	None	240+	++++	0

[a]Test system: 1.0 ml fresh blue shark blood at 20°C + 0.1 ml CaCl$_2$ (molarity as given).

fibrinolytic activity in the recalcified blood clot was inhibited by EACA, and the clots retracted instead of dissolving.

3.8. Cellular Elements

All the blood cells of Chondrichthyes including the RBCs and thrombocytes were nucleated. Table 5.10 shows the RBC cell counts, hemoglobin (HGB), and hematocrit (HCT). Elasmobranchs had relatively high HCTs, low HGB, and very low RBC counts. This resulted in high mean corpuscular volumes (MCVs) and mean corpuscular hemoglobins (MCHs) and relatively low mean corpuscular hemoglobin concentrations (MCHCs). The WBCs were high with variable differentials. Thrombocyte counts were sometimes greater than WBC counts.

3.9. Thrombocyte Aggregation

Crude suspensions of thrombocytes were prepared either by resuspending buffy coat in plasma or by brief low-speed centrifugation to sediment erythrocytes. In any case, the suspensions contained other white cells, particularly lymphocytes. In one experiment, dogfish connective tissue was snipped into very fine pieces and washed repeatedly with 0.85% saline to remove any thrombin or blood products. When added to a suspension of dogfish thrombocytes, some cells appeared (microscopically) to adhere to the fibers, but gross aggregation

TABLE 5.10
Cellular Elements of the Peripheral Blood in Chondrichthyes

	Dogfish		Sharks				
Element	New (7)	Old (4)	Mako (1)	Blue (2)	BT (1)	Skates (2)	Guitarfish (1)
HCT (%)	30	26	24	20	34	23	28
HGB (g/dl)	5.1	4.4	5.7	5.4	4.7	3.4	4.5
RBC (× 10^6/mm^3)	0.40	0.39	0.33	0.25	0.46	0.21	—
MCV (fl)	750	667	727	800	739	1095	—
MCH (pg)	128	113	173	216	102	162	—
MCHC (g/dl)	17	17	24	27	14	15	16
WBC (× 10^3/mm^3)	—	84	91	22	—	88	16
Hetero (%)	—	35	42	46	19	69	—
Lymph (%)	—	53	42	40	57	8	—
Mono (%)	—	7	0	10	5	5	—
Eos (%)	—	5	16	0	19	13	—
Bas (%)	—	0	0	4	0	5	—
Thromb (× 10^3/mm^3)	180	138	73	184	24	39	—

was not seen. No aggregation or clumping was detected when ADP (1×10^{-3} to 1×10^{-8} M) was added. Figure 5.2 is a transmission electron micrograph (TEM) of skate thrombocytes that were exposed to crude skate thrombin. The cells appear to be degranulating and the cytoplasm of one cell fusing with that of another. These cells are suspended in plasma, but fibrin is not evident at this thrombin concentration. Both skate and bovine thrombin, in low concentrations, produced thrombocyte agglutinates that could barely be seen without magnification.

3.10. Thrombocyte Ultrastructure

Figure 5.3 shows a TEM of a dogfish thrombocyte. A group of peripheral microtubules, sectioned lengthwise, can be seen in the left upper portion. There are some dark, circular (fried-egg) granules or inclusions that resemble myelin bodies. There are quite a few open vacuoles that may be part of an open canalicular system. There are distinct small electron-opaque bodies, probably glycogen, scattered in the cytosome. The nucleus has very dense peripheral chromatin.

4. SUMMARY

Hemostasis in Chondrichthyes appeared to be a relatively simple process—primarily an extrinsic mechanism. An injury released tissue factor, which, in the

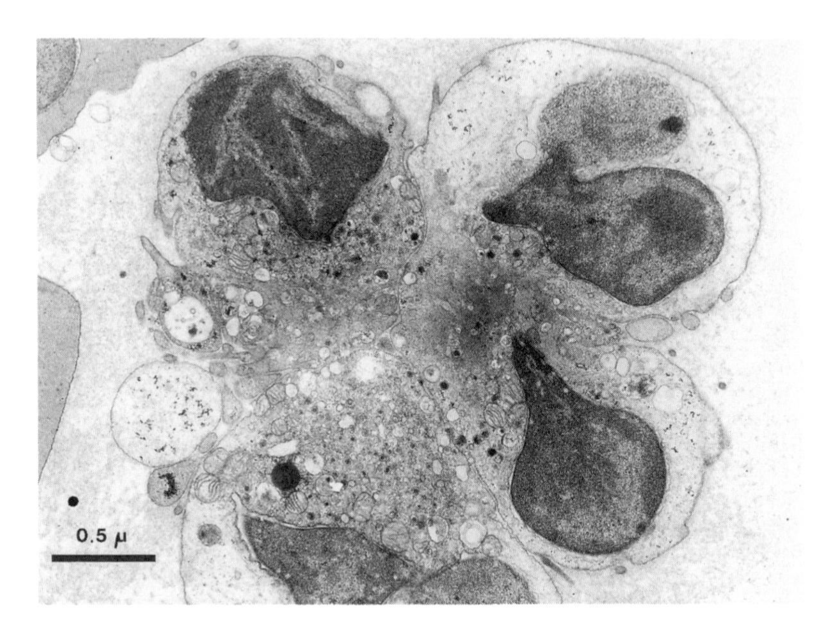

FIGURE 5.2. TEM of skate thrombocytes exposed to skate thrombin. Note the loss of distinct cell boundaries.

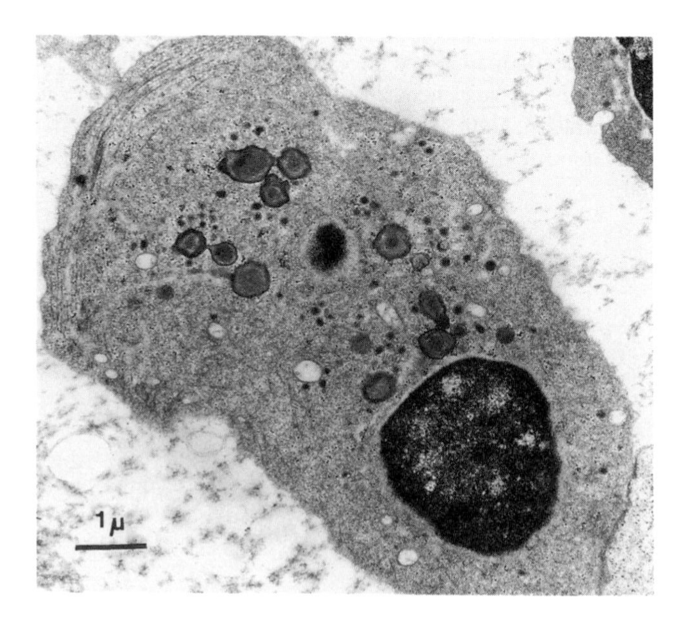

FIGURE 5.3. TEM of a dogfish thrombocyte.

presence of calcium, converted prothrombin to thrombin. Whether or not other factors were involved is unknown. Thrombin had at least two functions. The primary one was the agglutination of thrombocytes to form a hemostatic plug. Thrombin also converted fibrinogen to fibrin. In this function, it was readily inhibited by fibrinogen itself or a heat-labile derivative.

Blood obtained by nontraumatic techniques did not clot unless tissue factor or hypertonic solution ($CaCl_2$ or sea water) was added. The calcium optimum for clotting was variable and somewhat dependent on the presence of tissue factor. Fibrinolysis of whole blood or plasma clots caused by added calcium was frequently seen. These was no plasminogen-activating effect of streptokinase (SK), urokinase (UK), or staphylokinase (Staph K).

Nucleated thrombocytes were necessary for clot retraction. They have some ultrastructural similarities to mammalian platelets. Microtubules were visible, as were an open canalicular system and some glycogen particles.

REFERENCES

Belamarich, F. A., Fusari, M. H., Shepro, D., and Kren, M., 1966, *In vitro* studies of aggregation of non-mammalian thrombocytes, *Nature (London)* **212:**1579–1580.

Belamarich, F. A., Shepro, D., and Kien, M., 1968, ADP is not involved in thrombin-induced aggregation of thrombocytes of a non-mammalian vertebrate, *Nature (London)* **220:**509–510.

Doolittle, R. F., 1963, Further studies on clotting and fibrinolysis in plasma from the smooth dogfish (*Mustelis canis*), *Br. J. Haematol.* **9:**464–470.

Doolittle, R. F., and Surgenor, D. M., 1962, Blood coagulation in fish, *Am. J. Physiol.* **203:**964–940.

Lewis, J. H., 1972, Comparative hemostasis: Studies on elasmobranchs, *Comp. Biochem. Physiol.* **42A:**233–240.

SUGGESTED READINGS

Hyder, S. L., Cayer, M. L., and Pettey, C. L., 1983, Cell types in peripheral blood of the nurse shark: An approach to structure and function, *Tissue Cell* **15:**437.

Irisawa, H., and Irisawa, A. F., 1954, Blood serum protein of the marine elasmobranchii, *Science* **120:**849.

Johansson-Sjobeck, M. L., and Stevens, J. D., 1976, Haematological studies on the blue shark, *Prionace glauca* L., *J. Mar. Biol. Assoc. U. K.* **56:**237.

Mainwarning, G., and Rowley, A. F., 1985, Studies on granulocyte heterogeneity in elasmobranchs, in: *Fish Immunology* (M. J. Manning and M. F. Tatner, eds.), p. 57, Academic Press, Orlando, FL.

Morrow, W. J. W., and Pulsford, A., 1980, Identification of peripheral blood leucocytes of the dogfish (*Scyliorhinus canicula* L.) by electron microscopy, *J. Fish Biol.* **17:**461.

Parish, N., Wrathmell, A., and Harris, J. E., 1985, Phagocytic cells in the dogfish (*Scyliorhinus canicula* L.), in: *Fish Immunology* (M. J. Manning and M. F. Tatner, eds.), p. 71, Academic Press, Orlando, FL.

Parish, N., Wrathmell, A., Hart, S., and Harris, J. E., 1986, The leucocytes of the elasmobranch *Scyliorhinus canicula* L.—a morphological study, *J. Fish Biol.* **28:**545.

Reznikoff, P., and Reznikoff, D. G., 1934, Hematological studies in dogfish (*Mustelus canis*), *Biol. Bull. (Woods Hole, MA)* **66:**115.

Saunders, D. C., 1966, Elasmobranch blood cells, *Copeia* **1966:**348.

Shepro, D., Belamarich, F. A., and Branson, R., 1966, The fine structure of the thrombocyte in the dogfish (*Mustelus canis*) with special reference to microtubule orientation, *Anat. Rec.* **156:**203.

Sherburne, S. W., 1974, Occurrence of both heterophils and neutrophils in the blood of the spiny dogfish, *Squalus acanthias, Copeia* **1974:**259.

Stokes, E. E., and Firkin, B. G., 1971, Studies on the peripheral blood of the Port Jackson shark (*Heterodontus portusjacksoni*) with particular reference to the thrombocyte, *Br. J. Haematol.* **20:**427.

Zapata, A., and Carrato, A., 1980, Ultrastructure of elasmobranch and teleost thrombocytes, *Acta Zool.* **61:**179.

CHAPTER 6

THE BONY FISH

Class: Osteichthyes

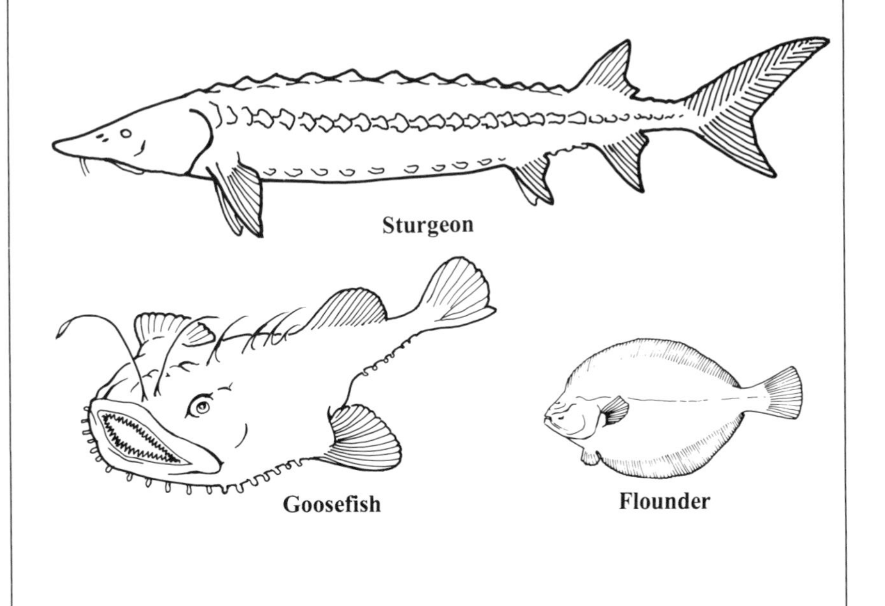

1. INTRODUCTION

The bony fish evolved in the late Silurian–early Devonian Age, about 300 MYA. Their ossified fossil remains show incredible diversity that continues today in the 35,000–40,000 species extant. There are more species of living fish than of all other vertebrates combined. With a rare peculiar exception, each species is subtly different from the next and can breed only within its own natural group.

Ostheichthyes are classified into two subclasses and six orders:

- Acanthopterygii (spiny-rayed fish)
  - Polypterii [polypherus and Reed fish (Africa)]
  - Chondrostei (sturgeons and paddlefish)
  - Holostei (gars and bowfish)
  - Teleostei (many orders and thousands of species)
- Sarcopterygii (soft-rayed fish)
  - Crossopterygii (coelacanths)
  - Dipnoi (lungfish)

Sturgeons are very primitive fish belonging to the order Chondrostei. All other fish in this study were Teleostei.

Teleosts live worldwide in freshwater lakes and streams, in saltwater seas, in deep oceans and shallow bays. Some are euryhaline (e.g., salmon, trout) and can go from sea water to fresh and back again.

Most fish have torpedo-shaped, flexible bodies with powerful fins and tails to aid in swimming. They come in all sizes, shapes, and colors. Some are flattened from top to bottom (flounder), some from side to side (orange filefish). Some have large teeth (wolffish) and some have horny spines (sculpin). Skin covered with protective, overlapping scales is characteristic of nearly all teleosts.

With a rare exception (lungfish), fish exchange O_2 and CO_2 through gills containing many small surface vessels over which water is forced in a countercurrent manner. Most Osteichthyes have a swim bladder or air sac that serves as a hydrostatic organ, changing size as the water pressure varies. This provision allows the animal to float at its optimal depth without sinking to the bottom. In contrast, as noted in Chapter 5, Chondrichthyes have no such air bladders and must swim continuously to prevent sinking.

A majority of fish are oviparious, spawning eggs, externally fertilized with milt (spermatazoa). These eggs hatch without nurturing, and a few grow to maturity. The others are lost to the vicissitudes of life. A few species are viviparous or ovoviviparous or even hermaphroditic. In a rare species such as the popularly known sea horse, the male parent cares for the eggs or young.

Although few bony fish, except the barracuda, will attack man, others may inflict injuries when they are handled. Fish with sharp teeth, pointed spines, or poisonous skin are hazardous to humans.

Whether caught or raised, fish are of great economic value, supplying over half the world's protein foodstuffs. They are also important for animal food, fish oil, fertilizer, and aquarium pets. Because of this high monetary value, their identification, life patterns, and diet are described in many publications. Little mention is made of hematology or blood coagulation.

Blood counts and hemoglobin studies were made by Hall and Gray (1929), Callegarin (1966), Eisler (1965), and Blaxhall and Daisley (1973). Early blood coagulation studies have been very limited (Jara, 1957; Katz *et al.*, 1950; Langdell *et al.*, 1965; Macnab and Ronald, 1965; Nolf, 1906. Zunz (1933) has done preliminary tests and some studies with species-specific thromboplastins.

2. SOURCE

Table 6.1 lists the sources of the fish, their weights, and the number of each species studied. The personnel at the research laboratories (Nos. 1 and 2) were very helpful in supplying fish and helping with the phlebotomies. Arctic grayling (*Thymallus arcticus*) and trout (*Salvelinus namaycush*) were caught by two medical sportsmen who did the clotting times and processed the samples for return to Pittsburgh. The blood from the sturgeons (*Acipenser oxyrhynchus*) was obtained through the cooperation of the U.S. Fish and Wildlife Service. The fish were gill-netted on the Suwannee River in Florida. After routine measurements and venipuncture, the fish were returned to the river. The author wishes to thank the many individuals who helped with blood samples from the various Osteichthyes.

3. TESTS CONDUCTED

3.1. Bleeding Times

Bleeding times were done on many of the fish using a Simplate or by making a 5- to 6-mm shallow cut with a new rounded knife blade. For goosefish (*Lophius americanus/piscatorius*), four fish received two cuts each, and the mean bleeding time was 75 sec. One striped mullet (*Mugil cephalus*), one porgy (*Stenotomus chrysops*), one sea robin (*Prionotus carolinus*), and one tautog (*Tautoga onitus*) each received one Simplate bleeding time. Results were 60–90 sec. Five sculpin (*Myoxocephalus octodecemspinosus*) each received one cut, which bled 3¼, 0, 0, 0, and 3¾ min, respectively (mean: 1.4 min).

TABLE 6.1
Whole Blood or Plasma Clotting Times, Clot Retraction, and Clot Lysis in Osteichthyes[a]

	Source	Wt		Whole blood[b]						Plasma Recal T		
				Clot T (min)		Clot Retr		Clot Lys				
	No.[c]	(kg)	N	G	S	G	S	G	S	Clot T (sec)	MCA lysis	Lys
Amberjack	1	9	1	4	20	4	4	0	+	[d]	0	0
Arctic grayling	3	2	5	1	1	4	4	0	0	—	—	—
Arctic trout	3	8	4	7	27	4	4	1/3	0	—	—	—
Cobia	1	10	1	12	120	4	4	0	0	[d]	0	0
Flounder	2	1	3	6	60	2	3	0	0	[d]	0	0
Goosefish	2	2	4	44	120+	2	2	0	0	[d]	0	0
Grouper	1	4	2	4	240	4	4	0	0	[d]	0	0
Mullet	1	<1	6	5	10	3	3	0	0	—	—	—
Orange filefish	1	2	1	8	40	4	4	0	1	—	—	—
Porgy	1	<1	3	4	12	4	4	0	0	—	—	—
Sculpin	2	<1	4	6	22	4	4	0	0	75	0	0
Sea bass	1	2	3	6	30	3	3	0	0	45	0	0
Sea robin	1	1	4	2	2	2	2	0	0	—	—	—
Sturgeon	4	15	7	52	456	2	2	3/7	4/7	[d]	0	0
Tautog	1	1.5	2	30	120	4	4	0	0	43	0	0
Toadfish	1	1.5	1	30	240	4	4	0	0	[d]	0	0
Triggerfish	1	2	2	3	6	2	2	0	0	[d]	0	0

[a]The tests were conducted at ambient temperature.
[b]Tubes: (G) glass; (S) siliconized.
[c]Source: (1) Marineland Research Laboratory, St. Augustine, FL; (2) Mt. Desert Island Biological Laboratory, Salsbury Cove, ME; (3) Great Bear Lake, Canada (fresh water); (4) U.S. Fish and Wildlife Service, Gainesville, FL, Suwannee River (fresh water).
[d]More than 120 sec.

3.2. Coagulation Tests and Factors

Table 6.1 shows that the clotting times for Osteichthyes were variable. Goosefish, sturgeon, tautog, and toadfish (*Opsanus tau*) clotted very slowly in both glass and siliconized tubes. Most of the fish blood clotted in times close to the human range (see Table 2.1). A few clotted very rapidly, and this did not seem to relate to phlebotomy technique. For example, all samples from five Arctic grayling clotted very rapidly, while those from four Arctic trout, caught in the same cold (2°C) lake, clotted much more slowly. Most fish blood clots retracted well, and only a rare sample lysed. Recal Ts were done when the sample size was sufficient. Most of these plasma samples clotted slowly and did not lyse alone or in 1% MCA. The slow clotting may be related to removal of thrombocytes by centrifugation.

Sturgeon blood was unique among the fishes studied. Samples anticoagulated with the usual calcium chelators (citrate, oxalate, or EDTA) showed marked hemolysis. Samples collected into heparin did not hemolyze. The clear heparinized plasma was frozen and transported to Pittsburgh, where the heparin was then removed with Hepabsorb and citrate added.

Table 6.2 shows the Simplastin, homologous tissue, RVV, thrombin, and Atroxin times for the fish from which sufficient plasma was available. The mean

TABLE 6.2
Prothrombin and Thrombin Times in Osteichthyes

	Prothrombin time (sec)					Thrombin time (sec)	
		Homologous tissues[b]					
Fish	Simplastin	Brain	Gill	Skin	RVV	Bov Thr	Atroxin
Amberjack	60.8	—	—	—	—	83.2	53.0
Cobia	57.4	20.4	[a]	[a]	45.3	[a]	40.0
Flounder	87.0	68.8	14.8	18.8	37.4	[a]	21.0
Goosefish	89.0	16.5	22.0	13.3	[a]	59.4	[a]
Porgy	15.4	8.1	7.6	13.6	17.8	12.3	—
Sculpin	18.9	19.2	22.8	14.6	31.0	28.0	24.6
Sea bass	91.2	18.0	15.6	21.0	31.8	13.8	—
Sturgeon	[a]	—	—	—	[a]	[a]	[a]
Tautog	15.8	15.0	10.2	16.5	13.8	8.4	—
Toadfish	31.5	10.9	18.0	54.0	[a]	9.4	—
Triggerfish	[a]	13.4	18.0	14.5	—	[a]	—
Human (mean)	12.1	12.2	—	—	15.1	14.8	13.6

[a]More than 120 sec.
[b]Prothrombin times of human plasma plus homologous fish tissue = more than 120 seconds.

TABLE 6.3
Thromboplastin Generation Test in Osteichthyes

Generating mixture[a]					Substrate clotting time (sec)[a,b]				
H	C	Fl	OF	P	H	C	Fl	OF	P
H (mean of 4)	—	—	—	—	**9.9**	—	40+	40+	—
—	C	—	—	—	40+	**27.0**	—	—	—
—	—	Fl	—	—	26.2	—	**25.4**	—	—
—	—	—	OF	—	40+	—	—	**40+**	—
—	—	—	—	P	18.6	—	—	—	**18.0**

[a]Plasmas: (H) human; (C) cobia; (Fl) flounder; (OF) orange filefish; (P) porgy.
[b]Homologous systems are in **boldface**.

Simplastin time for human controls was 12.1 sec, and for the fish it varied from 15.4 to over 120 sec. Homologous teleost tissue also had quite variable thromboplastic effects. None was active on human plasma. Diverse effects were also found with RVV, thrombin, and Atroxin.

The TGT (Table 6.3) was done on cobia (*Rachycentron canadus*), flounder (*Paralychthys lethostigmus*), orange filefish (*Alutera schoepfii*), and porgy. Moderate activity was found in porgy and low activity in cobia and flounder.

Coagulation factors are shown in Table 6.4. Most of the fibrinogen (factor I) levels fell within the human range. Factors II, V, VII, and X were at very low levels, probably due to ineffective species reactivity of tissue factor and substrate. Both porgy and tautog showed moderate factors VIII, IX, and XI activities. Most of the fish showed traces of factors XII and XIII.

3.2.1. Effects of Heparin

To determine whether or not heparin would prolong fish clotting, one small triggerfish (*Balistes carolinensis*) (284 g) was given an injection of 25 U. Three postinjection samples showed marked prolongation of the clotting time (Table 6.5).

3.3. Fibrinolytic Enzyme System

Plasmas from goosefish Nos. 1, 2, and 3 were tested with streptokinase (500 U/ml) and urokinase (250 U/ml) and showed no fibrinolytic activity in the ^{125}I fibrin test system. Goosefish plasmas No. 1 and 3 inhibited the fibrinolytic activity of 1% trypsin to about the same extent that human plasma did.

TABLE 6.4
Coagulation Factors in Osteichthyes

Animal	I (mg/dl)	II (U/ml)	V (U/ml)	VII (U/ml)	X (U/ml)	VIII (U/ml)	IX (U/ml)	XI (U/ml)	XII (U/ml)	XIII (R)
Amberjack	230	0.02	0.02	[a]	[a]	0.05	0.05	0.30	0.20	4
Cobia	131	[a]	[a]	[a]	[a]	[a]	[a]	[a]	0.02	—
Flounder	188	[a]	[a]	[a]	[a]	[a]	[a]	—	0.90	4
Goosefish (mean of 4)	204	[a]	[a]	[a]	[a]	[a]	[a]	[a]	[a]	9
Grouper	66	[a]	[a]	[a]	[a]	[a]	[a]	—	0.02	—
Mullet	110	[a]	[a]	[a]	[a]	[a]	[a]	[a]	[a]	—
Orange filefish	225	[a]	[a]	[a]	[a]	[a]	[a]	—	0.02	2
Porgy	225	[a]	[a]	0.05	0.05	1.00	1.00	1.00	0.20	—
Sculpin (mean of 4)	135	0.05	0.34	[a]	0.04	0.02	0.02	0.01	0.03	4
Sea bass	—	[a]	[a]	[a]	[a]	[a]	[a]	[a]	0.05	2
Sea robin	—	[a]	[a]	[a]	—	0.05	0.05	—	—	—
Sturgeon (mean of 5)	212	[a]	0.04	0.03	[a]	[a]	[a]	[a]	[a]	—
Tautog	319	[a]	[a]	0.05	0.05	1.00	0.20	1.00	0.05	4
Toadfish	563	[a]	[a]	[a]	[a]	[a]	[a]	0.15	0.05	4
Triggerfish	385	[a]	[a]	[a]	[a]	[a]	[a]	[a]	[a]	4

[a]Less than 0.01 U/ml.

TABLE 6.5
Effects of Intramuscular Heparin on Clotting of Triggerfish Blood[a]

	Clott T (min)	Clot Retr
Preinjection	2½	4+
Postinjection		
2 hr	2880	Liquid
24 hr	360	4+
48 hr	15	5+

[a]The heparin dose was 25 U/284 g fish (88 U/kg). At 48 hr, the blood appeared anemic, and the fish was returned to the sea.

3.4. Miscellaneous Tests

The four goosefish plasmas clumped the standard staphylococci preparation in dilutions of 1:8 or 1:16. The human control was active to a dilution of 1:1024.

When goosefish plasma was heated to 56°C for 5 min, no precipitate was formed, indicating that goosefish fibrinogen did not have the same heat-precipitable characteristic that mammalian plasma had.

3.5. Biochemical Tests

Table 6.6 shows results of biochemical tests on amberjack (*Seriola dumerili*), goosefish, and triggerfish sera. The Ca, P, chloride, K, and Na were higher than human. Protein, albumin, uric acid, blood urea nitrogen (BUN), and CO_2 were lower.

3.6. Cellular Elements

Table 6.7 shows the parameters of the large nucleated RBCs found in the various fish. There was a marked variation from species to species. The red cell sizes varied from slightly larger than human (MCV = 102) to about three times that size (MCV = 263). WBC results are shown in Table 6.8. The counts of both WBCs and thrombocytes were also quite variable. In general, the lymphocytes predominated.

TABLE 6.6
Biochemical Tests in Osteichthyes

Substance	Amberjack[a] (1)	Goosefish[a] (4)	Triggerfish[a] (1)	Normal human range
TP (g/dl)	—	*2.9*	*5.6*	6.0–8.0
Albumin (g/dl)	*2.3*	*1.2*	*1.4*	3.5–5.0
Ca (mg/dl)	**>15.0**	10.5	**13.4**	8.5–10.5
P (mg/dl)	**>10.0**	**9.3**	**8.6**	2.5–4.5
Cholesterol (mg/dl)	**325**	*81*	*118*	150–200
Glucose (mg/dl)	**137**	*46*	*25*	65–110
Uric acid (mg/dl)	*<2.0*	*1.0*	*<2.0*	2.5–8.0
Creatinine (mg/dl)	*0.1*	**2.9**	*0.1*	0.7–1.4
T bilirubin (mg/dl)	0.2	1.1	0.3	0.1–1.4
BUN (mg/dl)	*6.5*	*1.3*	*8.0*	10–20
Chloride (meq/liter)	**182**	**176**	**127**	95–105
CO_2 (meq/liter)	*8*	*6.6*	*7*	24–32
K (meq/liter)	5	**5.8**	**10**	3.5–5.0
Na (meq/liter)	**206**	**189**	**191**	135–145
Alk Phos (IU/liter)	37	11	77	30–85
SGOT (IU/liter)	**92**	14	8	0–41

[a] Values **higher** than human are in **boldface**, those *lower* than human in *italics*.

TABLE 6.7
Red Blood Cell Elements of the Peripheral Blood in Osteichthyes

Fish	*N*	HCT (%)	HGB (g/dl)	RBC ($\times 10^3/mm^3$)	MCV (fl)	MCH (pg)	MCHC (g/dl)
Amberjack	1	48	12.0	1.82	263	66	25
Cobia	1	35	—	1.35	259	—	—
Flounder	4	24	8.3	2.00	120	41	35
Goosefish	4	21	4.5	0.96	215	47	22
Grouper	2	24	—	2.10	114	—	—
Orange filefish	1	35	4.4	1.93	181	23	13
Porgy	3	37	6.1	1.76	210	35	16
Red snapper	2	22	3.4	1.04	211	33	15
Sculpin	4	36	5.8	2.00	—	—	—
Sea bass	1	33	5.9	1.90	173	31	18
Sturgeon	4	27	10.3	1.31	203	78	39
Tautog	2	49	9.5	4.80	102	20	19
Triggerfish	2	47	4.3	3.69	127	12	9

TABLE 6.8
White Blood Cell Elements of the Blood in Osteichthyes

Fish	*N*	WBC (× 10^3/mm^3)	Hetero (%)	Lymph (%)	Mono (%)	Eos (%)	Bas (%)	Thromb (× 10^3/mm^3)
Amberjack	1	23	11	69	20	0	0	43
Arctic grayling	5	—	28	48	20	4	0	—
Arctic trout	4	—	28	62	10	0	0	—
Flounder	4	18	46	46	8	0	0	38
Goosefish	4	10	12	27	60	0	1	41
Orange filefish	1	54	20	77	30	0	0	21
Porgy	3	14	50	20	29	0	1	55
Red snapper	2	20	12	69	18	1	0	12
Sculpin	4	87	84	8	8	0	0	119
Sea bass	1	14	34	54	12	0	0	67
Sea robin	4	10	53	29	23	0	0	10
Tautog	2	19	56	38	6	6	0	24
Toadfish	1	83	24	69	6	1	0	90
Triggerfish	2	122	21	66	13	0	0	18

3.7. Thrombocyte Aggregation

Thrombocyte-rich plasma (TRP) was prepared from amberjack and triggerfish blood by very slow-speed centrifugation. The TRPs together with human platelet-rich plasma (PRP) were tested for gross and microscopic aggregation by adding 0.05 ml reagent to 0.4 ml plasma and gently rocking for 5 min and then examining. When the reagent was ADP (10 μM), bovine collagen (0.2 mg/ml), or ristocetin (9 mg/dl), human PRP formed heavy clumps in 10–24 sec. With triggerfish collagen (50 mg chopped tendon/5 ml), human PRP clumped in 180 sec. Amberjack and triggerfish TRPs did not clump grossly or microscopically with any reagent.

3.8. Thrombocyte Ultrastructure

TEMS of the thrombocytes of sculpin (Fig. 6.1) and sturgeon (Fig. 6.2) were very similar. Each showed a large, slightly spindle-shaped cell with a central nucleus in which the chromatin was heavily distributed along the periphery of the nucleus. An open canalicular system, some microtubules, and an occasional granule were apparent.

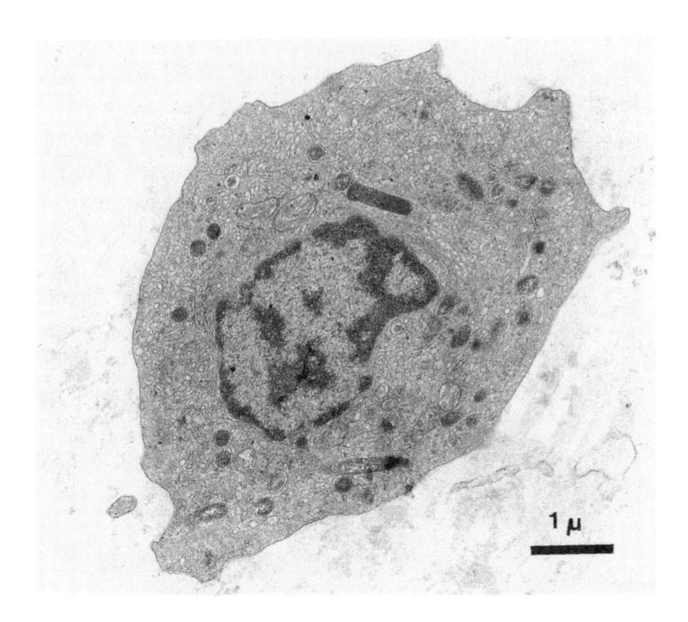

FIGURE 6.1. TEM of a sculpin thrombocyte.

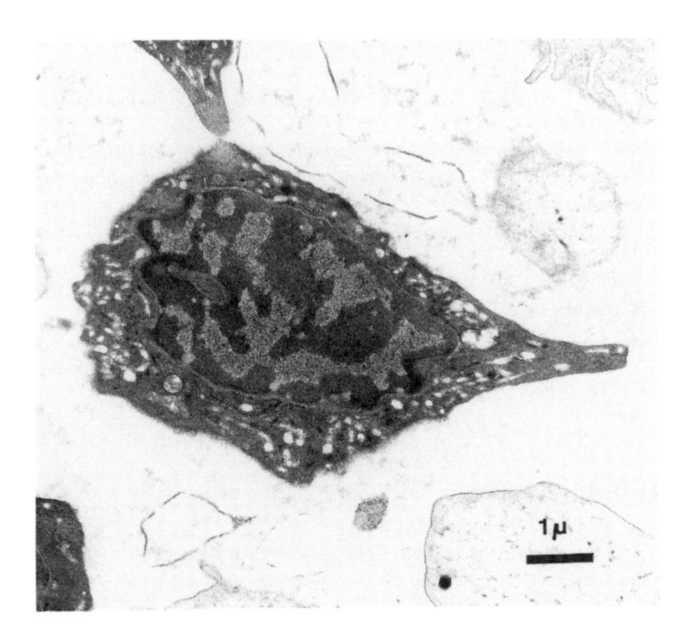

FIGURE 6.2. TEM of a sturgeon thrombocyte.

4. SUMMARY

The findings in the sturgeons and the teleosts showed many variations from species to species. The very primitive sturgeon, covered with protective bony plates or scutes, is claimed to be the world's largest and longest-lived fish. It was the largest fish we studied and the only nonteleost. Fish clotting times varied from very long to very short among the diverse species. Clot retraction was usually good, and lysis was rare. A number of fish tissues were tested, and their thromboplastic effects showed no uniformity. Fish coagulation factors, assayed in human assay systems, were often present in trace or very low amounts. RBC and WBC parameters were also quite variable. Thrombocytes of amberjack, goosefish, and triggerfish did not aggregate with common human aggregating agents. The thrombocytes of sculpin and sturgeon were nucleated cells with dense chromatin at the periphery of the nucleus, an open canalicular system, and some microtubules and granules.

REFERENCES

Blaxhall, P. C., and Daisley, K. W., 1973, Routine haematological methods for use with fish blood, *J. Fish Biol.* **5**:771.

Eisler, R., 1965, Erythrocyte counts and emoglobin content in nine species of marine teleosts, *Chesapeake Sci.* **6**:119.

Hall, F. G., and Gray, I. E., 1929, The haemoglobin concentration of the blood of marine fishes, *J. Biol. Chem.* **81**:589.

Jara, Z., 1957, The blood coagulation system in carp (*Cyprinus carpio* L.), *Zool. Pol.* **8**:113.

Katz, M., and Southward, M., 1950, Blood clotting time in spent silver salmon, *Oncarhyncus kisutch* (Walbaum), *Copeia* **2**:150.

Langdell, R. D., Bryan, F. T., and Gibson, W. S., Jr., 1965, Coagulation of catfish blood, *Proc. Soc. Exp. Biol. Med.* **118**:439.

Macnab, H. C., and Ronald, K., 1965, Blood clotting time of the Atlantic cod, *Gadus morhua, J. Fish. Res. Board Can.* **22**:1299.

Nolf, P., 1906, La coagulation du sang des poissons, *Arch Int. Physiol.* **4**:216.

Zunz, E., 1933, Contribution a l'étude de la coagulation du sang chez les poissons, *Arch. Int. Physiol.* **37**:282.

SUGGESTED READINGS

Cannon, M. S., Mollenhauer, M. H., Eurell, T. E., Lewis, D. H., Cannon, A. M., and Tompkins, C., 1980, An ultrastructural study of the leukocytes of the channel catfish, *Ictalurus punctatus, J. Morphol.* **164**:1.

Catton, W. T., 1951, Blood cell formation in certain teleost fishes, *Blood* **6**:39.

Conroy, D. A., and Rodriguez, L., 1966, Erythrocyte measurements of some Argentine fishes, *Prog. Fish Cult.* **28**:46.

Doolittle, R. F., and Surgenor, D. M., 1962, Blood coagulation in fish, *Am. J. Physiol.* **203:**964.

Engle, R. L., Jr., Woods, K. R., Paulsen, E. C., and Pert, J. H., 1958, Plasma cells and serum proteins in marine fish, *Proc. Soc. Exp. Biol. Med.* **98:**905.

Field, J. B., Elvehjem, C. A., and Juday, C., 1943, A study of blood constituents of carp and trout, *J. Biol. Chem.* **148:**261.

Hougie, C., 1972, Coagulation changes in healthy and sick Pacific salmon, in: *Advances in Experimental Medicine and Biology,* Vol. 22, *Comparative Pathophysiology of Circulatory Disturbances* (L. M. Bloor, ed.), p. 89, Plenum Press, New York.

Kaname, S., 1954, Biochemical studies on fish blood. I. On elements of blood coagulation and coagulation time, *Bull. Jpn. Soc. Fish* **19:**1139.

Kisch, B., 1949, Hemoglobin content, size, and amount of erythrocytes in fishes, *Exp. Med. Surg.* **7:**118.

Lepkovsky, S., 1930, Distribution of serum and plasma proteins in fish, *J. Biol. Chem.* **85:**667.

Murtaugh, P. A., Halver, J. E., and Gladner, J. A., 1973, Cross-linking of salmon fibrinogen and fibrin by factor XIII and transglutaminase, *Biochem. Biophys. Res. Commun.* **54:**849.

Pilos, C., and Guise, J. V., 1962, A contribution to the investigation on coagulation of blood of carp, *Acta Hydrobiol.* **4:**413.

Ronald, K., 1965, Blood clotting time of the Atlantic cod *Gadus morhua*, *J. Fish. Res. Board Can.* **22:**1299.

Saunders, D. C., 1968, Variations in thrombocytes and small lymphocytes found in the circulating blood of marine fishes, *Trans. Am. Microscoc. Soc.* **87:**39.

Smith, A. C., 1980, Formation of lethal blood clots in fishes, *J. Fish Biol.* **16:**1.

Smith, H. W., 1929, The composition of the body fluids of goosefish (*Lophius piscatorius*), *J. Biol. Chem.* **82:**71.

Srivastava, A. K., 1969, Studies of the hematology of certain fresh water teleosts. 5. Thrombocytes and the clotting of blood, *Anat. Anz.* **124:**368.

Weinreb, E. L., 1963, Studies on the fine structure of teleost blood cells. I. Peripheral blood, *Anat. Rec.* **147:**219.

Wolf, K., 1959, Plasmaptysis and gelation of erythrocytes in coagulation of blood of freshwater bony fishes, *Blood* **14:**1339.

CHAPTER 7

THE AMPHIBIANS
Class: Amphibia

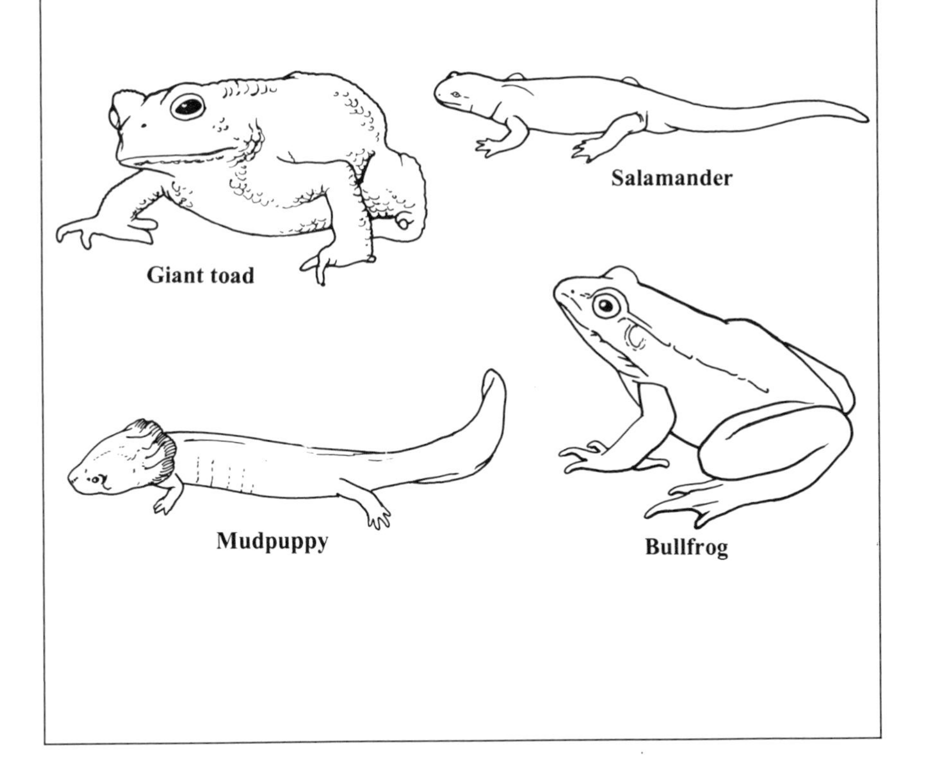

1. INTRODUCTION

Amphibia include two major groups: those with tails (subclass Urodela or Caudata) and those without (subclass Anura or Salientia). The name Amphibia derives from the Greek words *amphi* ("both") and *bios* ("mode of life"). The name is appropriate to most Amphibia, which live partly in water and partly on land. They probably differentiated from early fish during the period of great species expansion that occurred about 300 MYA. They are poikilothermic, with body temperatures dependent on the temperature of the environment. Amphibia are intermediate between fishes and reptiles, with many characteristics of each. Respiration is by gills, lungs, or skin, separately or in combination. Gills are present in the early stages of life or throughout life. All Amphibia reproduce in the water. Fertilization is internal or external. Most Amphibia are oviparous, laying large numbers of eggs enclosed in gelatinous material that sticks the eggs together, forming a handball size mass that can often be seen floating in a pond in the early spring.

Amphibians live near or in water but rarely in salt water. The Congo eel (*Amphiuma means*), hellbender (*Cryptobranchus alleganiensis*), and mud puppy (*Necturus maculosus*) live entirely in water and have external gills. The salamander used in this study was *Ambystoma Tigrinum,* which is terrestrial and in adult life has lungs. Anurans in the early tadpole stages have tails, gills, and so forth and go through a complex metamorphosis to become frogs and toads. Adult anurans have lungs but live in damp places or in the water, hibernating deep in lakes or burrowing in pondside mud in the winter. Most adult Amphibia eat only live, moving creatures, such as insects, worms, small mollusks, or fish.

Scientific literature dealing with the clotting of amphibian blood or aggregation of thrombocytes is uncommon. Early studies by Hackett and LePage (1961a,b) and by Rex and Freytag (1964) noted the long clotting times and species specificity of tissue factors. Hackett and Hann (1964), Blofield (1965), and Srivastava *et al.* (1981) described the frequency of spontaneous fibrinolysis. The relationship of thrombocyte appearance and aggregation with clotting was suggested by Tait and Green (1926) and investigated further by Stiller *et al.* (1974).

2. SOURCE

Most of the animals were ordered from a laboratory animal supplier (Carolina Biological Supply Company or W. A. Lemberger Company). Cryptobranchus specimens were caught in tributaries of the Allegheny River. A large new plastic laundry cart with bottom drain was converted to a tank after many rinses with distilled water. A mesh cover was constructed. Pond water and a few large rocks were collected and put in the tanks just before the animals were put in. Most of the Amphibia would not eat dried bugs. When live insects or worms could not be obtained, the blood collection commenced as soon as the animals were rehydrated. The anurans were anesthetized, usually with ether, in closed plastic containers.

Blood from anurans was obtained by rapid, surgical exposure of the heart and cardiac puncture or vena cava puncture with a No. 22 needle and small syringe containing anticoagulant or, for clotting times, a dry syringe. Blood from urodeles was obtained by external puncture into the ventral postcaval vein.

3. TESTS CONDUCTED

3.1. Bleeding Times

The skin on most of the Amphibia was very loose, and no subcutaneous tissue was apparent. The bleeding times on amphiuma, necturus, grass frog (*Rana pipiens*), and bullfrog (*Rana catesbeiana*) were done by pulling up a piece of the loose skin and making a small cut (6 mm) with very sharp scissors. These cuts bled for 30 sec in two amphiuma, not at all in three necturus, 0–1 min (mean: 30 sec) in four frogs, and 0–2 min (mean: 90 sec) in six bullfrogs. In four giant toads (*Bufo marinus*), two knife cuts were made on opposite legs. These bled $^1/_2$, $^1/_2$, $2^1/_2$, $2^1/_2$, 1, and 1 min (mean: 1.4 min). The bleeding cuts were so large in proportion to the animal size that the same animals were not used for coagulation tests or blood counts. However, their blood was used for electron micrographs, serum proteins, or cell measurements.

3.2. General Coagulation Tests and Factors

The striking observation in the clotting tests was the frequency of whole blood or Recal T clot lysis (Clot Lys). Lysis was positive in amphiuma, cryptobranchus, frog, bullfrog, and giant toad. Necturus and salamander whole blood clots did not lyse, but did retract well. One in three necturus Recal clots did lyse. Also, clots from the common toads (*Bufo terrestris*) did not lyse. Seasonal, nutritional, or other unknown factors may have played a role in the presence or absence of clot lysis.

Table 7.1 shows that the urodeles had short clotting times and the anurans long clotting times. Prothrombin times with Simplastin were moderately long

TABLE 7.1
General Coagulation Tests in Amphibia[a]

Test		A	C	N	S	F	BF	T	GT	Normal human range
	N:	3	2–4	3	1	1	6–14	3	4, 5	
	Wt. (kg):	2	1	<1	<1	<1	<1	<1	<1	
Clot T (min)										
Glass		4	6	5	3	32	19	20	14	6–12
Silicone		9	18	7	3	94	96	26	120+	20–59
Clot Retr		[b]	2–3	4	4	2–4	[b]	2–3	[b]	3–4+
Clot Lys		+	1/4[e]	0	0	+	+	0	4/5[e]	0
PT (sec)										
Simplastin		29.0	38.0	17.8	37.8	[c]	[c]	24.8	24.6	10–13
Homologous										
Brain		26.5	28.2	21.0	51.8	15.7	42.9	35.5	32.1	—
Gill/lung		17.4	15.2	10.6	31.6	17.3	17.4	11.3	15.4	—
Skin		16.3	21.6	9.8	23.8	12.9	12.8	14.2	15.6	—
RVV		15.4	14.0	14.1	13.4	18.0	16.4	12.1	33.4	13–18
APTT (sec)		87.0	[c]	[c]	68.0	[c]	66.6	[c]	[c]	45–55
Recal T (sec)		117	103	120	72	185	146	53	128	90–180
Lysis		+	1/3[e]	1/3[e]	0	2/5[e]	+	0	4/5[e]	0
Lys MCA		?[d]	0	0	0	0	?[d]	0	0	0
Thromb T (sec)										
Bov Thr		28.2	63.8	37.3	—	53.0	[c]	55.0	[c]	11–18
Atroxin		—	—	—	—	105	[c]	—	[c]	11–18

[a]The amphibian tests were conducted at a temperature of 22–25°C. Animals: (A) *Amphiuma means* (Congo eel); (C) *Cryptobranchus alleganiensis* (hellbender); (N) *Necturus maculosus* (mud puppy); (S) *Ambystoma tigirinum* (salamander); (F) *Rana pipiens* (grass frog); (BF) *Rana catesbeiana* (bullfrog); (T) *Bufo terrestris* (toad); (GT) *Bufo marinus* (giant toad). Human tested at 37°C.
[b]The clot lysed.
[c]The time was more than 120 sec or the concentration less than 0.01 U/ml.
[d]The clot lysed before the 1-hr observation period for MCA lysis.
[e]Indicates number lysed/number observed.

The animal tissues had no clot-accelerating activity on human plasma, but did on homologous animal plasma. Skin was usually the most potent. RVV was also an effective clotting accelerator in the amphibians. APTTs and Recal T were similar to those of human plasma. Thrombin times and Atroxin times (done on frog, bullfrog, and giant toad) were long.

Table 7.2 shows the lack of thromboplastin generating ability in the amphibian mixtures. It should also be noted that the human generating mixtures, which clotted human plasma in about 10.5 sec, did not clot the amphibian plasmas.

Coagulation factor assays are shown in Table 7.3. Fibrinogen (factor I) assayed at normal human levels, although the consistency of the formed fibrin was more gelatinous. Other coagulation factors (II–XIII) were found in very low to trace amounts. Only factor XII in the toad was found in the low human range (0.5 U/ml).

Table 7.4 shows that the presence of buffy coat (containing thrombocytes) shortens the recalcification but does not cause rapid clot retraction. Recal clot retraction was not observed in bullfrog clots because they lysed.

TABLE 7.2
Thromboplastin Generation Test in Amphibia

Generating mixture					Substrate clotting time (sec)[a]				
Urodela[b]									
H (mean)	A	C	N	S	H	A	C	N	S
H (4)	—	—	—	—	**10.5**	40+	40+	34.6	32.4
—	A	—	—	—	40+	**37.0**	—	—	—
—	—	C	—	—	40+	—	**40+**	—	—
—	—	—	N	—	40+	—	—	**24.6**	—
—	—	—	—	S	40+	—	—	—	**40+**
Anura[b]									
H (mean)	F	BF	T	GT	H	F	BF	T	GT
H (4)	—	—	—	—	**10.3**	40+	40+	40+	40+
—	F	—	—	—	40+	**40+**	—	—	—
—	—	BF	—	—	40+	—	**40+**	—	—
—	—	—	T	—	40+	—	—	**40+**	—
—	—	—	—	GT	40+	—	—	—	**40+**

[a]Homologous systems are in **boldface**.
[b]Plasmas. (H) human; animals: See footnote *a* in Table 7.1.

TABLE 7.3
Coagulation Factors in Amphibia[a]

Factor	A (3)	C (4)	N (3, 4)	S (1)	F (5)	BF (7–14)	T (2, 3)	GT (4)	Normal human range
I (mg/dl)	213	257	284	363	337	293	152	179	150–450
II (U/ml)	0.03	0.06	0.06	0.11	0.04	0.02	0.17	0.06	0.70–1.30
V (U/ml)	0.02	0.06	0.13	0.20	0.04	0.11	0.03	0.18	0.65–1.45
VII (U/ml)	[b]	[b]	0.02	[b]	[b]	0.04	[b]	[b]	0.50–1.30
X (U/ml)	[b]	[b]	[b]	[b]	[b]	0.02	[b]	0.03	0.75–1.25
VIII (U/ml)	0.04	0.04	0.01	[b]	0.04	[b]	0.04	[b]	0.75–1.40
IX (U/ml)	0.18	[b]	0.02	[b]	0.16	[b]	0.02	[b]	0.65–1.70
XI (U/ml)	0.27	0.48	0.02	[b]	[b]	[b]	0.02	[b]	0.75–1.40
XII (U/ml)	0.15	0.19	0.20	[b]	0.18	0.04	0.5	[b]	0.50–1.45
XIII (R)	2	2	10	2	2	12	2	2	4–16

[a]Animals: See footnote *a* in Table 7.1. [b]Less than 0.01 U/ml.

TABLE 7.4
Effect of Buffy Coat on Recalcification Time in Bullfrog No. 5[a]

Procedure	Clot T (sec)	Clot Retr (4 hr)	Clot Lys (24 hr)
Plasma			
1,500g 10 min, decanted	288	0	+
20,000g 20 min, decanted	1560	0	+
20,000g 20 min, remixed	151	0	+

[a]Test system: 0.2 ml plasma + 0.025 M $CaCl_2$.

3.3. Serum Proteins

Table 7.5 shows total proteins (TPs) and paper electrophoretic distribution. The TPs and albumins are all very low. For more details, see Chapter 29.

3.4. Cellular Elements

Table 7.6 shows the RBC, WBC, and thrombocyte counts of the Amphibia. RBCs of the urodeles were huge, amphiuma and necturus being the largest, with MCVs over 7000. The anuran RBCs were about 1/10th the size of those of urodeles, but still very large. The MCHs were very high, but the MCHCs were very much like those of mammals and all other vertebrates. WBC and thrombocyte counts were low and the cells were large.

TABLE 7.5
Serum Proteins by Paper Electrophoresis in Amphibia[a]

Protein	A (2)	N (1)	S (1)	F (2)	BF (4)	T (2)	GT (6)	Normal human range
TP (g/dl)	2.6	1.9	1.81	3.6	3.9	2.8	2.8	6.3–7.9
Albumin (g/dl)	0.6	0.4	0.5	1.0	1.1	0.6	0.9	3.2–4.4
α_1-Globulin (g/dl)	1.2	0.4	0.7	1.6	0.4	0.3	0.3	0.2–0.4
α_2-Globulin (g/dl)	[b]	0.4	[b]	[b]	1.5	1.4	0.6	0.4–1.0
β-Globulin (g/dl)	0.7	0.4	0.5	0.5	0.5	1.4	0.5	0.5–1.0
γ-Globulin (g/dl)	0.2	0.3	0.1	0.2	0.4	0.5	0.5	0.6–1.8

[a]Animals: See footnote *a* in Table 7.1.
[b]α_1-Globulin and α_2-globulin were assayed together.

TABLE 7.6
Cellular Elements of the Blood in Amphibia[a]

Element	A (4)	C (3)	N (5)	S (2)	F (7)	BF (13)	T (3)	GT (6)	Normal human range
HCT (%)	41	34	27	33	31	30	24	39	37–52
HGB (g/dl)	14.8	8.6	8.2	8.8	7.0	9.2	6.8	12.2	12–18
RBC (× 10^6/mm^3)	0.056	0.091	0.038	0.124	0.50	0.44	0.33	0.81	4.2–6.2
MCV (fl)	7391	3938	7325	2698	654	714	755	504	79–97
MCH (pg)	2653	889	2193	764	146	221	230	159	27–31
MCHC (g/dl)	36	26	31	26	22	31	30	31	32–36
WBC (× 10^3/mm^3)	1.2	—	1.1	—	15.2	18.8	—	—	5–10
Hetero (%)	63	—	49	—	57	38	4	—	55–75
Lymph (%)	31	—	36	—	27	56	56	—	20–40
Mono (%)	0	—	0	—	5	2	6	—	2–10
Eos (%)	3	—	10	—	7	3	0	—	0–3
Bas (%)	3	—	5	—	4	1	34	—	0–3
Thromb (× 10^3/mm^3)	6.6	—	6.3	—	77	41	—	—	150–450

[a]Animals: See footnote *a* in Table 7.1.

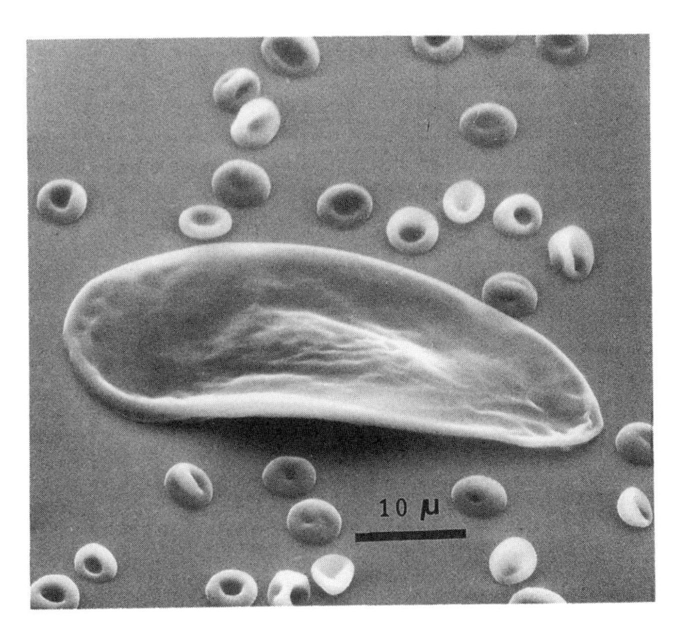

FIGURE 7.1. SEM of amphiuma and human erythrocytes.

Figure 7.1 shows a scanning electron micrograph (SEM) of an amphiuma RBC mixed with doughnut-shaped relatively small human RBCs. The enormous size of the amphiuma cells is easy to appreciate. Human RBCs appear to be about 5–6 μm in diameter and the amphiuma cells about 50 × 15 μm. The central bump is the layer of cytoplasm over the nucleus. The entire surface appears uneven and pockmarked. Figure 7.2 is an SEM of three bullfrog erythrocytes mixed with human RBCs. The central nuclear mounds are prominent, as are the rounded, rolled cytoplasmic edges. Figure 7.3 is an SEM of frog erythrocytes with clumped, clotted thrombocytes taken from a buffy coat preparation that stood in a glass tube without anticoagulant for 30 min. This figure shows thrombocyte processes that appear to be connected to some early fibrin. Figure 7.4 shows a single floating frog erythrocyte, our flying saucer.

3.5. Thrombocyte Aggregation

Thrombocyte aggregation (Table 7.7) was studied by mixing slow-centrifuged giant toad or human plasmas with the aggregating agent on a siliconized slide, gently swirling the mixture about every minute, and then observing for 10 min at low power under a phase-contrast microscope. The control

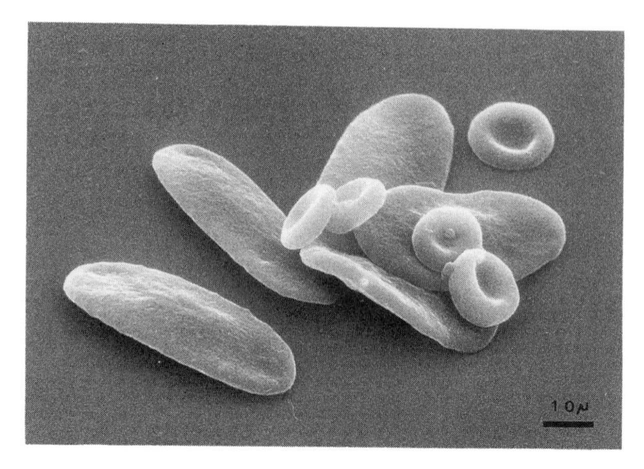

FIGURE 7.2. SEM of bullfrog and human erythrocytes. Note that the human RBCs are much larger than in Fig. 7.1 (the magnification is almost twice as great). The frog cell is smaller than that of amphiuma. ×3425.

FIGURE 7.3. SEM of bullfrog blood-clotting. There is a clump of thrombocytes loosely attached to an erythrocyte. ×3550.

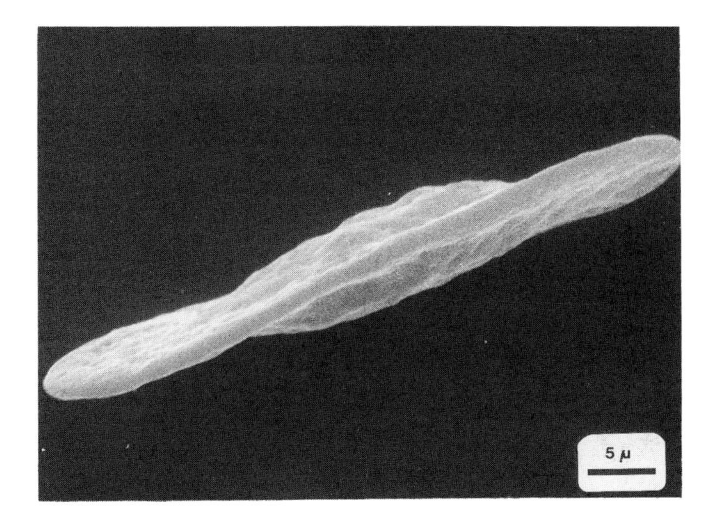

FIGURE 7.4. SEM of a bullfrog erythrocyte in an edgewise view (a "flying saucer"). Note the nuclear swelling, rough surface, and rolled edge. ×7520.

(buffer) showed individual cells of various types in toad plasma or platelets in human platelet-rich plasma (PRP) and did not change in the 10-min observation period. In the human PRP, aggregates were seen with ADP, bovine collagen, and ristocetin (Risto) within seconds. The thrombin preparations clotted within 1 min. Toad collagen caused very little aggregation in human PRP. Toad collagen aggregated thrombocytes of toad No. 2, but not those of toad No. 5. Trace (+)

TABLE 7.7
Aggregation of Human Platelets and Giant Toad Thrombocytes by Low-Power Phase-Contrast Microscopy

	Aggregation		
		Toad	
Reagent	Human control	No. 2	No. 5
ADP (20 μM)	++++	0	+
Collagen (0.5 mg/ml)	+++	+	+
Toad tendon	+	+++	0
Risto (0.9 mg/ml)	++++	+	+
Bov Thr (0.5 U/ml)	+++ → Clot	+	+
(0.75 U/ml)	Clot	0	Clump
Buffer	0	0	0

aggregation was seen with bovine collagen, Risto, and bovine thrombin in both toad thrombocyte-rich plasmas. ADP caused trace aggregation in toad No. 5.

3.6. Thrombocyte Ultrastructure

Figures 7.5 and 7.6 are TEMS of sectioned bullfrog thrombocytes. The nuclei were large, with dense peripheral chromatin. The cytoplasm contained many vacuoles that did not seem to be connected but could be a convoluted open canalicular system. The lower cell in Fig. 7.5 showed a nondescript inclusion body that probably was a phagocytosed RBC.

4. SUMMARY

Amphibian blood clotted slowly, and the clots lysed frequently. Tissues and RVV accelerated clotting slightly. Coagulation factors, except for fibrinogen, were in the low–trace human range. Fibrinogen was present at human levels, but its consistency was more gelatinous. RBCs were very large; the largest were found in amphiuma and necturus. These were almost 100 times the size of a

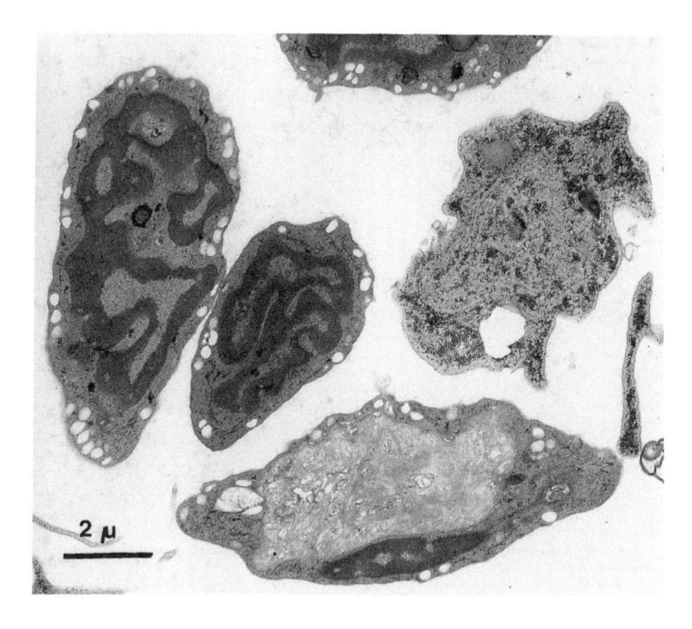

FIGURE 7.5. TEM of bullfrog thrombocytes. Note that the cell at the bottom of the photo may contain a phagocytosed RBC.

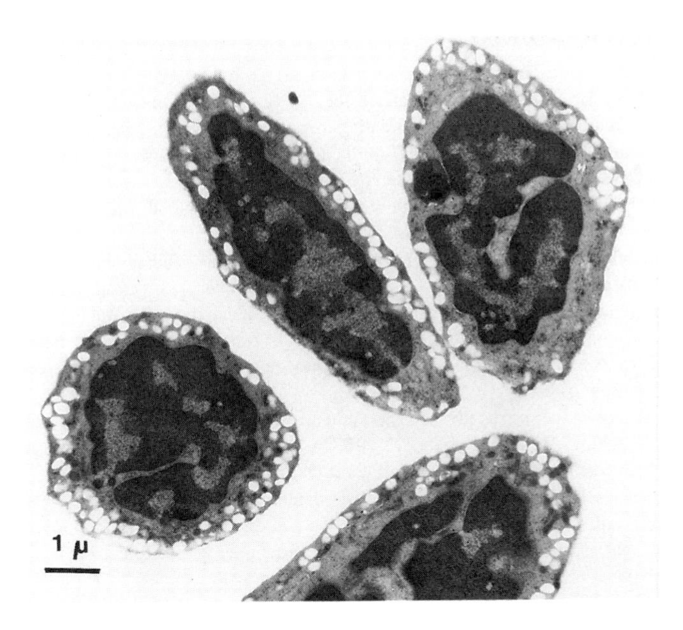

FIGURE 7.6. TEM of bullfrog thrombocytes.

human RBC. Thrombocytes appeared similar to those of other premammalian vertebrates.

REFERENCES

Blofield, A., 1965, A spontaneously active fibrinolytic system in *Xenopus laevis* which is further activated by human urokinase, *Nature (London)* **206:**736.

Hackett, E., and Hann, C., 1964, Erythrocytes and the liquefying of clotted amphibian blood in vitro, *Nature (London)* **204:**590.

Hackett, E., and LePage, R., 1961a, The clotting of the blood of an amphibian, *Bufo marinus* Linn. 1. Prothrombin–thrombin and "fibrinogen–fibrin" stages, *Aust. J. Exp. Biol.* **39:**57.

Hackett, E., and LePage, R., 1961b, The clotting of the blood of an amphibian, *Bufo marinus* Linn. 2. Blood thromboplastic activity, *Aust. J. Exp. Biol.* **39:**67.

Rex, J. O., and Freytag, G. E., 1964, Comparative investigation of the blood coagulation of amphibians, *Acta Biol. Med. Ger.* **13:**168.

Srivastava, V. M., Dube, B., Dube, R. K., Agarwad, G. P., and Ahmad, N., 1981, Blood fibrinolytic system in *Rana tigrina, Thromb. Diath. Haemorrh.* **45:**252.

Stiller, R. A., Belamirich, F. A., and Shepro, D., 1974, Frog thrombocytes: Aggregation and the release reaction, *Thromb. Diath. Haemorrh.* **32:**685.

Tait, J., and Green, F., 1926, The spindle-cells in relation to coagulation of frog's blood, *Q. J. Exp. Physiol.* **16:**141.

SUGGESTED READINGS

Ahmad, N., Dube, B., Agarwal, G. P., and Dube, R. K., 1979, Comparative studies of blood coagulation in hibernating and non-hibernating frogs (*Rana tigrina*), *Thromb. Haemost.* **42:**959.

Anstall, H. B., and Huntsman, R. G., 1960, Influence of temperature upon blood coagulation in a cold- and a warm-blooded animal, *Nature (London)* **166:**726.

Baitsell, G. A., 1917, A study of the clotting of the plasma of frog's blood and the transformation of the clot into a fibrous tissue, *Am. Jour. Physiol.* **44:**109.

Bertolini, B., and Monaco, G., 1976, The microtubule marginal band of the newt erythrocyte: Observations on the isolated band, *J. Ultrastruct. Res.* **54:**59.

Broyles, R. H., Dorn, A. R., Maples, P. B., Johnson, G. M., Kindell, G. R., and Parkinson, A. M., 1981, Choice of hemoglobin type in erythroid cells of *Rana catesbeiana,* in: *Hemoglobins in Development and Differentiation* (G. Stamatoyannopoulos, and A. W. Nienhius, eds.), p. 179, A. R. Liss, New York.

Campbell, F. R., 1970, Ultrastructure of the bone marrow of the frog, *Am. J. Anat.* **129:**329.

Daimon, T., Mizuhira, V., and Uchida, K., 1979, Fine structural distribution of the surface-connected canalicular system in frog thrombocytes, *Cell Tissue Res.* **201:**431.

Dent, J. H., and Schuellein, R. J., 1930, A consideration of the prothrombin times of several amphibians with notes on effects of parasitization and disease, *Physiol. Zool.* **23:**23.

Fey, F., 1966, Vergleichende Hämozytologie niederer Vertebräten. II. Thrombozyten, *Folia Haematol.* **85:**205.

Foxon, G. E. H., 1964, Blood and respiration in: *Physiology of the Amphibia* (J. A. Moore, ed.), p. 151, Academic Press, London.

Gorzula, A., and Arocha-Pinango, C. L., 1975, A coagulation study of *Amphisbaena alba linnaeus, Br. J. Herpetol.* **5:**629.

Gouchi, H., 1982, Ultrastructure of eosinophil granules of bullfrogs, *Rana catesbeiana, J. Med. Soc. Toho Univ.* **29:**9.

Harris, J. A., 1972, Seasonal variation in some hematological characteristics of *Rana pipiens, Comp. Biochem. Physiol. A* **43:**975.

Jordan, R. E., 1983, Antithrombin in vertebrate species: Conservation of the heparine-dependent anticoagulant mechanism, *Arch. Biochem. Biophys.* **227:**587.

Kase, F., 1978, Letter to the editor: Thrombocytes or spindle cells?, *Thromb. Haemost.* **39:**775.

Kuramoto, M., 1981, Relationships between number, size and shape of red blood cells in amphibians, *Comp. Biochem. Physiol. A* **69:**771.

Pringle, H., and Tait, J., 1910, Natural arrest of hemorrhage in the tadpole, *J. Physiol.* **40:**lvi.

Setoguti, T., Fujii, H., and Isono, H., 1970, An electron microscopic study on neutrophil leukocytes of the toad, *Bufo vulgaris japonicus, Arch. Histol. Jpn.* **32:**87.

Sinclair, G. D., and Brasch, K., 1975, The nucleated erythrocyte: A model of cell differentiation, *Rev. Can. Biol.* **34:**287.

Spitzer, J. J., and Spitzer, J. A., 1952, The blood coagulation mechanism of frogs, with respect to the species specificity of thromboplastin, to intracardial thrombin injection, and to the effect of seasonal changes, *Can. J. Med. Sci.* **30:**420.

Tait, J., and Burke, H. E., 1926, Platelets and blood coagulation, *Q. J. Exp. Physiol.* **16:**129.

CHAPTER 8

THE REPTILES
Class: Reptilia

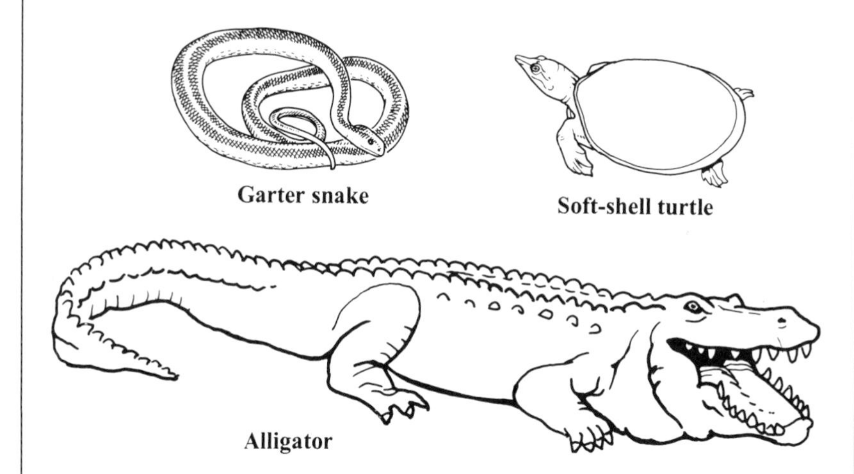
Garter snake

Soft-shell turtle

Alligator

1. INTRODUCTION

Reptilia is the first class of vertebrates completely adapted to life on land. They have dry, cornified skin, often scaly or horny, fore and hindlimbs (lost in snakes), a bony skeleton, and lungs. In general, reptiles are omnivorous, eating whatever is available. Some eat only insects or small animals. Reptiles may live for many years. Most lay eggs, but a few produce living offspring. Nurturing is not a reptilian characteristic, and the young are "on their own" and are often prey for other animals. Some ancient reptiles developed into dinosaurs, which flourished from 200 to 65 MYA. The cause of their slow extinction is still controversial.

Hematological studies on reptiles are limited. Early observers measured the sizes of erythrocytes and their nuclei. Coagulation studies are few. Absence of a number of coagulation factor activities has been described by Fantl (1961), Denson (1976), and others. Hackett and Hann (1967) and Nahas *et al.* (1973) have described potent inhibitors. Ratnoff *et al.* (1990), studying the Burmese python, suggest the existence of analogues of the mammalian intrinsic and extrinsic pathways (see Chapter 3).

2. SOURCE

2.1. Chelonia

One green turtle (*Chelonia mydas*) and two loggerhead marine turtles (*Caretta caretta caretta*) were studied at the Marineland Research Laboratory, St. Augustine, Florida. The green turtle weighed 50 kg and the loggerheads 44 and 41 kg, respectively. These large animals were placed on their backs on a surgical table equipped with running sea water. The plastron (ventral shell) was drilled in the cardiac area, and blood was drawn with a 6-inch No. 17 needle and multiple syringes. Animals were bathed with sea water every 15 to 20 min. To close the sternal openings, small steel plates were cemented to the cleaned, dried plastron.

Land turtles were purchased from an animal supplier (W. A. Lemberger and Company or Carolina Biological Company). A hole was drilled in the plastron of the 2.3–6.8 kg animals, and blood was obtained from heart or aorta with syringes and No. 19 needles. Most of the studies were done on the pond turtle (*Pseudemys scripta elegans*) (PSE). Other small turtles are listed in the tables.

2.2. Squamata

One Mexican iguana (*Basilicus americanus*) and six garter snakes (*Aglypha thamnophis*) were purchased. The iguana weighed 4.5 kg and the garter snakes 3–4 kg. Blood was obtained from the caudal sinus by venipuncture.

2.3. Crocodilia

Six alligators (*Alligator mississippiensis*) (weighing 24–32 kg) were studied in 1992. Blood and brain were purchased from Gatorland, Orlando, Florida. The blood was easily obtained from the dorsal neck vessels. The blood counts and chemistries were done at the Sea World Research Laboratory in Orlando through the courtesy of Drs. Odell, Walsh, and Campbell. Clotting times and clot retraction were done at room temperature. Citrated plasma was separated, labeled, and frozen in Florida and returned in dry ice to our Pittsburgh Blood Bank laboratory for coagulation factor assays and other studies. Buffy coat was fixed before transport to Pittsburgh.

3. TESTS CONDUCTED

3.1. Bleeding Times

Adequate cuts for measuring bleeding time were almost impossible on these animals with thick, hard skin. On four snakes, knife-blade cuts bled for 1½, 2, 0, and 1½ min (mean: 1½ min).

3.2. General Coagulation Tests and Factors

Table 8.1 shows that clotting times were very long in the animals from which blood was obtained without difficulty. In the marine turtles, the clots were often still soft after 24 hr in glass tubes, and clot retraction was poor. The PSE turtle blood clotted slowly in glass and siliconized tubes, retracted poorly, and did not lyse. The small turtles' blood clotted more rapidly and retracted well. This may have reflected technical difficulties in obtaining the blood. Snake blood clotted slowly, but retracted well. Alligator blood clotted in somewhat longer times than human, retracted well, and did not lyse.

The Simplastin times were variable, but always longer than human times. Homologous brain was not a very effective clot accelerator in these reptiles. Homologous lung extracts clotted green and PSE turtle plasmas quite rapidly.

TABLE 8.1
General Coagulation Tests in Reptilia[a]

Test	G (1)	L (2)	P (11)	M (4)	S (5)	A (6)	Normal human range
Clot T (min)							
Glass	420	120	33	9	61	20	6–12
Silicone	1440	1440	141	13	120+	135	20–59
Clot Retr	[b]	[b]	0–2	4	3–4	4	3–4+
Clot Lys	0	0	0	0	0	0	0
PT (sec)							
Simplastin	21.8	75.0	19.6	30.5	[c]	44.9	10–13
Homologous							
Brain	24.2	86.2	83.7	93.0	26.2	[c]	—
Lung	11.4	—	16.1	61.0	28.3	—	—
RVV	13.4	25.1	28.9	21.0	24.8	14.0	13–18
APTT (sec)	[c]	[c]	80.7	55.0	[c]	71.0	24–34
Recal T (sec)	600+	600+	546	400	300+	363	90–180
Lys	0	0	0	0	0	0	0
Lys MCA	0	0	0	0	0	0	0
Thromb T (sec)							
Bov Thr	9.4	11.3	69.6	28.0	[c]	[c]	11–18
CHH	—	—	64.4	—	—	[c]	11–18
Atroxin	—	—	78.4	—	—	[c]	11–18

[a]Animals: (G) green marine turtle; (L) loggerhead marine turtle; (P) *Pseudemys scripta elegans*; (M) miscellaneous small land turtles; (S) garter snake; (A) alligator.
[b]Not clotted. [c]More than 120 sec.

RVV was the overall best agent for reptiles in the prothrombin time test, but these times were often much longer than human. The APTT test was always long, as was the Recal T. The Recal clots did not lyse alone or in 1% MCA. Thrombin times were quite variable, from shorter than human to over 120 sec.

Table 8.2 shows the results with thromboplastin generating mixtures. Human mixture contained all human materials including platelet suspension. Its shortest clotting time on human plasma was 10.5 sec. The human mixture clotted the animal plasmas in somewhat longer times, but in less than 40 sec. The animal mixtures using their own buffy coat concentrates did not clot human or homologous plasmas.

Table 8.3 shows the results of coagulation factor assays in the reptiles. Fibrinogen (factor I) always fell within the normal human range. Most of the other coagulation factors assayed in the trace to very low range. One exception was the high level of factor V found in the PSE turtle. Factor V assayed at high human levels in all six of the PSE turtles tested. ATIII appeared in the normal human range in alligators.

TABLE 8.2
Thromboplastin Generation Test in Reptilia[a]

Generating mixture[b]						Substrate clotting time (sec)[b,c]					
H	G	L	P	S	A	H	G	L	P	S	A
H (4 mean)	—	—	—	—	—	**10.5**	13.0	26.0	14.8	—	38.4
—	G	—	—	—	—	40+	**40+**	—	—	—	—
—	—	L	—	—	—	40+	—	**40+**	—	—	—
—	—	—	P	—	—	40+	—	—	**40+**	—	—
—	—	—	—	S	—	40+	—	—	—	**40+**	—
—	—	—	—	—	A	40+	—	—	—	—	**40+**

[a]Human (H) plasmas were at 37°C, reptilian plasmas at ambient temperature.
[b]Animals: See footnote *a* in Table 8.1.
[c]Homologous systems are in **boldface** type.

TABLE 8.3
Coagulation Factors in Reptilia[a]

Factor	G (1)	L (2)	P (6)	M (4)	I (1)	S (4)	A (6)	Normal human range
I (mg/dl)	267	388	329	267	281	162	178	150–450
II (U/ml)	[b]	0.05	0.12	[b]	[b]	0	0.21	0.70–1.30
V (U/ml)	0.20	0.22	3.47	0.28	[b]	[b]	0.02	0.65–1.45
VII (U/ml)	[b]	[b]	0.04	[b]	[b]	0	[b]	0.50–1.30
X (U/ml)	[b]	[b]	0.24	[b]	[b]	0	[b]	0.75–1.25
VIII (U/ml)	[b]	[b]	0.01	[b]	0.05	0	[b]	0.75–1.40
IX (U/ml)	[b]	[b]	[b]	[b]	[b]	0	0.02	0.65–1.70
XI (U/ml)	0.10	[b]	0.03	0.10	0.02	0	0.02	0.75–1.40
XII (U/ml)	0.05	[b]	0.03	0.01	0.02	[b]	0.02	0.50–1.45
Flet (U/ml)	—	—	[b]	—	—	—	[b]	0.50–1.50
Fitz (U/ml)	—	—	—	—	—	—	[b]	0.50–1.50
ATIII (U/ml)	—	—	[b]	—	—	—	1.05	0.80–1.20
RCF (U/ml)	—	—	0.22	—	—	—	[b]	0.50–1.50
Prot C (A) (U/ml)	—	—	—	—	—	—	0.08	0.60–1.40
Prot C (I) (U/ml)	—	—	—	—	—	—	[b]	0.60–1.40
Plgn (U/ml)	—	—	—	—	—	—	0.06	0.80–1.20
Antipl (U/ml)	—	—	—	—	—	—	0.42	0.80–1.20
XIII (R)	2	4	2	—	—	—	—	4–16

[a]Animals: See footnote *a* in Table 8.1; (I) iguana.
[b]Less than 0.01 U/ml.

3.3. Effects of Calcium Molarity, Buffy Coat Content, and Temperature

Table 8.4 shows that there was no effect of increasing calcium strength on the clotting time or clot retraction of loggerhead turtle blood. Table 8.5 shows the effects of increasing calcium strength on the clotting of plasma from the PSE turtle. The optimum calcium level was the same as that for human plasma, which was just sufficient to match the citrate concentration of the plasma and leave some free calcium to support clotting. Table 8.5 also shows the effects of temperature on clotting. Both human and turtle clotting times were slightly longer at 25°C than at 37°C. Table 8.6 also shows effects of temperature, this time on alligator whole blood clotting in glass or siliconized tubes. The glass clotting was faster at 37°C, but in siliconized tubes it was faster at 25°C. Clot retraction occurred at any temperature, and these clots did not lyse.

Tables 8.7 and 8.8 show effects of cell (thrombocyte) content. PSE turtle blood was centrifuged at various speeds, and the resultant plasmas were recalcified and clotting and clot retraction observed. Very slow or very short centrifu-

TABLE 8.4
Effects of Calcium on Clotting of Loggerhead Turtle Blood[a]

$CaCl_2$ (M)	Clott T (min)	Appearance in 18 hr
1.0	60+	Solid (no retraction)
0.1	60+	
0.05	60+	
0.04	60+	
0.025	60+	
Sea water	60+	
Distilled water	60+	
None	60+	

[a]Test system: 0.9 ml fresh whole blood + 0.1 ml $CaCl_2$ or other agent at 28°C.

gation yielded plasmas that clotted well and retracted slightly. Longer or faster centrifugation resulted in slow clotting and no retraction. In Table 8.8, green turtle fast-centrifuged clear plasma was mixed with its buffy coat resuspended in plasma and recalcified. Clotting times were clearly dependent on buffy coat content. Retraction was not seen here or in whole green turtle blood.

3.4. Biochemical Tests in Alligators

Table 8.9 shows that the total protein and albumin levels were lower than human, as were the cholesterol, uric acid, creatinine, BUN, CO_2, and alkaline

TABLE 8.5
Effects of Temperature and Calcium Strength on Recalcification Time[a]

	Recalcification time (sec)			
	Human		PSE turtle	
$CaCl_2$ (M)	37°C	24°C	37°C	25°C
0.01	165	240	250	375
0.02	120	240	195	375
0.035	195	270	390	900+
0.05	270	300	390	900+
0.075	390	900+	900+	900+
0.1	480	900+	900+	900+

[a]Test system: 0.1 ml plasma + 0.1 ml buffer + 0.1 ml $CaCl_2$.

TABLE 8.6
Effect of Temperature on Clotting Time, Clot Retraction, and Lysis of Blood of Alligators No. 1 and 2

Test	25°C		37°C		5°C	
	No. 1	No. 2	No. 1	No. 2	No. 1	No. 2
Clotting time (sec)						
Glass (G)	12	16	10¼	14½	13	35
Silicone (S)	30	34	46	58	120	120
Clot Retr (G/S)	3/3	4/4	4/2	4/4	4/2	4/4
Clot Lys	0	0	0	0	0	0

TABLE 8.7
Effects of Speed and Time of Centrifugation on Clotting Time, Clot Retraction, and Clot Lysis of *Pseudemys scripta elegans* Turtle Plasma[a]

Centrifugation		Clotting time (min)		Clot retraction		
Speed	Time	Glass	Silicone	Glass	Silicone	Clot lysis
300*g*	10 min	7½	120	2	2	0
800*g*	3 min	7	120	1	1	0
800*g*	10 min	8½	120	0	0	0
2,000*g*	15 min	20	120	0	0	0
20,000*g*	15 min	41	240+	0	[b]	0

[a]Technique: 0.2 ml PSE citrate plasma + 0.2 ml p.025 M $CaCl_2$ at 28°C.
[b]Clot too soft to detect retraction or lysis.

TABLE 8.8
Effect of Buffy Coat on Recalcification Time and Clot Retraction of Green Turtle Plasma Mixtures[a]

Mixture (ml)						
Clear plasma	0.25	0.20	0.15	0.1	0.05	0
Buffy coat in plasma	0	0.05	0.10	0.15	0.20	0.25
Clot (min)						
Soft	360+	60	40	20	10½	10½
Hard	1440+	1440+	360	60	60	60
Retraction	[b]	[b]	0	0	0	0

[a]Tests were conducted in glass tubes at 28°C.
[b]Clot too soft to detect retraction or lysis.

phosphatase (Alk Phos). Ca, triglyceride, glucose, bilirubin, K, Na, and serum glutamic-pyruvic transaminase (SGPT) were within the human range. P, chloride, creatine phosphokinase (CPK), lactic dehydrogenase (LDH), and serum glutamic-oxalacetic transaminase (SGOT) were high.

3.5. Erythrocyte Parameters

Table 8.10 shows the erythrocyte parameters observed in the Reptilia. The HCTs were variable—from 21 in PSE turtles to 40 in the very large green marine turtle. HGBs and RBCs also varied among the species. The cells were moderately large, with MCVs from 254 to 739 fl, and hemoglobin content (MCH) from 48 to 185 pg.

3.6. Thrombocyte Ultrastructure

Figure 8.1 shows three alligator thrombocytes and a fragment. The cells are characterized by spindle shape, large nuclei with heavy peripheral chromatin, a predominant fine open canalicular system and a few tubular structures (? microtubules). Specific granules were not apparent.

4. SUMMARY

If obtained cleanly, without tissue contamination, reptilian blood clotted slowly in glass or siliconized tubes. The blood of the large marine turtles clotted very slowly, and clot retraction was not seen. Blood from the smaller turtles, iguana, snakes, and alligators retracted fully.

Clot retraction in the latter group related to numbers of thrombocytes present. Prothrombin times with Simplastin, homologous brain or lung, and RVV were variable, but usually showed some accelerating activity. Reptilian tissues had no accelerating activity on human plasma prothrombin times. APTTs and Recal Ts were long. Recal clots did not lyse alone or in 1% MCA. Thrombin times in the marine turtle plasmas were fast, but in miscellaneous turtles they were moderately long. In snake and alligator plasmas, the thrombin times were over 2 min. TGT showed no activity generated in the reptilian mixtures. The

TABLE 8.9
Biochemical Tests in Alligators

Substance	Alligator No.[a] 1	2	3	4	5	6	*N*	Mean[a]	SD	Normal human range
TP (g/dl)	*5.8*	*2.7*	*3.5*	*4.0*	*5.0*	*6.1*	6	*4.5*	1.3	6.0–8.0
A/G[b] (g/dl)	*2.1/3.7*	*1.0/1.7*	*1.2/2.3*	*1.5/2.5*	*1.7/3.3*	*2.2/3.9*	6	*1.6/2.9*	0.48/2.9	3.5–5.0[c]
Ca (mg/dl)	11.2	8.5	8.9	9.8	10.5	10.5	6	9.9	1.0	8.5–10.5
P (mg/dl)	**4.6**	4.1	4.3	**4.9**	**6.8**	**5.3**	6	**5.0**	1.0	2.5–4.5
Cholesterol (mg/dl)	*92*	*34*	*41*	*60*	*68*	*68*	6	*60*	21	150–200
Triglyceride (mg/dl)	184	—	20	56	160	98	5	104	69	45–200
Glucose (mg/dl)	73	66	63	72	99	89	6	77	14	65–110
Uric acid (mg/dl)	*1.3*	*0.3*	*0.7*	*0.5*	*1.2*	*1.1*	6	*0.85*	0.4	2.5–8.0
Creatinine (mg/dl)	*0.2*	*0.2*	*0.2*	*0.2*	*0.3*	*0.3*	6	*0.23*	0.05	0.7–1.4
T bilirubin (mg/dl)	0.1	0.1	0.1	0.2	0.2	0.1	6	0.13	0.05	0.1–1.4
BUN (mg/dl)	*1.0*	*0.9*	*0.8*	*0.8*	*1.1*	*0.8*	6	*0.9*	0.13	10–20
Chloride (meq/liter)	**112**	**115**	**115**	**111**	**109**	**108**	6	**112**	2.9	95–105
CO_2 (meq/liter)	*15.3*	*19*	*17.4*	*16.7*	*14.1*	*15.8*	6	*16.4*	1.7	24–32
K (meq/liter)	4.9	3.6	3.7	3.5	2.8	2.9	6	3.6	0.8	3.5–5.0
Na (meq/liter)	150	141	143	140	143	141	6	143	2.2	135–145
Alk Phos (IU/liter)	*24*	*16*	*21*	*18*	*23*	*24*	6	*21*	3.3	30–85
CPK (IU/liter)	**470**	**548**	**480**	**829**	**796**	**809**	6	**655**	173	0–110
LDH (IU/liter)	**244**	186	**324**	**333**	**309**	**314**	6	**285**	58	60–200
SGOT (IU/liter)	**224**	**219**	**233**	**331**	**252**	**396**	6	**276**	72	0–41
SGPT (IU/liter)	37	21	22	31	36	25	6	28.7	7.0	10–50

[a]Values **higher** than human are in **boldface**, those *lower* than human in *italics*.
[b]Albumin/globulin ratio. [c]For albumin.

TABLE 8.10
Red Blood Cell Elements of the Blood in Reptilia[a]

Element	G (1)	L (2)	P (6)	M (4)	I (1)	S (5)	A (6)
HCT (%)	40	34	21	27	36	31	33
HGB (g/dl)	5.9	8.5	5.6	7.1	6.3	7.9	13.8
RBC ($\times 10^6/mm^3$)	0.84	0.46	0.63	0.58	1.31	1.22	0.77
MCV (fl)	476	739	333	466	275	254	433
MCH (pg)	70	185	89	122	48	65	179
MCHC (g/dl)	13	25	27	26	18	25	42

[a]The values are means. Animals: See footnote *a* in Table 8.1; (I) iguana.

human mixture clotted human substrate in the usual time (10.5 sec). It also clotted the animal plasmas in 38 sec or less. Coagulation factors in reptiles were within the normal human range only for fibrinogen. Other factors were low to trace or undetectable with two exceptions: Factor V was high in the PSE turtle, and ATIII was in the normal range for alligators. RBCs were moderately large, nucleated cells. TEMS of alligator thrombocytes resembled those from the other classes of premmalian vertebrates.

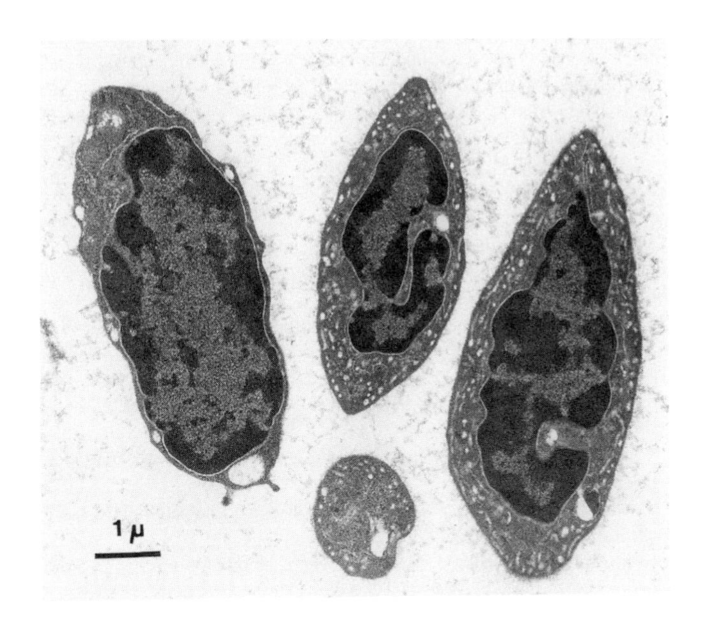

FIGURE 8.1. TEM of alligator thrombocytes.

REFERENCES

Denson, K. W. E., 1976, The clotting of a snake (*Crotalus viridis helleri*) plasma and its interaction with various snake venoms, *Thromb. Haemost.* **35:**314.

Fantl, P., 1961, A comparative study of blood coagulation in vertebrates, *Aust. J. Exp. Biol.* **39:** 403.

Hackett, E., and Hann, C., 1967, Slow clotting of reptile blood, *J. Comp. Pathol.* **77:**175.

Nahas, L., Betti, F., Kamiguti, A. S., and Sato, H., 1973, Blood coagulation inhibitor in snake plasma (*Bothrops jararaca*), *Thromb. Diath. Haemorrh.* **30:**106.

Ratnoff, O. D., Rosenberg, J. J., Everson, B., Emanuelson, M., and Tulodziecki, N., 1990, Notes on clotting in Burmese python (*Python molurus bivittatus*), *J. Lab. Clin Med.* **115:**629.

SUGGESTED READINGS

Arocha-Pinango, C. L., Gorzula, S. J., and Ojeda, A., 1982, The blood clotting mechanism of spectacled *Caiman crocodilus, Mol. Physiol.* **22:**161.

Baril, E. F., Palmer, J. L., and Bartel, A. H., 1961, Electrophoretic analysis of young alligator serum, *Science* **133:**278.

Belamarich, F. A., and Eskridge, R. W., 1963, Some observations on clotting in alligators, *Thromb. Diath. Haemorrh.* **9:**223.

Bramble, C. E., 1941, Prothrombin activity of turtle blood and the effect of a synthetic vitamin K derivative, *J. Cell. Comp. Physiol.* **18:**221.

Dessauer, H. C., 1970, Blood chemistry of reptiles: Physiological and evolutionary aspects, in: *Biology of the Reptilia,* Vol. 3, *Morphology* (C. Gans and T. S. Parsons, eds.), p. 1, Academic Press, New York.

Duguy, R., 1970, Numbers of blood cells and their variation, in: *Biology of the Reptilia,* Vol. 3, *Morphology* (C. Gans and T. S. Parsons, eds.), p. 93, Academic Press, New York.

Heady, J., and Rogers, M., 1962, Turtle blood cell morphology, *Proc. Iowa Acad. Sci.* **69:**587.

Jaques, F. A., 1963, Blood coagulation and anticoagulant mechanisms in the turtle *Pseudemys elegans, Comp. Biochem. Physiol.* **9:**241.

Lavres, A. A. C., Fichman, M., Hiraichi, E., Bocault, M. A., Tobo, T., Schmuziger, P., Nanas, L., and Picarelli, Z. P., 1979, Deficiency of kallikrein–kinin system and the presence of potent kinanase activity in the plasma of *Bothrops jararaca (Serpentes crotaline), Cienc. Cult.* **31:**168.

Nahas, L., Kamiguti, A. S., Betti, F., Sano Martins, I. S., and Rodrigues, M. I., 1981, Blood coagulation mechanisms in the snakes *Waglerophis merremii* and *Bothrops jararaca, Comp. Biochem. Physiol.* **69:**739.

Ryerson, D. L., 1949, A preliminary survey of reptilian blood, *Ent. Zool.* **41:**49.

St. Girons, M. C., 1970, Morphology of the circulating blood cells, in: *Biology of the Reptilia,* Vol. 3, *Morphology* (C. Gans and T. S. Parsons, eds.), p. 73, Academic Press, New York.

Taylor, K., and Kaplan, H. M., 1961, Light microscopy of the blood cells of pseudemys turtles, *Herpetologica* **17:**186.

Taylor, K. W., Kaplan, H. M., and Hirano, T., 1963, Electron microscope study of turtle blood cells, *Cytologia* **28:**248.

Wood, F. E., and Ebanks, G. K., 1984, Blood cytology and hematology of the green sea turtle, *Chelonia mydas, Herpetologica* **40:**331.

Zain-Ul-abedin, M., and Katorski, B., 1966, Increased blood clotting time in a hibernating lizard, *Can. J. Physiol. Pharmacol.* **44:**505.

Zweig, G., 1957, Differentiation of species by paper electrophoresis of serum proteins of pseudemys turtles, *Science* **126:**1065.

CHAPTER 9

DOMESTIC FOWL

Class: Aves

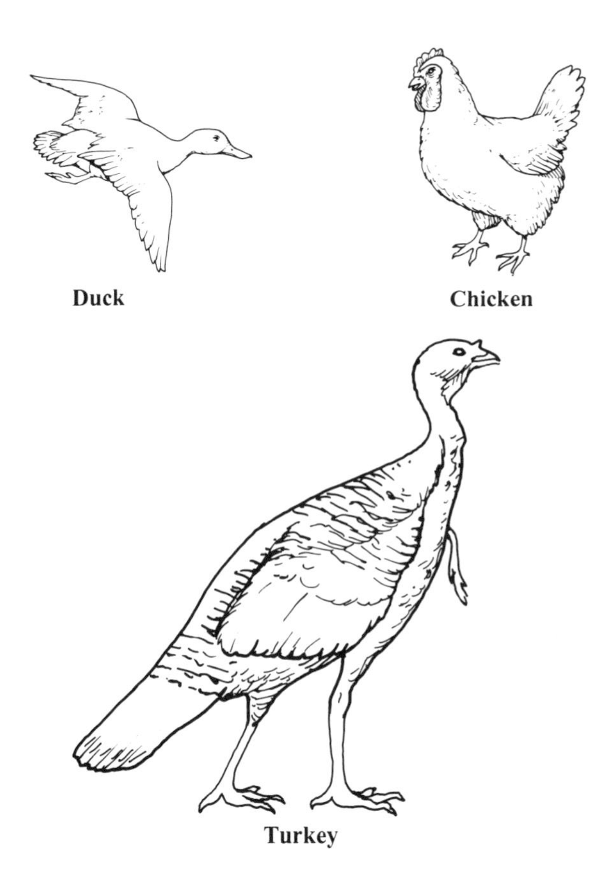

1. INTRODUCTION

It is thought that birds evolved from early reptiles, perhaps dinosaurs, in the upper Jurassic Age (150 MYA). Aves have spread worldwide and developed some 8000 species. They are unique in being covered with feathers, having a warm body temperature (40–43°C), laying hard-shelled eggs, and nurturing their young. Their bodies are streamlined, with beaked heads, flexible necks, two forelimbs that are wings, and two weight-supporting hindlimbs. Most birds have four toes that end in horny claws. Birds also have tails composed primarily of feathers, which, together with the wings and light skeletons, make flying possible. To obtain energy for flying, birds must eat almost continually. Their diet varies with their habitat—seeds, berries, insects, and small fishes or animals are common food.

Fowl became domesticated as man became a farmer. They provide man with eggs, food, and feathers for pillows or insulation. A previous publication concerning turkeys (Lewis et al., 1979) is reprinted in part with the kind permission of the editor of *Comparative Biochemistry and Physiology.*

2. SOURCE

All animals were obtained from or bled at poultry farms. After their eyes were covered and their abdomens gently rubbed, they were quiet and did not need anesthesia. Most blood samples were collected by venipuncture from an anterior axillary (wing) vein. Chickens [*Gallus gallus (domesticus)*] weighed 3–5 kg; ducks (*Anas platyrhynchos*), 2–3 kg; geese (*Anser albifrons*), 4.5–7 kg; pigeons (*Columba livia*), 2 kg; and turkeys (*Meleagris gallopavo*), 7–10 kg.

3. TESTS CONDUCTED

3.1 Bleeding Times

On six chickens, Simplate bleeding times were performed in a breast area from which the feathers had been plucked. These bleeding times were all shorter than 90 sec, with a mean of 48 sec. On three turkeys, bleeding times were done without plucking, by separating feathers on the thigh and making a short cut (6 mm) with a new rounded knife blade. The bleeding times were 2, 2 1/2, and 2 1/2 min (mean 2.3 min).

3.2 General Coagulation Tests and Factors

Table 9.1 shows that with the exception of most of the chickens, the clotting times of the birds' blood samples were borderline-long to long compared to the human range. One chicken (No. 7) was not clotted at 6 hr, so it is possible that the others were short because technical problems introduced tissue factor. This conjecture is supported by the long Recal Ts. There was little difference between clotting times in glass or siliconized tubes. Clots retracted poorly or not at all. Prothrombin times with Simplastin were very long. Using homologous animal brain suspensions, they were much shorter. Bird brain suspensions did not accelerate human clotting. RVV was thromboplastic to both bird and human plasmas. Thrombin times on Aves with bovine thrombin were much longer than on human plasma. Thrombin times with the venoms, CHH and Atroxin, did not clot chicken or turkey plasmas in less than 120 sec.

The prolonged thrombin and prothrombin times (with Simplastin) found in turkey plasma were studied. The results are shown in Table 9.2. Eqimixtures of human and turkey plasma gave longer times than with human alone. This inhibitory effect was greatly reduced after the turkey plasma was heated to 56°C for 5 min and centrifuged to remove the precipitate. Barium-treated plasma (Ba Pl) was still inhibitory. The inhibitor was heat-labile, not heparinlike, and may have been the fibrinogen itself.

TGT results are shown in Tables 9.3 and 9.4. Generating mixtures contained homologous buffy coat, serum, and Ba Pl. The human mixture clotted all substrates. Chicken, duck, goose, and turkey clotted neither human nor homologous plasma in less than 40 sec. Pigeon mixture clotted both human and pigeon plasma, but only after long times. Table 9.4 shows the results of an experiment in which goose generating components were substituted individually for human components. None supported thromboplastin generating activity.

Coagulation factor assays on Aves are shown in Table 9.5. All birds had factor I (fibrinogen) and factor XIII levels that fell within the human range. Factors VIII, V, and II were also detectable, but the activity was lower than human. To determine whether or not the avian VIII activity resembled that of human plasma, an attempt was made to inhibit it. Plasma from a patient with an acquired inhibition to human factor VIII was incubated for 2 hr at 37°C with normal human or turkey plasma. Against human plasma, there were over 500, and against turkey plasma, there were only 2 Bethesda units of anti-VIII.

To see whether the factor V in turkey plasma was labile, as it is in human plasma, both plasmas were incubated at 37°C for 4 hr, as shown in Table 9.6. The incubation-prolonged prothrombin time of human plasma can be shortened by the addition of fresh Ba Pl, which contains factor V but is free of factors II, VII, and X. The observations shown in Table 9.6 suggest that the turkey plasma

TABLE 9.1
General Coagulation Tests in Aves

	Chicken[a]			Duck			Goose			Pigeon			Turkey			Human			
																Controls			Normal range
	N	Mean	SD	*N*	Mean	SD	*N*	Mean	SD	*N*	Mean	SD	*N*	Mean	SD	*N*	Mean	SD	
Clot T (min)																			
Glass	9	8	1.4	4	16	9.1	6	12	4.1	2	25	6.4	8	61	37	6	7.2	1.2	6–12
Silicone	9	8	2.1	4	19	12	6	22	15	2	32	11	8	69	38	6	24	4.0	20–59
Clot Retr	10	0–2	—	4	0–0	—	6	2–3	—	2	0–3	—	8	0–2	—	6	4–4	0	3–4+
Clot Lys	10	0	0	4	0	0	6	0	0	2	0	0	8	0	0	6	0	0	0
SPT (sec)																			
Human	10	[b]	—	4	[b]	—	6	[b]	—	2	[b]	—	—	—	—	7	36.7	3.4	20+
Homolog	6	13.7	1.4	4	12.9	0.7	6	[b]	—	2	[b]	—	—	—	—	7	[b]	—	—
PT (sec)																			
Simplastin	10	[b]	—	4	40+	—	6	[b]	—	2	46.8	4.2	8	60+	—	8	13.5	1.3	10–13
Homolog BR	6	13.9	1.5	4	12.2	0.2	6	12.9	0.4	2	22.5	5.5	6	13.6	1.2	8	[b]	—	—
RVV	10	15.7	2.8	4	19.3	2.1	6	14.5	1.9	2	12.0	0.8	8	18.2	2.4	8	17.8	3.0	13–18
APTT (sec)	10	[b]	—	4	50+	—	6	90+	—	2	[b]	—	8	40+	—	6	32.6	8.0	24–34
Recal T (sec)	6	355	35	4	453	210	6	1510	450	2	600+	—	8	600+	—	8	128	5.6	90–180
Lys	6	0	0	4	0	0	5	0	0	2	0	0	8	0	0	8	0	0	0
Lys MCA	6	0	0	4	0	0	5	0	0	2	0	0	8	0	0	8	0	0	0
Thromb T (sec)																			
Bov Thr	10	[b]	—	4	24.9	2.8	6	19.2	2.0	2	[b]	—	6	39.5	31.4	8	16.0	1.9	11–18
CHH	7	[b]	—	—	—	—	—	—	—	—	—	—	6	[b]	—	4	16.4	0.8	11–18
Atroxin	7	[b]	—	—	—	—	—	—	—	—	—	—	6	[b]	—	5	14.3	0.4	11–18

[a]The Clot T for chicken No. 7 was more than 360 min and was omitted from the mean.
[b]More than 120 sec.

TABLE 9.2
A Heat-Labile Inhibitor in Turkey Plasma

Equimixture	Time (sec) Thrombin	Prothrombin
T + T	66.0	150.0
H + H	15.2	11.7
H + T	44.8	18.7
H + T (56°C)	15.8	14.5
H + T (Ba)	36.0	18.5
H + H (56°C)	21.0	15.6
H + H (Ba)	16.4	15.1

TABLE 9.3
Thromboplastin Generation Test in Aves

Generating mixture[a]						Substrate clotting time (sec)[a,b]					
H	C	D	G	P	T	H	C	D	G	P	T
H (mean 4)	—	—	—	—	—	**9.2**	22.8	18.1	14.2	17.8	13.8
—	C	—	—	—	—	40+	**40+**	—	—	—	—
—	—	D	—	—	—	40+	—	**40+**	—	—	—
—	—	—	G	—	—	40+	—	—	**40+**	—	—
—	—	—	—	P	—	30.2	—	—	—	**24.2**	—
—	—	—	—	—	T	40+	—	—	—		**40+**

[a]Plasmas: (H) human; (C) chicken; (D) duck; (G) goose; (P) pigeon; (T) turkey.
[b]Homologous systems are in **boldface.**

TABLE 9.4
Thromboplastin Generation Test in Goose

Generating mixture			Substrate Clot T (sec)[a]	
Buffy	Serum	Ba Pl	Human	Goose
H	H	H	**9.6**	14.2
G	H	H	40+	40+
H	G	H	40+	40+
H	H	G	40+	40+
G	G	G	40+	**40+**

[a]Homologous systems are in **boldface.**

TABLE 9.5
Coagulation Factors in Aves

Factor	Chicken			Duck			Goose			Pigeon			Turkey			Normal human range
	N	Mean	SD	*N*	Mean	SD	*N*	Mean	SD	*N*	Mean	SD	*N*	Mean	SD	
I (mg/dl)	8	371	94	4	358	19	6	223	21	2	243	54	8	441	94	150–450
II (U/ml)	8	0.14	0.1	4	0.08	0.03	6	0.08	0.4	2	0.21	0.01	6	0.30	0.17	0.70–1.30
V (U/ml)	8	0.06	0.1	4	0.30	0.2	6	0.16	0.5	2	0.07	0.01	8	0.19	0.18	0.65–1.45
VII (U/ml)	8	[a]	—	4	[a]	—	6	[a]	—	2	0.03	0.03	8	[a]	—	0.50–1.30
X (U/ml)	8	[a]	—	4	[a]	—	6	[a]	—	2	[a]	—	8	[a]	—	0.75–1.25
VIII (U/ml)	8	0.31	0.2	4	0.33	0.15	6	0.18	0.7	2	0.55	0.64	8	0.57	0.27	0.75–1.40
IX (U/ml)	8	[a]	—	4	[a]	—	6	0.06	0.2	2	0.03	0.03	8	[a]	—	0.65–1.70
XI (U/ml)	5	[a]	—	4	[a]	—	6	0.02	0.3	2	0.08	0.03	8	[a]	—	0.75–1.40
XII (U/ml)	8	[a]	—	4	[a]	—	6	[a]	—	2	0.03	0.03	8	[a]	—	0.50–1.45
XIII (R)	8	6.5	2.1	4	9.0	5.0	6	5.7	2.7	0	—	—	8	33	15	4–16

[a]Less than 0.01 U/ml.

TABLE 9.6
Effects of Incubation or Barium Treatment or Both on Turkey Prothrombin Time

	Prothrombin time (sec)	
	Tissue factor	
	Simplastin	Turkey Br.
	Plasma	
Treatment	Human	Turkey
Fresh	12.3	12.0
4 hr 4°C	12.3	11.0
4 hr 37°C	16.8	20.7
Ba	120+	120+
3/4 37°C + 1/4 Ba	14.8	15.2

did contain a heat-labile factor that was present in the Ba Pl. Turkey brain suspension was used in the turkey system.

3.3. Fibrinolytic Enzyme System

Goose plasminogen was not activated with streptokinase (250 U/ml) or staphylokinase (2%). One of the five geese tested showed partial activation with urokinase (250 U/ml). The antiplasmin activity was 48% of the human control.

3.4. Serum Proteins

Table 9.7 shows the total protein assays and distribution by paper electrophoresis. Total protein values for chickens fell within the normal human range, but for the other fowl, it was lower. The albumin assayed below human for chicken, duck, and turkey but was normal for goose. α_1 globulin was in the normal human range in all four birds. α_2 globulin was normal for chicken but low for the other three. β globulin was normal for chicken and goose, high for duck, and low for turkey. γ globulin assayed above the human range for chicken but was normal for the other three.

3.5. Biochemical Tests

Results of standard human biochemical tests on chicken and turkey sera are shown in Table 9.8. Tests done on both species showed cholesterol levels and BUN

TABLE 9.7
Serum Proteins by Paper Electrophoresis in Aves

Protein	Chicken			Duck			Goose			Turkey			Normal human range
	N	Mean	SD	*N*	Mean	SD	*N*	Mean	SD	*N*	Mean	SD	
TP (g/dl)	4	6.4	1.5	4	3.6	0.4	6	5.7	0.2	3	3.8	0.4	6.3–7.9
Albumin (g/dl)	4	1.7	0.3	4	1.3	0.2	6	3.7	0.4	3	2.1	0.3	3.2–4.4
α_1-Globulin (g/dl)	4	0.4	0.2	4	0.4	0.1	6	0.3	0.1	3	0.2	0.3	0.2–0.4
α_2-Globulin (g/dl)	4	0.8	0.3	4	0.2	0.2	6	0.3	0.1	3	0.3	0.2	0.4–1.0
β-Globulin (g/dl)	4	1.0	0.4	4	1.1	0.1	6	0.8	0.3	3	0.4	0.1	0.5–1.0
γ-Globulin (g/dl)	4	2.5	0.9	4	0.6	0.2	6	0.6	0.2	3	0.8	0.3	0.6–1.8

TABLE 9.8
Biochemical Tests in Aves

Substance	Chicken			Turkey			Normal human range
	N	Mean[a]	SD	*N*	Mean[a]	SD	
TP (g/dl)	6	*3.9*	0.6	—	—	—	6.0–8.0
Albumin (g/dl)	—	—	—	4	*1.8*	0.2	3.5–5.0
Ca (mg/dl)	6	9.9	1.5	4	**11.7**	0.7	8.5–10.5
P (mg/dl)	—	—	—	2	**4.6**	0.8	2.5–4.5
Cholesterol (mg/dl)	6	*113*	19	2	*88*	3.5	150–200
Glucose (mg/dl)	6	**150**	45	4	**325**	17	65–110
Uric acid (mg/dl)	—	—	—	2	6.7	1.0	2.5–8.0
Creatinine (mg/dl)	—	—	—	6	*0.5*	0.2	0.7–1.4
T bilirubin (mg/dl)	—	—	—	5	0.2	0.0	0.1–1.4
BUN (mg/dl)	6	*<2.0*	—	6	*2.2*	1.4	10–20
Chloride (meq/liter)	6	**117**	2.5	6	**117**	2.8	95–105
CO_2 (meq/liter)	—	—	—	6	26	2.8	24–32
K (meq/liter)	6	**6.6**	2.5	5	*2.5*	0.4	3.5–5.0
Na (meq/liter)	6	**153**	1.4	6	**161+**	—	135–145
Alk Phos (IU/liter)	6	**300**	62	4	**298+**	—	30–85
SGOT (IU/liter)	6	**238**	32	4	**300+**	—	0–41

[a]Values **higher** than human are in **boldface**, those *lower* than human in *italics*.

to be low and glucose, chloride, Na, Alk Phos, and SGOT (AST) to be high in both. Ca was normal in chicken and high in turkey, and K was high in chicken and low in turkey.

3.6. Miscellaneous Tests

Plasma from chickens, ducks, and turkeys either did not clump the standard staphylococcal preparation or did so only at full strength.

3.7. Cellular Elements

Table 9.9 shows that the HCT in pigeon and the HGB in chicken were below the human range. All RBCs were low, and the calculated MCVs and MCHs, were high. MCHC was high for pigeon but within the normal range for the other fowl. The erythrocytes had roundish nuclei and a mean cell length × width of 12.4 μm × 6.8 μm. The white blood counts were high, and on Wright's stained smears at least five types of cells could be differentiated: Heterophils showed

TABLE 9.9
Cellular Elements of the Blood in Aves

Element	Chicken			Duck			Goose			Pigeon			Turkey			Normal human range
	N	Mean	SD	*N*	Mean	SD	*N*	Mean	SD	*N*	Mean	SD	*N*	Mean	SD	
HCT (%)	11	37	5.2	4	40	1.7	6	45	3.2	4	34	4.6	8	40	2.5	37–52
HGB (g/dl)	11	11.6	1.8	4	13.2	0.9	6	14.5	1.1	4	13.9	8.8	8	12.8	0.7	12–18
RBC (× 10^6/mm^3)	11	2.7	0.5	4	2.6	0.4	6	2.9	4.7	4	3.3	2.1	6	2.3	0.4	4.2–6.2
MCV (fl)	11	135	6.4	4	158	32	6	153	21	4	106	7.3	6	171	21	79–97
MCH (pg)	11	44	7.5	4	52	9.4	6	51	7.5	4	43	1.2	6	56	8.5	27–31
MCHC (g/dl)	11	32	5.7	4	33	3.1	6	33	0.8	4	42	3.1	6	32	2.2	32–36
WBC (× 10^3/mm^3)	11	10.4	62	4	35	18	6	38	20	2	14	13	6	51	31	5–10
Hetero (%)	11	18	6.6	4	35	17	6	29	6.0	1	71	—	6	39	10	55–75
Lymph (%)	11	59	10	4	49	19	6	65	7.1	1	18	—	6	40	13	20–40
Mono (%)	11	15	7.3	4	8	5.8	6	5	3.6	1	6	—	6	4	2.3	2–10
Eos (%)	11	6	8.2	4	4	1.4	6	1	2.5	1	5	—	6	14	9.3	0–3
Bas (%)	11	2	2.2	4	4	2.5	6	0.2	0.4	1	0	—	6	3	2.2	0–3
Thromb (× 10^3/mm^3)	11	62	24	4	40	9.2	6	58	19	1	72	—	6	45	18	150–450

multilobed nuclei and large pale red granules, lymphocytes had roundish single-lobe nuclei and a few granules, monocytes (rare) were large cells with dark nuclei and pale cytoplasm, eosinophils had multilobed nuclei and large red round or sickle-shaped granules, and basophils had blue granules. Thrombocytes were spindle-shaped nucleated cells with fine granular cytoplasm and oval nuclei in which the chromatin was dense at the periphery.

3.8. Thrombocyte Aggregation

Slightly hazy "thrombocyte-rich" plasmas were prepared by the same differential centrifugation used in preparation of human platelet-rich plasma. Smears from this plasma showed rare RBCs, but the WBCs were of all types. Only minor aggregation or clearing of this cell-rich plasma was found after addition of ADP, collagen, ristocetin, arachidonic acid (Arach A), CHH, or thrombin (Table 9.10).

3.9. Thrombocyte Ultrastructure

Figures 9.1 and 9.2 show TEMs of chicken thrombocytes, and Figure 9.3 shows a turkey thrombocyte. They are spindle-shaped with a large nuclei characterized by dark-staining chromatin material on the periphery. The cytoplasm shows numerous small vacuoles, probably part of an open canalicular system.

TABLE 9.10
Thrombocyte Aggregation in Aves

	Aggregation (%)						
	Chicken No.			Turkey No.		Human	
Reagent	2	5	6	1	3	Control	Normal range
ADP (20 μM)	5	0	0	0	4	100	70–100
ADP (5 μM)	0	0	0	0	0	80	50–100
Collagen							
Bov (0.19 mg/ml)	8	5	8	0	5	93	80–100
Turkey (0.2 mg/ml)	0	0	0	10	0	75	—
Risto (0.9 mg/ml)	3	0	0	0	3	100	80–100
Arach A (0.5 mg/ml)	23	5	5	—	3	68	60–100
CHH (0.05 mg/ml)	5	3	7	—	10	92	60–100
Bov Thr (0.4 U/ml)	3	3	8	10	10	82	60–100

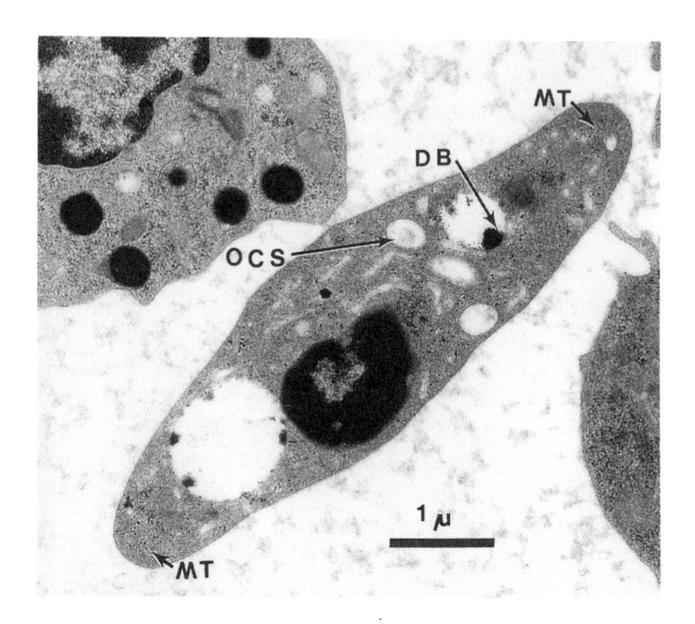

FIGURE 9.1. TEM of chicken thrombocytes. Internal structures: (MT) microtubule; (DB) dense body; (OCS) open canalicular system.

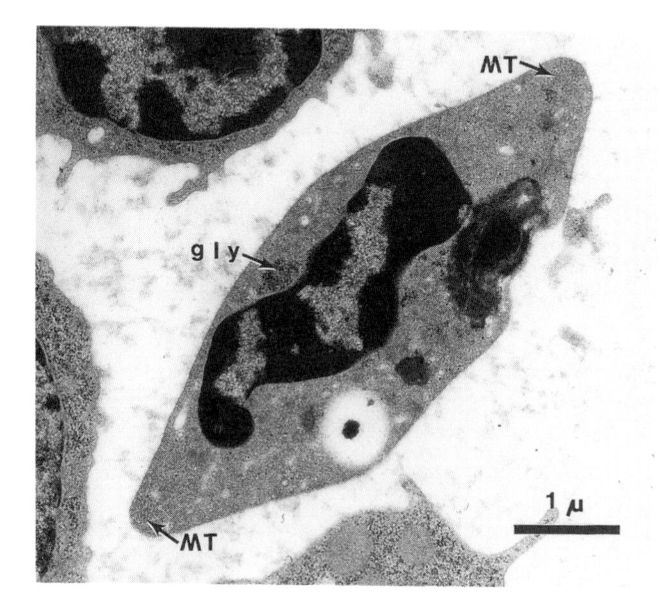

FIGURE 9.2. TEM of chicken thrombocytes. Internal structures: (MT) microtubule; (gly) glycogen.

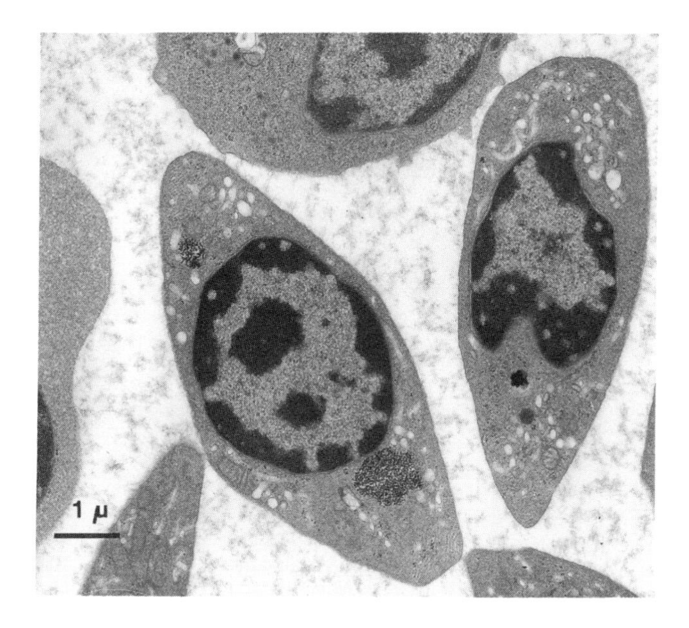

FIGURE 9.3. TEM of turkey thrombocytes.

There are a few clumps of glycogen particles, but granules are not apparent. Patches of microtubules are barely visible at the ends of the cells.

4. SUMMARY

The blood of most of the fowl studied clotted slowly and retracted poorly. Prothrombin times with Simplastin were long, but prothrombin times were much shorter with homologous brain extract or RVV. Animal brain extracts did not accelerate human clotting. APTT and Recal Ts were long. Thrombin times were longer than on human plasma. CHH and Atroxin did not clot Aves plasma in less than 120 sec. Turkey plasma contained a heat-labile thrombin inhibitor. TGT generation mixtures from all the five avian species were inactive. Coagulation factor assays showed normal humanlike amounts of fibrinogen and factor XIII. Factor VIII activity (ability to clot hemophilic plasma) was 0.18–0.57 U/ml, but this activity differed from human in the lack of inhibition by human anti-VIII. Trace amounts of factors II and V and in some species VII, IX, and XI were found. Factor X was always undetectable.

RBCs were moderately large nucleated cells. Thrombocytes were nucleated cells almost as large as RBCs. Aggregation with the usual agents was absent or minimal.

REFERENCE

Lewis, J. H., Hasiba, U., and Spero, J. A., 1979, Comparative hematology: Studies on class Aves, domestic turkey (*Meleagris gallopavo*), *Comp. Biochem. Physiol.* **62A:**735.

SUGGESTED READINGS

Aves: General

Barrett, L. A., and Dawson, R. B., 1974, Avian erythrocyte development: Microtubules and the formation of the disk shape, *Dev. Biol.* **36:**72.

Barrett, L. A., and Scheinberg, S. L., 1972, The development of the avian red cell shape, *J. Exp. Zool.* **182:**1.

Belamarich, F. A., 1975, The effect of reserpine on aggregation of avian thrombocytes, *Thromb. Diath. Haemorrh.* **34:**360.

Bierer, B. W., Eleazer, T. H., and Roebuck, D. E., 1964, Sedimentation rate, packed cell volume, buffy coat value, and rectal temperature of chickens and turkeys at various ages, *J. Am. Vet. Med. Assoc.* **144:**727.

Bigland, C. H., 1964, Blood clotting times of five avian species, *Poult. Sci.* **43:**1035.

Bigland, C. H., 1964, Poor serum yields from avian blood as a result of delayed clotting, *Poult. Sci.* **43:**1387.

Bigland, C. H., and Starr, R. M., 1965, Comparison of simple blood coagulation tests in birds, *Can. Vet. J.* **6:**233.

Blain, D., 1928, A direct method for making total white blood counts on avian blood, *Proc. Soc. Exp. Biol. Med.* **25:**594.

Campbell, F., 1967, Fine structure of the bone marrow of the chicken and pigeon, *J. Morphol.* **123:**405.

Dein, F. J., 1982, Avian clinical hematology, *Proc. Assoc. Avian Vet.* **1982:**5–29.

Edmonds, R. H., 1968, Electron microscope studies on the hemostatic process in bird embryos. 1. The initial plug, *J. Ultrastruct. Res.* **24:**295.

Grant, R. A., and Zucker, M. B., 1973, Avian thrombocyte aggregation and shape change in vitro, *Am. J. Physiol.* **225:**340.

Hann, C. S., 1966. Fibrinopeptides in one lizard and four bird species, *Biochim. Biophys. Acta* **124:**398.

Harris, J. R., and Brown, J. N., 1971, Fractionation of the avian erythrocyte: An ultrastructural study, *J. Ulstrastruct. Res.* **36:**8.

Hodges, R. D., 1977, Normal avian (poultry) haematology, in: *Comparative Clinical Haematology* (R. K. Archer and L. B. Jeffcot, eds.), p. 483, Blackwell, Oxford.

Kelly, J. W., and Dearstyne, R. S., 1935, Haematology of the fowl. A. Studies on normal chick and normal adult blood, *N. C. Agric. Exp. Stn. Tech. Bull.*, No. 50.

Kuruma, I., Okada, T., Kataoka, K., and Sorimachi, M., 1970, Ultrastructural observation of

5-hydroxytryptamine-storing granules in the domestic fowl thrombocytes, *Z. Zellforsch. Mikrosk. Anat.* **108:**268.
Lucas, A. M., and Jamroz, C., 1961, *Atlas of Avian Hematology,* U.S. Department of Agriculture Monograph 25, pp. 41–45, U.S. Government Printing Office, Washington, DC.
Maxwell, M. H., 1973, Comparison of heterophil and basophil ultrastructure in six species of domestic bird, *J. Anat.* **115:**187.
Maxwell, M. H., 1974, An ultrastructural comparison of the mononuclear leucocytes and thrombocytes in six species of domestic bird, *J. Anat.* **117:**69.
Olson, C., 1959, Avian hematology, in: *Diseases of Poultry,* 4th ed. (H. E. Biester and L. H. Schwarte, eds.), pp. 53–69, Iowa State University Press, Ames.
Olson, C., 1965, Avian hematology, in: *Diseases of Poultry,* 5th ed. (H. E. Biester and L. H. Schwarte, eds.), pp. 100–119, Iowa State University Press, Ames.
Rowsell, H. C., 1968, The hemostatic mechanism of mammals and birds in health and disease, *Adv. Vet. Sci.* **12:**337.
Rowsell, H. C., 1968, Hemostasis in the chicken and turkey, *Fed. Proc. Fed. Am. Soc. Exp. Biol.* **27:**627.
Shum, Y., and Griminger, P., 1972, Prothrombin time stability of avian and mammalian plasma: Effect of anticoagulant, *Lab. Anim. Sci.* **22:**384.
Sturkie, P. D., 1965, *Avian Physiology,* 2nd ed., Comstock, Ithaca, NY.
Topp, R. C., and Carlson, H. C., 1972, Studies on avian heterophils. III. Histochemistry, *Avian Dis.* **16:**369.
Topp, R. C., and Carlson, H. C., 1972, Studies on avian heterophils. IV. Phagocytic properties, *Avian Dis.* **16:**374.
Traill, K. N., Bock, G., Boyd, R., and Wick, G., 1983, Chicken thrombocytes: Isolation, serological and functional characterisation using the fluorescence activated cell sorter, *Dev. Comp. Immunol.* **7:**111.

Chicken

Allen, B. V., 1972, Fibrinolysis in the domestic chicken (*Gallus domesticus*), *Thromb. Diath. Haemorrh.* **27:**644.
Bell, D. J., and Freeman, B. M., 1971, *Physiology and Biochemistry of the Domestic Fowl,* Vol. 2, Academic Press, New York.
Bigland, C. H., and Triantaphyllopoulos, D. C., 1960, Chicken prothrombin, thrombin, and fibrinogen, *Am. J. Physiol.* **200:**1013.
Bigland, C. H., and Triantaphyllopoulos, D. C., 1960, A re-evaluation of the clotting time of chicken blood, *Nature (London)* **186:**644.
Blount, W. P., 1939, Thrombocyte formation in the domestic hen, *Vet. J.* **95:**195.
Chang, C. F., and Hamilton, P. B., 1979, The thrombocyte as a primary phagocyte in chickens, *J. Reticuloendothel. Soc.* **25:**585.
Cury, G., Gassiole, D., and Dollander, A., 1965, Some morphological characteristics of fowl thrombocytes, *C. R. Seances Soc. Biol. Filiales* **159:**1410.
Daimon, T., and Caxton-Martins, A., 1977, Electron microscopic and enzyme cytochemical studies on granules of mature chicken granular leucocytes, *J. Anat.* **123:**553.
Dam, H., and Schonheyder, F., 1978, A deficiency disease in chicks resembling scurvy, *Biochemistry* **28:**1356.
Dhingra, L. D., Parrish, W. B., and Venzke, W. G., 1969, Electron microscopy of granular leucocytes of chicken (*Gallus domesticus*), *Am. J. Vet. Res.* **30:** 637.

Dhingra, L. D., Parrish, W. B., and Venzke, W. G., 1969, Electron microscopy of nongranular leucocytes and thrombocytes of chickens, *Am. J. Vet. Res.* **30:**1837.

Diesem, C. D., Venzke, W. G., and Moore, E. N., 1958, The hemogram of chickens, *Am. J. Vet. Res.* **19:**719.

Doerr, J. A., and Hamilton, P. B., 1981, New evidence for intrinsic blood coagulation in chickens, *Poult Sci.* **60:**237.

Doerr, J. A., and Hamilton, P. B., 1981, Aflatoxin and intrinsic coagulation function in broiler chickens, *Poult. Sci.* **60:**1406.

Doerr, J. A., Wyatt, R. D., and Hamilton, P. B., 1975, Investigation and standardization of prothrombin times in chickens, *Poult. Sci.* **54:**969.

Enbergs, H., and Kriesten, K., 1968, Die weissen Blutzellen des Haushuhns im elektronenmikroskopischen Bild, *Dtsch. Tierarztl. Wochenschr.* **75:**271.

Enbergs, H., and Kriesten, K., 1968, Zytoplasmatische Feinstrukturen der Thrombozyten des Haushuhns, *Experientia* **24:**597.

Forkner, C. E., 1929, Blood and bone marrow cells of the domestic fowl, *J. Exp. Med.* **50:**121.

Janzarik, H., 1981, Nucleated thrombocytoid cells. II. Phase and interference-contrast microscopic studies on blood cells of the domestic fowl, *Cell Tissue Res.* **219:**497.

Janzarik, H., and Morgenstern, E., 1979, The nucleated thrombocytoid cells. I. Electron microscopic studies on chicken blood cells, *Thromb. Haemost.* **41:**608.

Kriesten, K., and Enbergs, H., 1970, Zur Ultrastruktur der Lymphozyten des Haushuhns, *Blut* **20:**229.

Maxwell, M. H., and Trejo, F., 1970, The ultrastructure of white blood cells and thrombocytes of the domestic fowl, *Br. Vet. J.* **126:**583.

Medway, W., and Kare, M. R., 1959, Blood and plasma volume, hematocrit, blood specific gravity and serum protein electrophoresis of the chicken, *Poult. Sci.* **38:**624.

Menache, D., Cesbron, N., Guillin, M., and Schlegel, N., 1973, Chicken fibrinogen and human fibrinogen. Comparative immunological studies, *Thromb. Diath. Haemorrh.* **15:**72.

Planas, J., and Recio, J. M., 1961, Las proteinas plasmaticas de la gallina, *Rev. Esp. Fisiol.* **17:**95.

Sankaranarayanaw, G., and Nambiar, K. T. K., 1971, Studies in blood coagulation in the domestic fowl, *Indian Vet. J.* **48:**30.

Schonheyder, F., 1938, Prothrombin in chickens, *Am. J. Physiol.* **123:**349.

Schumacher, A., 1965, Zur submikroskopischen Struktur der Thrombozyten, Lymphozyten und Monozyten des Haushuhnes (*Gallus domesticus*), *Z. Zellforsch. Mikrosk. Anat.* **66:**219.

Skjorten, F., and Evensen, S. A., 1973, Induction of disseminated intravascular coagulation in the factor XII–deficient fowl: Morphological effects of liquoid, bacterial endotoxin and tissue thromboplastin in the normal and anticoagulated fowl, *Thromb. Diath. Haemorrh.* **15:**25.

Smith, N., and Engelbert, V. E., 1969, Erythropoiesis in chicken peripheral blood, *Can. J. Zool.* **47:**1269.

Sorbye, O., 1962, Studies on the coagulation of chicken blood. III. Differentiation of x-, o_1- and o_2-factors by adsorption, *Acta Chem. Scand.* **16:**799.

Sorbye, O., 1962, Studies on the coagulation of chicken blood. IV. Adsorption of o-factor activity, *Acta Chem. Scand.* **16:**903.

Sorbye, O., 1962, Studies on the coagulation of chicken blood. IX. Labile factor activity of fresh oxalated plasma: The combined activity of sixteen discrete labile coagulation factors, *Acta Chem. Scand.* **16:**2411.

Sorbye, O., 1963, Studies on the coagulation of chicken blood. X. Adsorption analysis of the x- and o-factor activities, *Acta Chem. Scand.* **17:**123.

Sorbye, O., 1963, Studies on the coagulation of chicken blood. XI. Determination of the o_1- and o_2-factor concentrations in plasma, and detection of unidentified dietary factors essential for maximum level of o_2-factor, *Acta Chem. Scand.* **17:**1015.

Sorbye, O., and Kruse, I., 1962, Studies on the coagulation of chicken blood. V. Differentiation and assay of labile factors, adsorbable by barium carbonate, *Acta Chem. Scand.* **16:**1221.
Sorimachi, M., Kataoka, K., and Inouye, A., 1970, Release in vitro of 5-hydroxytryptamine from spindle cells of the domestic fowl, *Eur. J. Pharmacol.* **10:**243.
Stalsberg, H., and Prydz, H., 1963, Studies of chick embryo thrombocytes. I. Morphology and development, *Thromb. Diath. Haemorrh.* **9:**279.
Stalsberg, H., and Prydz, H., 1963, Studies of chick embryo thrombocytes. II. Function in primary hemostasis, *Thromb. Diath. Haemorrh.* **9:**291.
Stopforth, A., 1970, A study of coagulation mechanisms in domestic chickens, *J. Comp. Pathol.* **80:**525.
Stratil, A., 1972, Low-molecular-weight proteins in chicken serum, *Comp. Biochem. Physiol. B* **43:**257.
Sweeny, P. R., and Carlson, H. C., 1968, Electron microscopy and histochemical demonstration of lysosomal structures in chicken thrombocytes, *Avian Dis.* **12:**636.
Wachter, H., 1965, Serum proteins of hen with regard to race, sex and age, *Acta Biol. Med.* **15:**138.
Wartelle, O., 1957, Mecanisme de la coagulation chez la poule. I. Étude des elements du complexe prothrombique et de la thromboplastino-formation, *Rev. Hematol.* **12:**351.
Weller, E. M., and Schechtman, A. M., 1962, Ontogeny of serum proteins in the chicken. I. Paper electrophoresis studies, *Dev. Biol.* **4:**517.

Duck

Belamarich, F. A., and Simoneit, L. W., 1972, Duck thrombocyte aggregation by 5-hydroxytryptamine, paper presented at the Third Congress of the International Society on Thrombosis and Hemostasis, Washington, DC, p. 166 (abstract).
Belamarich, F. A., and Simoneit, L. W., 1973, Aggregation of duck thrombocytes by 5-hydroxytryptamine, *Microvasc. Res.* **6:**229.
Magath, T. B., and Higgins, G. M., 1934, The blood of the normal duck, *Folia Haematol.* **51:**230.
Murdock, H. R., and Lewis, J. O. D., 1964, A simple method for obtaining blood from ducks, *Proc. Soc. Exp. Biol. Med.* **116:**51.
Stiller, R. A., Belamari, F. A., and Shepro, D., 1975, Aggregation and release in thrombocytes of duck, *Am. J. Physiol.* **229:**206.

Goose

Hunsaker, W. G., Hunt, J. R., and Aitken, J. R., 1964, Physiology of the growing and adult goose. I. Physical characteristics of blood, *Br. Poult. Sci.* **5:**249.
Schram, A. C., 1970, Serum proteins of the domestic goose, *Anser anser, Comp. Biochem. Physiol.* **32:**81.

Pigeon

De Eds, F., 1927, Normal blood counts in pigeons, *J. Lab. Clin. Med.* **12:**437.
Kennedy, W. P., and Climenko, D. R., 1929, Studies on the blood of birds. I. The corpuscles of the pigeon, *Q. J. Exp. Physiol.* **19:**43.

Turkey

Boulanger, P., 1952, Seasonal changes in the coagulability of the blood of turkeys, *Can. J. Comp. Med.* **16:**222.

Ferguson, T. M., Golan, F. A., Trammel, J., Smith, E., Omar, E., and Couch, J. R., 1964, Hematological data and blood pressure for male broad breasted bronze turkeys, *Poult. Sci.* **43:**1318.

McCartney, M. G., 1952, Total blood and corpuscular volume in turkey hens, *Poult. Sci.* **31:**184.

Simpson, C. F., 1968, Ultrastructural features of the turkey thrombocyte and lymphocyte, *Poult. Sci.* **47:**848.

Venkataratnam, A., and Clarkson, M. J., 1962, The blood cells of the turkey, *Res. Vet. Sci.* **3:**455.

Wachowicz, B., 1982, Binding of adenosine diphosphate to turkey thrombocytes, *Haemostasis* **11:**139.

CHAPTER 10

PRIMITIVE AUSTRALIAN MAMMALS

Echidna

Quokka

Wallaby

1. INTRODUCTION

Monotremata is the first or most primitive order of living mammals. They are found in the Australian area. The two major species are: *Ornithorhynchus anatinus,* the duck-billed platypus, and *Tachyglossus aculeatus,* the echidna or spiny anteater. Both lay one to three eggs with large yolks and soft shells. When the young hatch, they cling to the mother's abdomen and lap up the milk that aexudes from primitive teats. The echidna is covered with spiny, scruffy hair and has a long beak suited to catching ants and termites. They are thought to have arisen about 100 MYA from basic mammalian stock or a small mammal-like reptile.

Marsupialia is the next most primitive order, its members delivering live young after a 2-week gestation period. These infants must make their way to the marsupium (pouch) on the mother's abdomen that contains ten or more teats and protects the newborn. Animals of the order Marsupialia are found in the Australian region and in North and South America.

Hematological studies on Monotremata are very limited. Those on Marsupialia are somewhat more common. Data from a previous study (Lewis *et al.*, 1968) are reprinted here with the kind permission of the editor of *Comparative Biochemistry and Physiology.*

2. SOURCE

The animals were studied in the laboratory of Dr. C. Hann at the Institute of Medical and Veterinarian Science, Adelaide, Australia. After anesthesia with intraperitoneal Nembutal, blood was obtained by cardiac puncture from three echidnas, one quokka (*Setonix brachyurus*), and one wallaby (*Thylogale eugenii*). These are "cat-sized" animals; the echidnas weighed 5–6 kg, the quokka 8 kg, and the wallaby 14 kg. Most studies were carried out on fresh blood, but a few samples of plasma from the animals and human controls were lyophilized and studied later after our return to Pittsburgh.

3. TESTS CONDUCTED

3.1. Bleeding Times

Bleeding times were not done on these animals.

3.2. General Coagulation Tests and Factors

Table 10.1 illustrates the results of general coagulation tests. The blood coagulation times were short in glass tubes and slightly longer in siliconized tubes. Clot retraction was less in the animal blood than in human blood. The clots did not lyse. SPTs were long, as were those of the normal humans. The prothrombin times with Simplastin were longer in the animals than in humans. RVV and marsupial brain extracts were active in clotting both animal and human plasmas. The APTTs were shorter on animal than on human plasma. The Recal Ts were similar to those of human plasma, and the clots did not lyse by themselves or in 1% MCA. Thrombin times with human or bovine thrombin were much longer in the animals than in the humans.

TABLE 10.1
General Coagulation Tests in Primitive Australian Mammals[a]

						Human		
	Echidna No.					Control No.		Normal
Test	1	2	3	Quokka	Wallaby	1	2	range
Clot T (min)								
Glass	4	5	1	6	5	7	8	6–12
Silicone	18	7	3	14	14	37	—	20–59
Clot Retr	2/2	2/2	3/3	4/4	3/3	4/4	4/4	3–4+
Clot Lys	0	0	0	0	0	0	0	0
SPT (sec)								
Human	42.0	[b]	47.0	[b]	[b]	37.0	43.0	20+
PT (sec)								
Simplastin	28.0	37.0	33.8	20.0	22.9	13.2	13.8	10–13
Echidna Brain	—	—	29.4	28.7	50.3	—	58.5	—
Wallaby Brain	—	—	23.2	9.5	13.9	17.9	—	—
Quokka Brain	—	—	14.2	7.4	9.6	—	13.4	—
RVV	14.7	13.7	16.8	12.5	21.8	13.1	9.5	13–18
APTT (sec)	46.0	31.5	37.0	19.5	22.5	52.8	42.0	45–55
Recal T (sec)	150	150	140	56	66	130	140	90–180
Lys	0	0	0	0	0	0	0	0
Lys MCA	0	0	0	0	0	0	0	0
Thromb T (sec)								
Hu Thr	100+	47.6	49.2	28.5	49.2	13.1	10.2	11–18
Bov Thr	89	80	360	48.0	115	14.8	14.1	11–18

[a]Echidna No. 1, wallaby, and human control No. 1 were studied on the first day; echidnas No. 2 and 3, quokka, and human control were studied on the next day.
[b]More than 120 sec.

TABLE 10.2
Thromboplastin Generation Test in Primitive Australian Mammals

Generating mixture[a]				Substrate Clot T (sec)[a,b]			
H	E (mean)	Q	W	H	E	Q	W
H	—	—	—	**11.1**	21.3	21.3	23.6
—	E (2)	—	—	18.1	**19.9**	—	—
—	—	Q	—	21.5	—	**18.1**	—
—	—	—	W	19.3	—	—	**21.5**

[a]Plasmas: (H) human; (E) echidna; (Q) quokka; (W) wallaby.
[b]Homologous systems are in **boldface.**

TGTs (Table 10.2) were performed on each species employing its own platelet suspension, serum, and barium-treated plasma (Ba Pl) using platelet-poor plasma (PPP) from each species as substrate. The clotting time of the human generating mixture tested on human substrate was in the usual range: 9.8–11.2 sec. The animal generating mixtures were all less active than human, and species specificity was not prominent. Animal platelets, serum, and Ba Pl were tested in otherwise human component tests. Only echidna serum showed poor activity.

The results of factor assays carried out on the animal plasmas are shown in Table 10.3. A lyophilized sample of echidna No. 3 was tested in Pittsburgh. Factors I, V, VIII, XII, and XIII appear to fall in the normal human range,

TABLE 10.3
Coagulation Factors in Primitive Australian Mammals

Factor	Echidna 1	Echidna 2	Echidna 3	Mean	Quokka	Wallaby	Normal Human Range
I (mg/dl)	281	667	—	474	208	105	150–450
II (U/ml)	0.42	0.79	0.44	0.55	0.46	0.26	0.70–1.30
V (U/ml)	0.60	1.20	1.20	1.00	1.28	1.40	0.65–1.45
VII (U/ml)	[a]	[a]	0.05	0.02	0.05	[a]	0.50–1.30
X (U/ml)	[a]	[a]	0.05	0.02	0.05	[a]	0.75–1.25
VIII (U/ml)	1.40	0.90	1.10	1.13	0.90	0.75	0.75–1.40
IX (U/ml)	0.80	0.15	0.15	0.37	0.08	0.10	0.65–1.70
XI (U/ml)	0.20	0.50	0.50	0.40	1.0+	1.0+	0.75–1.40
XII (U/ml)	1.0+	1.0+	1.0+	1.0+	1.0+	1.0+	0.50–1.45
XIII (R)	4	4	4	4	4	8	4–16

[a]Less than 0.01 U/ml.

although high values were not tested to their full dilutions. Factor XI was moderately lower in the echidna and normal or high in the marsupials. Factors II and IX appear in the low human range. Factors VII and X were very low. These two factors were assayed on fresh plasma in the old proconvertin assay, which tests for both activities. Lyophilized samples were retested a few months later when factor VII and factor X substrates became available, and the results still showed these factors to be almost undetectable. Human plasma lyophilized in Australia gave the same results when retested as it did when fresh.

3.3. Fibrinolytic Enzyme System

No casein hydrolysis was observed with any of the plasmas alone or after acid treatment (to remove inhibitor). Table 10.4 shows that echidna was activated with urokinase, but not with streptokinase. Quokka and wallaby plasmas were activated with both kinases, but to a lesser extent than human plasma.

3.4. Serum Proteins

Serum protein distribution after paper electrophoresis is shown in Table 10.5. Echidna total protein (TP) was below that of the marsupials or the human range. All show distribution similar to human.

3.5. Miscellaneous Tests

Echidna plasma in dilutions up to 1:2048 and wallaby plasma at 1:512 clumped the standard staphylococcal preparation.

TABLE 10.4
Plasminogen Activation with Streptokinase and Urokinase in the Plasmas of Primitive Australian Mammals by the Caseinolytic Method[a]

Plasma	I/S	SK (250 U/ml)	UK (250 U/ml)	Antiplasmin
Echidna	0%	2%	60%	100%
Quokka	0%	40%	16%	43%
Wallaby	0%	80%	31%	97%
Human	0%	100%	100%	100%

[a]Caseinolytic method results are calculated as the percentage of the activity of a standard human plasma panel.

TABLE 10.5
Serum Proteins by Paper Electrophoresis in Primitive Australian Mammals

Protein	Echidna (pool)	Quokka	Wallaby	Normal human range
TP (g/dl)	5.7	7.3	8.0	6.3–7.9
Albumin (g/dl)	3.6	5.6	4.6	3.2–4.4
α_1-Globulin (g/dl)	0.1	0.2	0.2	0.2–0.4
α_2-Globulin (g/dl)	0.1	0.1	0.8	0.4–1.0
β-Globulin (g/dl)	0.8	0.8	0.8	0.5–1.0
γ-Globulin (g/dl)	1.1	0.6	1.6	0.6–1.8

3.6. Cellular Elements

Table 10.6 shows the cell counts for the three echidnas, the wallaby, and the quokka. These primitive animals have blood counts that are remarkably like those of humans. The HCTs and MCVs of the echidnas and quokka were only slightly below the human range. The WBC counts were 11.0 × 10^3/mm^3 for echidna No. 3 and quokka and 4.5 × 10^3/mm^3 for the wallaby. The differential counts show no standard patterns. The platelets were numerous.

TABLE 10.6
Cellular Elements of the Blood in Primitive Australian Mammals

Element	Echidna No.			Quokka	Wallaby	Normal human range
	1	2	3			
HCT (%)	37	34	36	36	47	37–52
HGB (g/dl)	14.2	13.2	13.8	13.9	17.2	12–18
RBC (× 10^6/mm^3)	6.1	4.9	5.6	5.3	6.6	4.2–6.2
MCV (fl)	61	69	64	70	71	79–97
MCH (pg)	23	27	25	26	26	27–31
MCHC (g/dl)	38	39	38	39	37	32–36
WBC (× 10^3/mm^3)	6.7	10.1	11.0	11.0	4.5	5–10
Neutr (%)	88	50	40	43	19	55–75
Lymph (%)	9	46	50	52	74	20–40
Mono (%)	3	2	3	2	4	2–10
Eos (%)	0	2	4	1	3	0–3
Bas (%)	0	0	3	0	0	0–3
Plat C (× 10^3/mm^3)	488	660	498	1180	390	150–450

4. SUMMARY

The primitive mammals studied obviously have effective hemostatic mechanisms that resemble those of other mammals.

Echidna and marsupial blood clotted rapidly in glass and siliconized tubes. Clot retraction was less evident in the echidna blood. No lysis was seen. Prothrombin times with Simplastin and RVV were longer than human. Animal brain extracts clotted animal plasmas more rapidly than human. In the TGTs, animal mixtures usually clotted homologous substrate most rapidly.

The echidna and the marsupials assayed low in factors VII and X, moderately low in II, and within the human range in factors I, V, VIII, XII, and XIII. Only factor XI was low in echidnas and normal in the marsupials.

Fibrinolytic studies showed lack of streptokinase activation in the animals. When a small amount of human proactivator was added, some activity was obtained in marsupial plasma. With urokinase, echidna was more active than the marsupials.

Echidna, wallaby, and quokka had moderately small (MCV = 61–71 fl) but plentiful biconcave erythrocytes. Their leukocytes resembled those of human blood. Platelets were small and numerous.

REFERENCE

Lewis, J. H., Phillips, L. L., and Hann, C., 1968, Coagulation and hematological studies in primitive Australian mammals, *Comp. Biochem. Physiol.* **25:**1129.

SUGGESTED READINGS

Barrets, H., Reigel, K., Kleinhauer, E., and Lang, E. M., 1966, Comparative studies of the respiratory function of mammalian blood. II. Marsupialia: Great grey kangaroo and Tasmanian devil, *Respir. Physiol.* **1:**145.

Fantl, P., and Ward, H. A., 1957, Comparison of blood clotting in marsupials and man, *Aust. J. Exp. Biol. Med. Sci.* **35:**209.

Maxwell, C. M., Elliott, R. B., and Kneebone, G. M., 1964, Haemodynamics of kangaroos and wallabies, *Am. J. Physiol.* **206:**967.

Parer, J. T., and Metcalfe, J., 1967, Respiratory studies of monotremes. II. Blood of the echidna (*Tachyglossus setosus*), *Respir. Physiol.* **3:**143.

Parsons, R. S., and Guiler, E. R., 1972, Observations on the blood of the marsupial tiger cat, *Dasyurops maculatus* (Kerr) (Dasyuridae), *Comp. Biochem. Physiol. A* **43:**935.

Parsons, R. S., Heddle, R. W. L., Flux, W. G., and Guiler, E. R., 1970, Studies on the blood of the Tasmanian devil, *Comp. Biochem. Physiol.* **32:**345.

Parsons, R. S., Atwood, J., Guiler, E. R., and Heddle, R. W. L., 1970, Comparative studies on the blood of monotremes and marsupials. I. Haematology, *Comp. Biochem. Physiol. B* **39:**203.

Parsons, R. S., Guiler, E. R., and Heddle, R. W. L., 1971, Comparative studies on the blood of monotremes and marsupials. I. Electrolyte organic constituents, proteins, gas analysis and enzymes, *Comp. Biochem. Physiol. B* **39:**209.

CHAPTER 11

THE OPOSSUM

1. INTRODUCTION

The opossum (*Didelphis marsupialis virginiana*) is the only mammal of order Marsupialia found in North America. The other marsupials live in or near Australia (see Chapter 10) or South America. Opossums are cat-sized with white, pointed faces, scraggly fur, and long, scaly, prehensile tails. They are nocturnal, omnivorous animals that eat insects, grubs, small animals, fruits and berries.

Marsupials are unique in that the gestation period is short—at 2–3 weeks. After birth, the tiny babies crawl to the marsupium (pouch) on the mother's abdomen, which contains the mammary teats (13 for the opossum). After 8–10 weeks, the young climb out. At 4 months later, they are independent.

Hematological studies on opossums are sparse. Data concerning North American opossums are found in Ponder *et al.* (1929), Didisheim *et al.* (1959), Youatt *et al.* (1961), Mays and Loew (1968), Timmons and Marques (1969), and Giacometti *et al.* (1972). Rothstein and Hunsaker (1972) studied the South American woolly opossum. Studies from this laboratory (Lewis, 1975) are reprinted here with the kind permission of the editor of *Comparative Biochemistry and Physiology.*

2. SOURCE

A total of 11 opossums were studied. Seven were purchased from a biological supply house; four were trapped in the wild. As listed in the tables, the first six were female and the next five were male. The animals were caged for a few days and fed raw eggs and lettuce. The opossums' average weight was 8 kg. Light Nembutal anesthesia was administered, and blood was obtained by cardiac puncture.

3. TESTS CONDUCTED

3.1. Bleeding Times

Skin bleeding times were performed on nine animals by making 6 × 2 mm cuts with new knife blades on the soft underbelly skin. Bleeding stopped promptly in all, giving a mean bleeding time of 1 min with a range of 45–75 sec.

3.2. General Coagulation Tests and Factors

Table 11.1 lists the results of general tests of coagulation. Clotting times in glass and siliconized tubes were within the human range. Clots retracted well and did not lyse. Prothrombin times with Simplastin were longer, with opossum brain shorter, and with RVV about the same for opossum plasma and human plasma. The APTT was much faster in opossum than in human plasma. The recalcified plasma clots did not lyse in 1% MCA, indicating the presence of factor XIII.

Thrombin times, using bovine thrombin, were very long. Thrombin times of opossums No. 6 and 7 were over 60 sec, while a human control was 14.8 sec and equimixtures of human and opossum were 35.4 and 37.4 sec, respectively. This inhibitory effect was not changed by heating opossum oxalated plasma to 56°C or by treating with $BaSO_4$.

Table 11.2 shows the results of TGTs in six opossums. The mean clotting times on human substrate were 10.4 sec and on opossum substrate 10.8 sec. These times are within the range usually found for humans: 9.8–11.2 sec.

Results of coagulation factor assays are shown in Table 11.3. Factors II, VII, and X were very low, but the others were within the human range or above.

3.3. Fibrinolytic Enzyme System

Fibrinolytic activity was not evident in the opossum euglobulin fraction in the presence of streptokinase (100 and 1000 U/ml), staphylokinase (2%), or urokinase (40–100 U/ml) by the ^{125}I fibrin method. The concentration of antifibrinolysin was about the same as in human plasma.

3.4. Serum Proteins

By paper electrophoresis, the albumin peak was prominent (Table 11.4). Frequently, the α_2- and β-globulins merged with the γ-globulin into a large, gradually sloped curve in which individual proteins could not be differentiated. In two of the animals, it was possible to distinguish peaks corresponding to those found in human serum.

3.5. Biochemical Tests

The striking findings in Table 11.5 were the very high levels of serum enzymes. It is not known why these enzymes assay at such high levels. There is no evidence that the values indicate liver disease.

TABLE 11.1
General Coagulation Tests in the Opossum

Test	Opossum No.													Human	
	1	2	3	4	5	6	7	8	9	10	*N*	Mean	SD	Controls (4) (mean)	Normal range
Clot T (min)															
Glass	4	6	6	4	14	6	5	4	3	5	10	6	3.1	8	6–12
Silicone	18	18	14	51	45	17	45	13	18	19	10	26	14	29	20–59
Clot Retr	4/4	4/4	4/4	3/4	4/4	4/4	3/3	3/3	3/3	4/4	10	3–4	—	4/4	3–4+
Clot Lys	0	0	0	0	0	0	0	0	0	0	10	0	0	0	0
SPT (sec)															
Human	33	63	44	56	26	44	90	54	53	62	10	53	18	34.2	20+
Opossum	32	60	—	—	—	22	29	30	32	30	7	34	1.2	45.0	—
PT (sec)															
Simplastin	27.2	25.0	21.3	26.2	33.0	40.4	28.0	33.6	37.2	36.9	10	30.9	6.2	11.2	10–13
Opos Br	11.8	—	8.8	—	18.4	9.8	13.2	11.4	13.0	12.2	8	12.3	2.9	18.7	—
RVV	—	—	—	—	14.4	19.4	16.8	16.8	15.4	16.0	6	16.5	1.7	15.5	13–18
APTT (sec)	—	—	24.2	19.4	26.2	27.8	26.4	16.4	24.8	20.2	8	23.2	4.0	32.8	24–34
Recal T (sec)	90	—	190	85	110	120	60	60	60	60	9	93	43	125	90–180
Lys	0	—	0	0	0	0	0	0	0	0	9	0	0	0	0
Lys MCA	0	—	0	0	0	0	0	0	0	0	9	0	0	0	0
Thromb T (sec)															
Bov Thr	60+	60+	60+	60+	60+	60+	60+	60+	60+	60+	10	60+	0	14.8	11–18

TABLE 11.2
Thromboplastin Generation Test in the Opossum

Generating mixture		Substrate Clot T (sec)[a]	
Human	Opossum		
Means		Human	Opossum
H (4)	—	**10.1**	9.8
—	O (6)	10.4	**10.8**

[a]Homologous systems are in **boldface.**

3.6. Miscellaneous Tests

Opossum plasma did not clump a standard preparation of killed staphylococci (Newman D_2C coagulase-negative) [see Chapter 29 (Section 3.3)] used in the staphylococcal clumping test.

3.7. Cellular Elements

Table 11.6 shows the cellular findings in nine opossums. The RBC parameters fell at the lower edge of the human range. WBCs were elevated in Nos. 6, 7, 8, and 9. The mean differential counts showed that lymphocytes were more common than neutrophils, and eosinophils averaged 7%. Both counts and sizes of opossum platelets were quite variable. The percentage of opossum platelets (average of 5 animals) of different diameters was compared to average normal human distribution (in parentheses): <2 μm = 6% (0%), 2–2.9 μm = 39% (90%), 3.0–3.9 μm = 41% (8%), 4.0–4.9 μm = 9% (2%), >5 μm = 5% (0%).

3.8. Platelet Activities

Table 11.7 shows that the platelet glass retention index was within the normal human range in the four opossums studied. Aggregation of platelets in platelet-rich plasma (PRP) was rapid and complete with ADP. On the other hand, aggregation with bovine collagen was absent in four, low in two, and over 50% in two of eight opossum PRPs. It took 5 times the concentration of thrombin to aggregate opossum platelets that it did for human platelets. At this thrombin strength, the human PRP clotted rapidly. The lack of thrombin aggregation may have been due to the thrombin inhibitor in opossum plasma.

TABLE 11.3
Coagulation Factors in the Opossum

Factor	Opossum No. 1	2	3	4	5	6	7	9	10	11	*N*	Mean	SD	Normal human range
I (mg/dl)	312	318	465	237	535	212	212	187	206	200	10	288	121	150–450
II (U/ml)	[a]	0.01	[a]	0.01	[a]	0.05	0.32	0.17	0.25	0.17	10	0.10	0.1	0.70–1.30
V (U/ml)	1.10	1.66	2.0+	2.0+	1.98	2.0+	2.34	3.20	2.90	1.91	10	2.0+	—	0.65–1.45
VII (U/ml)	[a]	[a]	[a]	[a]	[a]	[a]	0.26	[a]	0.03	0.03	10	0.03	0.1	0.50–1.30
X (U/ml)	[a]	[a]	[a]	[a]	[a]	[a]	[a]	[a]	[a]	[a]	10	[a]	—	0.75–1.25
VIII (U/ml)	2.0+	2.0+	0.75	2.0+	—	2.0+	2.0+	2.0+	2.0+	2.50	9	2.0+	—	0.75–1.40
IX (U/ml)	0.60	1.30	0.50	1.40	—	0.80	1.30	1.50	1.75	1.50	9	1.20	0.5	0.65–1.70
XI (U/ml)	—	—	0.75	—	—	—	—	1.75	2.00	2.50	4	1.80	0.7	0.75–1.40
XII (U/ml)	2.00	2.0+	—	—	—	2.0+	1.25	1.50	1.50	2.0+	7	1.5+	—	0.50–1.45
ATIII (U/ml)	—	—	—	—	1.00	1.08	1.10	1.20	1.25	1.15	6	1.10	0.1	0.80–1.20
XIII (R)	16	16	8	16	8	4	8	16	4	4	10	10	5.4	4–16

[a]Less than 0.01 U/ml.

TABLE 11.4
Serum Proteins by Paper Electrophoresis in the Opossum

Protein	Opossum No.					N	Mean	SD	Normal human range
	7	8	9	10	11				
TP (g/dl)	8.1	6.4	6.4	7.6	7.0	5	7.1	0.7	6.3–7.9
Albumin (g/dl)	2.9	2.5	2.5	2.2	2.2	5	2.5	0.3	3.2–4.4
α_1-Globulin (g/dl)	0.3	0.6	0.5	0.6	0.9	5	0.6	0.2	0.2–0.4
α_2-Globulin (g/dl)	1.5	1.9	[a]	[a]	1.1	3	[a]	—	0.4–1.0
β-Globulin (g/dl)	2.2	0.8	3.4	4.8	1.1	5	2.5	1.7	0.5–1.0
γ-Globulin (g/dl)	2.7	0.6	[a]	[a]	1.3	3	[a]	—	0.6–1.8

[a] Since electropheresis of some opossum sera did not differentiate α_2-, β-, and γ-globulins, no mean was calculated for these protein fractions.

3.9. Platelet Ultrastructure

TEMs of opossum platelets are shown in Figs. 11.1 and 11.2. In Fig. 11.1, the platelet was sectioned in the equatorial plane and a circumferential band of microtubules was evident. Internally, an open canalicular system, small glycogen particles, and granules were observed. The platelets in Fig. 11.2 each showed a discrete mass of fine fibrillar material. This fibrillar material was seen in about 10% of all platelets from three different opossums; its nature is unknown.

4. SUMMARY

Coagulation test results in opossums appeared to be fundamentally similar to those for humans, although factors II, VII, and X assayed well below the normal human range and V, VIII, XI, and XII were above this range. Fibrinogen was quantitatively similar to human, but different in its poor clottability with bovine thrombin and inability to clump staphylococci. A heat-stable, non-heparin-like thrombin inhibitor was present in plasma.

Erythrocytes were slightly smaller than human. Leukocytes were not very different and lymphocytes predominated.

Platelets were variable in number and size and adhered to glass and aggregated with ADP much as human platelets do. On the other hand, they aggregated poorly or not at all with collagen and thrombin. Platelet ultrastructure was similar to human except for two striking differences: Many platelets contained a circumscribed mass of wavy fine fibrillar material of unknown composition; also, striated elongated granules were quite common.

TABLE 11.5
Biochemical Tests in the Opossum

Substance	Opossum No.[a]									
	6	7	8	9	10	11	*N*	Mean[a]	SD	Normal human range
TP (g/dl)	6.5	8.2	6.6	—	7.0	8.0	5	7.2	0.8	6.0–8.0
Albumin (g/dl)	*0.5*	*0.8*	*0.7*	—	*0.6*	*0.4*	5	*0.6*	0.2	3.5–5.0
Ca (mg/dl)	8.9	10.3	10.5	—	9.5	9.6	5	9.7	0.6	8.5–10.5
P (mg/dl)	4.5	**6.3**	**6.5**	—	**5.7**	**5.4**	5	**5.6**	0.8	2.5–4.5
Cholesterol (mg/dl)	162	190	154	—	180	169	5	171	14	150–200
Glucose (mg/dl)	94	109	**163**	*59*	*57*	67	6	91	41	65–110
Uric acid (mg/dl)	*1.3*	2.9	*1.4*	—	*1.5*	*1.5*	5	*1.7*	0.7	2.5–8.0
Creatinine (mg/dl)	*0.4*	0.8	0.7	—	*0.6*	*0.6*	5	*0.6*	0.1	0.7–1.4
T bilirubin (mg/dl)	0.2	0.5	0.6	—	0.3	0.2	5	0.5	0.2	0.1–1.4
BUN (mg/dl)	**21**	**47**	**31**	**27**	**37**	**31**	6	**32**	8.9	10–20
Chloride (meq/liter)	103	103	107	106	105	101	6	104	2.1	95–105
CO_2 (meq/liter)	*19*	*8*	*14*	*17*	*20*	*23*	6	*17*	5.3	24–32
K (meq/liter)	4.6	**6.9**	4.7	5.0	**5.2**	**5.7**	6	**5.3**	0.8	3.5–5.0
Na (meq/liter)	138	141	141	146	141	141	6	141	2.7	135–145
Alk Phos (IU/liter)	85	**196**	**178**	—	**118**	**114**	6	**137**	42	30–85
HBD (IU/liter)	**290**	**1385**	134	**470**	**452**	**400**	6	**522**	441	14–185
CPK (IU/liter)	**2060**	**5490**	**4480**	**7400**	**891**	**978**	6	**3550**	2659	0–110
LDH (IU/liter)	**321**	**1680**	128	**400**	**519**	**405**	6	576	556	60–200
SGOT (IU/liter)	**390**	**575**	**373**	**171**	**196**	**329**	6	339	147	0–41
SGPT (IU/liter)	**54**	**144**	**65**	—	35	42	6	**68**	40	10–50

[a]Values **higher** than human are in **boldface**, those *lower* than human in *italics*.

TABLE 11.6
Cellular Elements of the Blood in the Opossum

Element	Opossum No.												Normal human
	1	2	3	4	5	6	7	8	9	*N*	Mean	SD	range
HCT (%)	39	41	44	36	46	47	39	45	41	9	42	3.7	37–52
HGB (g/dl)	12.4	12.9	14.2	15.1	14.9	14.4	12.5	14.7	12.9	9	13.8	1.0	12–18
RBC (× $10^6/mm^3$)	4.8	4.7	5.5	3.7	5.7	6.8	5.3	5.8	5.4	9	5.3	0.8	4.2–6.2
MCV (fl)	82	87	80	96	81	70	74	78	76	9	80	7.6	79–97
MCH (pg)	26	28	30	40	26	21	24	26	24	9	27	5.5	27–31
MCHC (g/dl)	32	31	32	42	32	30	32	33	32	9	33	3.5	32–36
WBC (× $10^3/mm^3$)	5.5	7.8	8.1	9.0	10.0	16.2	12.4	18.1	11.8	9	10.9	4.1	5–10
Neutr (%)	21	48	31	53	27	39	44	39	90	9	44	20	55–75
Lymph (%)	77	41	52	40	64	50	46	46	53	9	52	12	20–40
Mono (%)	0	8	9	7	0	1	2	0	3	9	6	4.7	2–10
Eos (%)	2	3	8	0	7	10	6	15	7	9	6	4.5	0–3
Plat C (× $10^3/mm^3$)	312	390	488	103	320	963	275	682	953	9	498	304	150–450

TABLE 11.7
Platelet Activities in the Opossum

Activity	Opossum No. 1	2	4	6	7	8	9	10	*N*	Mean	SD	Normal human range
Plat Glass Ret Ind (%)	—	—	35	58	34	80	—	—	4	52	22	>20
Aggregation (%)												
ADP (10 μM)	100	85	—	100	100	95	100	85	7	95	7.1	70–100
Collagen												
Bov (0.19 mg/ml)	10	0	55	62	45	0	0	0	8	22	28	80–100
Opossum (0.2 mg/ml)	5	0	0	68	0	0	0	0	8	9	23	—
Bov Thr (0.1 U/ml)	0	0	—	—	—	—	0	0	4	0	0	60–100
(0.25 U/ml)	—	—	—	—	—	—	—	0	1	0	0	60–100
(0.5 U/ml)	80	Clot	—	—	—	100	100	100	5	95	10	Clot

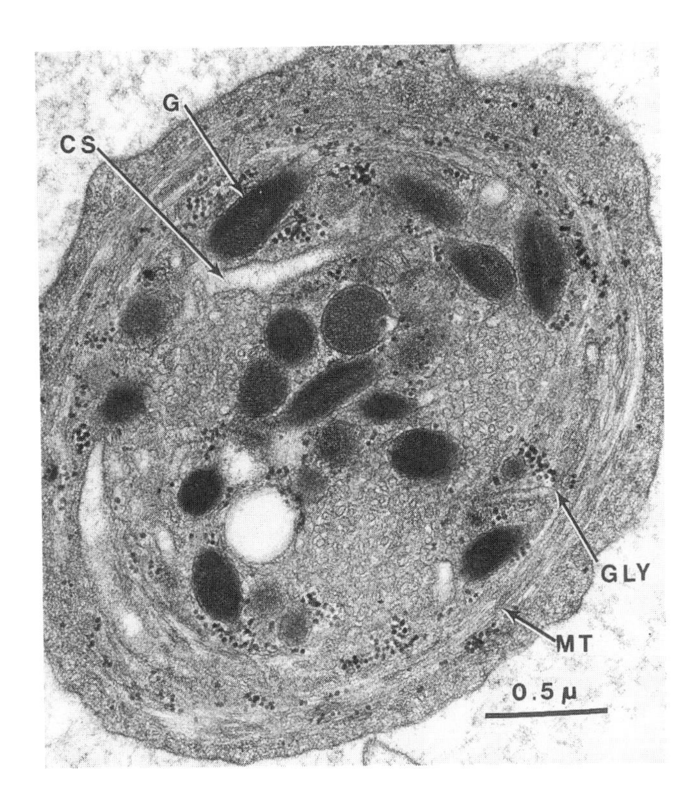

FIGURE 11.1. TEM of an opossum platelet. The platelet is sectioned in the equatorial plane and shows a circumferential band of microtubules (MT), open canalicular system (CS), a large granule (G), and sparse scattered glycogen particles (GLY).

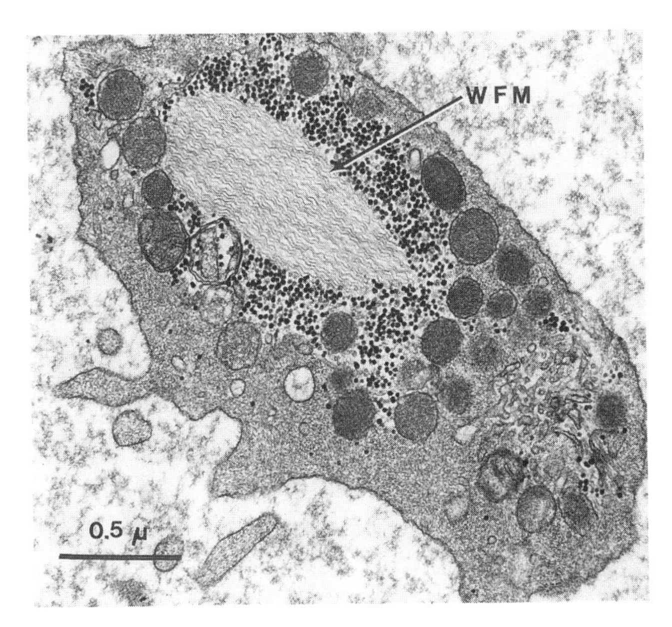

FIGURE 11.2. TEM of an opossum platelet. Note the large mass of wavy fibrillar material (WFM).

REFERENCES

Didisheim, P., Hattori, K., and Lewis, J. H., 1959, Hematologic and coagulation studies in various animal species, *J. Lab. Clin. Med.* **53:**866.

Giacometti, L., Berntzen, A. K., and Bliss, M. L., 1972, Hematologic parameters of the opossum (*Didelphis virginiana*), *Comp. Biochem. Physiol. A* **43:**287.

Lewis, J. H., 1975, Comparative hematology: Studies on opossums *Didelphis marsupialis (virginiana), Comp. Biochem. Physiol. A* **51:**275.

Mays, A., Jr., and Loew, F. M., 1968, Hemograms of laboratory-confined opossums (*Didelphis virginiana*), *J. Am. Vet. Med. Assoc.* **153:**800.

Ponder, E., Yeager, J. F., and Charipper, H. A., 1929, Studies in comparative haematology. III. Marsupialia, *Q. J. Exp. Physiol.* **19:**273.

Rothstein, R., and Hunsaker, D., II, 1972, Baseline hematology and blood chemistry of the South American woolly opossum, *Caluromys derbianus, Lab. Anim. Sci.* **22:**227.

Timmons, E. H., and Marques, P. A., 1969, Blood chemical and hematological studies in the laboratory-confined, unanesthetized opossum, *Didelphis virginiana, Lab. Anim. Care* **19:**342.

Youatt, W. G., Fay, L. D., Howe, D. L., and Harte, H. D., 1961, Hematologic data on some small mammals, *Blood* **18:**758.

CHAPTER 12

THE SPINY HEDGEHOG

1. INTRODUCTION

The order Insectivora consists of fewer than 400 species of insect-eating mammals: moles, shrews, and hedgehogs. Hedgehogs are large rat-sized animals weighing 0.4–1.1 kg. They live in the temperate climates of Europe and Asia. Their backs and sides are covered with a protective coat consisting of long hairs mixed with sharp striped spines. When the animal is frightened, circular muscles under the coat contract, erecting the spines every-which-way and rolling the animal into a protected ball.

Hedgehogs are chiefly nocturnal animals that rouse two or three times in the night to feed on insects, snails, mice, lizards, and bird eggs. Hedgehogs live in burrows under hedges, in woods, and in cultivated lands. They hibernate in the winter, producing heparin to prevent thrombosis during this period of inactivity. They have a maximum life expectancy of 10 years. Gestation is 5–6 weeks, and the female may have two litters of 5–7 young each year. The mother has eight nipples and nurses the tiny offspring for a few weeks. The hedgehog's natural enemies are parasites, a few aggressive large animals such as foxes and wolves, and the automobile.

References concerning the hematology of hedgehogs are scarce. Quilliam *et al.* (1971) describe counts of RBCs, WBCs, and platelets from three animals. Findings from this laboratory concerning coagulation and platelets were previously published (Lewis, 1976) and are reprinted here with the kind permission of the editor of *Comparative Biochemistry and Physiology.*

2. SOURCE

Two hedgehogs (*Erinaceus europaeus*) were imported from England through Charles P. Chase and Company. When we attempted to purchase more, we encountered import restrictions because these animals may be carriers of foot-and-mouth disease. The blood samples were obtained by cardiac puncture from the two animals after Nembutal anesthesia. Both animals were male and weighed 0.6–0.7 kg.

3. TESTS CONDUCTED

3.1. Bleeding Times

Bleeding times were done on the almost hairless underbelly skin by making a 5-mm cut with a new knife blade. Each animal received two such cuts. Bleeding ceased in 2 1/2 and 4 min in one animal and in 3 1/2 and 3 1/2 min. in the other, giving a mean bleeding time of 3 1/2 min.

3.2. General Coagulation Tests and Factors

Table 12.1 illustrates general coagulation tests. Blood clotted quickly and retracted poorly. The SPTs were long on both hedgehog and human substrates. Prothrombin times with Simplastin were about the same on hedgehog and human plasma. With hedgehog brain or RVV, hedgehog plasma clotted much more rapidly than human. The APTT and Recal Ts were much shorter than human. Thrombin times were long. The TGT using all-human components gave substrate clotting times within the expected range (9.8–11.2 sec). The hedgehog mixture was slightly slower, as shown in Table 12.2.

Results of coagulation factor assays are shown in Table 12.3. Hedgehog factors VII and X fell at the trace human level, and factor II was in the low range. Factors I, V, VIII, IX, XI, XII, and XIII were in the high to high–normal human range. ATIII was high.

TABLE 12.1
General Coagulation Tests in the Spiny Hedgehog

	Hedgehog No.			Human	
Test	1♂	2♂	Mean	Controls (2) (mean)	Normal range
Clot T (min)					
Glass	4	2	3	6	6–12
Silicone	20	10	15	29	20–59
Clot Retr	1/1	1/1	1–1	4/4	3–4+
Clot Lys	0	0	0	0	0
SPT (sec)					
Human	120+	120+	120+	34.0	20+
Hedgehog	28.4	32.4	30.4	120+	—
PT (sec)					
Simplastin	12.8	12.9	12.9	12.2	10–13
Hedgehog Br	10.1	10.4	10.3	19.8	—
RVV	14.5	16.4	15.5	20.4	13–18
APTT (sec)	19.7	19.3	19.6	49.0	45–55
Recal T (sec)	60	60	60	130	90–180
Lysis	0	0	0	0	0
Lys MCA	0	0	0	0	0
Thromb T (sec)					
Bov Thr	54.2	68.5	61.4	16.0	11–18

TABLE 12.2
Thromboplastin Generation Test in the Spiny Hedgehog

Generating mixture		Substrate Clot T (sec)[a]	
Human	Hedgehog	H	Hed
H	—	**10.4**	10.8
—	Hed	13.2	**11.0**

[a]Homologous systems are in **boldface.**

3.3. Fibrinolytic Enzyme System

Table 12.4 shows that hedgehog plasma clots lysed with streptokinase (330 U/ml) and urokinase (100 U/ml) in times slightly longer than human plasma clots.

3.4. Miscellaneous Tests

Table 12.4 includes some miscellaneous observations on hedgehogs. The ethanol gel test was positive, indicating partial clotting in the sample, probably due to difficulty in obtaining the blood sample. Staphylococcal clumping was positive to 1:1024 with hedgehog plasma.

TABLE 12.3
Coagulation Factors in the Spiny Hedgehog

	Hedgehog No.			Normal human range
Factor	1♂	2♂	Mean	
I (mg/dl)	280	296	288	150–450
II (U/ml)	0.16	0.26	0.21	0.70–1.30
V (U/ml)	1.96	2.40	2.18	0.65–1.45
VII (U/ml)	0.00	0.02	0.01	0.50–1.30
X (U/ml)	0.03	0.03	0.03	0.75–1.25
VIII (U/ml)	1.80	2.25	2.03	0.75–1.40
IX (U/ml)	2.00	2.25	2.13	0.65–1.70
XI (U/ml)	1.50	1.75	1.63	0.75–1.40
XII (U/ml)	1.50	1.50	1.50	0.50–1.45
ATIII (U/ml)	2.0+	2.0+	2.0+	0.80–1.20
XIII (R)	16	16	16	4–16

TABLE 12.4
Fibrinolytic and Miscellaneous Observations in the Spiny Hedgehog

Test	Hedgehog		Normal human range	
Euglobulin lysis time	3½	hr	3–6	hr
Plasma clot lysis	>120	min	>120	min
With UK (100 U/ml)	18	min	10½	min
With SK (330 U/ml)	20	min	10	min
Ethanol gel	4+		0	
Protamine gel	trace		0	
Cryoglobulin	0		0	
Sia test	0		0	
Staphylococcal clumping (titer) (on plasma)	1:1024		1:2048	

3.5. Cellular Elements

Table 12.5 shows the cellular elements in the two hedgehogs. HCT and HGB fell within the human range, but RBC counts were very high. The calculated indices showed the MCV (41 fl) and MCH (14 pg) to be much lower than humans, but the MCHC was in the human range. WBC counts were at the upper

TABLE 12.5
Cellular Elements of the Blood in the Spiny Hedgehog

Element	Hedgehog No. 1♂	Hedgehog No. 2♂	Mean	Normal human range
HCT (%)	40	42	41	37–52
HGB (g/dl)	13.5	13.5	13.5	12–18
RBC (× 10^6/mm^3)	9.9	9.6	9.8	4.2–6.2
MCV (fl)	39	42	41	79–97
MCH (pg)	13	14	14	27–31
MCHC (g/dl)	33	33	33	32–36
WBC (× 10^3/mm^3)	11.7	8.2	10.2	5–10
Neutr (%)	62	45	54	55–75
Lymph (%)	33	51	42	20–40
Mono (%)	3	2	3	2–10
Eos (%)	2	3	2	0–3
Bas (%)	0	0	0	0–3
Plat C[a] (× 10^3/mm^3)	93	132	113	150–450

[a]Hedgehog platelets appear to be larger than human.

TABLE 12.6
Platelet Aggregation in the Spiny Hedgehog

	Aggregation (%)				
	Hedgehog No.			Human	
Reagent	1♂	2♂	Mean	Control	Normal range
ADP (10 μM)	100	100	100	100	70–100
(5 μM)	100	90	95	100	50–100
Collagen Bov	100	—	100	95	80–100
Hedgehog	100	100	100	60	—

edge of the human range, with differential counts comparable to human. Platelets were low in number and appeared larger than their human counterparts.

3.6. Platelet Aggregations

Table 12.6 illustrates the results of aggregation tests in hedgehogs. Aggregation with ADP and collagen was similar to human.

3.7. Platelet Ultrastructure

Figures 12.1 and 12.2 show that TEMS of hedgehog platelets closely resemble human platelets. Circumferential microtubules, glycogen particles, dense bodies, an open canalicular system, a dense canalicular system, a Golgi apparatus, and mitochondria could be seen.

4. SUMMARY

Hedgehog blood clotted rapidly, and the clots retracted only slightly. The prothrombin time with Simplastin was like that on human plasma, but with hedgehog brain or RVV the times were much shorter. APTTs and Recal Ts were also short, but thrombin times were long. TGTs were much like human. Coagulation factor assays of factors VII and X were very low, and factor II was in the low human range. Factors I, V, VIII, IX, XI, XII, and XIII assayed in the high human range. RBCs were small and numerous. Leukocyte number and differential closely resembled their human counterparts. Platelets were low in count

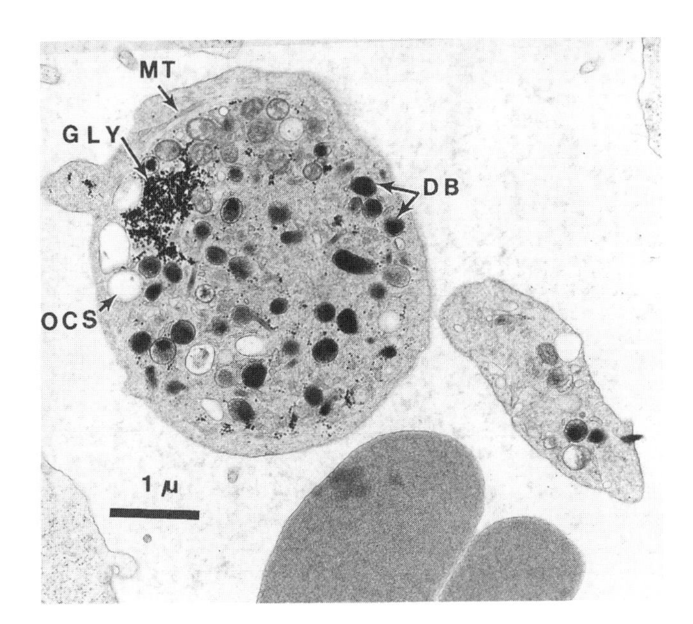

FIGURE 12.1 TEM of hedgehog platelets. Internal structures: (MT) circumferential microtubules; (GLY) glycogen particles; (DB) dense bodies; (OCS) open canalicular system.

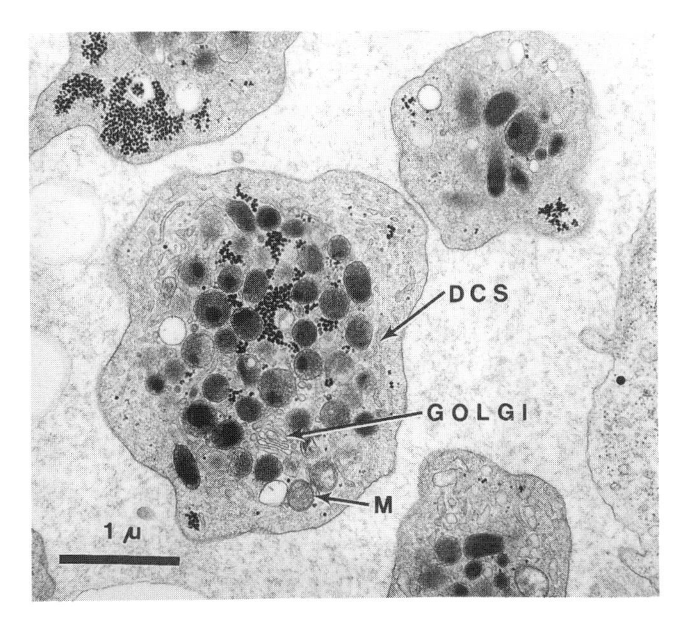

FIGURE 12.2. TEM of hedgehog platelets. Internal structures: (DCS) dense canalicular system; (GOLGI) Golgi apparatus; (M) mitochondrion.

but large. They reacted with ADP and collagen in a fashion similar to human platelets. Their ultrastructure was also similar to that of human platelets.

REFERENCES

Lewis, J. H., 1976, Comparative hematology—studies on hedgehogs (*Erinaceus europaeus*), *Comp. Biochem. Physiol. A* **53:**237.

Quilliam, T. A., Clarke, J. A., and Salsbury, A. J., 1971, The ecological significance of certain new haematological findings in the mole and hedgehog, *Comp. Biochem. Physiol. A* **40:**89.

SUGGESTED READINGS

Larsen, B., 1972, Seasonal variations in serum concentration and half-life of hedgehog IgG2, *Comp. Biochem. Physiol. A* **42:**115

Larsen, B., 1973, Properties of hedgehog immunoglobulin IgG2, *Comp. Biochem. Physiol. A* **44:**239.

Suomalainen, P., and Lehto, E., 1952, Prolongation of clotting time in hibernation, *Kurze Mitt.* **15:**65.

CHAPTER 13

THE FRUIT BAT

1. INTRODUCTION

Bats are the only mammals capable of true flight. Their wings are thin double membranes stretched between their fingers, arms, furry bodies, hindlimbs, and tails. The thumb and small hind feet are clawed. Bats have sharp teeth, although many eat only insects. Most are nocturnal, roosting by day in caves, trees, attics, and other sequestered places, hanging by their clawed feet with their heads downward. Poor vision is attributed to their small, beady eyes. Perhaps as a substitute, they have developed a sonar system called *echolocation.* They emit supersonic sound waves that bounce back from solid objects, enabling them to avoid head-on crashes in night flight and to locate their food: fruits or flying insects. Bats live in the temperate zones of the world. In frigid weather, some bats migrate to warmer climates, while others hibernate.

Bats constitute the order Chiroptera, which has been subdivided into two suborders, Megachiroptera and Microchiroptera, and almost a thousand species. Megachiroptera include the largest bat, *Pteropus giganteus,* which is found in Africa and Southern Asia. Great flocks of these bats may eat enormous amounts of fruit and are capable of destroying an entire orchard. Other bats eat only flying insects. Vampire bats consume blood and may harbor rabies or other disease vectors acquired from infected blood. Dried bat urine and feces, called *guano,* are valuable as fertilizer. Although bats usually produce only one offspring per year, they are not considered to be in danger of extinction. In fact, there are billions of bats. Their life expectancy is long, 20–25 years, and they are fertile most of this time. They have few natural enemies other than man, who frequently destroys their habitats.

Hematological studies of Chiroptera have been limited to cell counts (Krutzch and Wimsatt, 1963), capillary tube clotting times, or other tests that could be carried out on the small amounts of blood obtainable from these small mammals. Of interest have been the observations concerning the longer clotting times of brown bats (*Myotis lucifugus*) in winter compared to summer (Smith *et al.*, 1954a,b). In addition, these investigators showed that when bats were refrigerated, the clotting times became longer and the number of mast cells in the duodenum increased. Conversely, when dormant or hibernating bats were warmed, the clotting times became shorter and the number of mast cells decreased. These findings are suggestive of a release or increase of heparin during hibernation.

In order to obtain a sufficient volume of blood to perform coagulation studies, specimens of the large fruit bat of India (*Pteropus giganteus*) were utilized. Previously published coagulation and hematological studies performed in this laboratory (Lewis, 1977) are reprinted here with the kind permission of the editor of *Comparative Biochemistry and Physiology*.

2. SOURCE

Five fruit bats were imported from India through Charles P. Chase and Company. They weighed from 0.5 to 0.6 kg and had wingspans of approximately 1.4 m. They were caged for a few days before testing and were offered bananas, oranges, peaches, pears, apples, and water. When an attempt to acquire more fruit bats was made, importation was forbidden due to fear that these fruit-crop-destroying animals might escape and multiply. The blood samples were obtained by cardiac puncture.

3. TESTS CONDUCTED

3.1. Bleeding Times

Bleeding times were not done on bats.

3.2. General Coagulation Tests and Factors

Table 13.1 shows the results of general coagulation tests. Two of the bats, Nos. 4 and 5, had long glass clotting times. The shorter clotting times of bats No. 1–3 may simply reflect technical difficulties in obtaining the blood samples. The glass clotting times of bats No. 4 and 5 were longer than those of other mammals, except porpoises and some horses, and may reflect a true species deficiency. All bats had long APTTs, Recal Ts, and thrombin times. The prothrombin times with Simplastin, bat brain, and RVV were the same as or shorter than human. These findings suggested that the bat coagulation mechanism was similar to human in the extrinsic system but different in the intrinsic.

An inhibitor in the thrombin–fibrinogen reaction was explored in the tests in Table 13.2. This table shows thrombin times of mixtures of human and bat plasma after various treatments. These experiments were done with pooled frozen–thawed bat plasma. The thrombin time (0.1 ml thrombin + 0.2 ml plasma) for human plasma was 14.5 sec (a) and for bat plasma 120+ sec (h). Equimixture of bat and human plasmas gave a long thrombin time (b) that was shortened by treatment of the bat plasma with $BaCl_2$ (c) or by heating to 56°C (e). Simple addition of protamine shortened bat thrombin time from 120+ sec (h) to 17.5 sec (i) and human from 14.5 sec (a) to 11.3 sec (g). Shortening by $BaCl_2$,

TABLE 13.1
General Coagulation Tests in the Fruit Bat

	Bat No.								Human	
	1♂	2♂	3♀	4♀	5♀	N	Mean	SD	Control (2) (mean)	Normal range
Clot T (min)										
Glass	12	5	4	24	19	5	13	8.7	8	6–12
Silicone	—	5	—	36	33	3	25	17	32	20–59
Clot Retr	2/	3/3	3/	4/4	4/4	5	3–4	—	4/4	3–4+
Clot Lys	0	0	0	0	0	5	0	0	0	0
SPT (sec)										
Human	13.8	18.4	16.2	13.8	14.8	5	15.4	1.9	38.4	20+
PT (sec)										
Simplastin	10.2	10.4	11.4	10.8	9.9	5	10.5	0.6	11.4	10–13
Bat Br	13.8	19.2	14.4	19.2	20.4	5	17.0	3.3	21.8	—
RVV	17.4	19.2	13.7	15.0	14.8	5	16.4	2.2	17.4	13–18
APTT (sec)	180	189	209	354	161	5	233	82	48.0	45–55
Recal T (sec)	515	252	272	330	165	5	348	105	130	90–180
Lys	0	0	0	0	0	5	0	0	0	0
Lys MCA	0	0	0	0	0	5	0	0	0	0
Thromb T (sec)										
Bov Thr	44.4	73.0	68.0	45.5	40.4	5	57.7	18	16.1	11–18

which forms insoluble barium citrate, and by protamine suggested a heparin-type inhibitor.

Table 13.3 shows striking differences from human in the activities of bats' clotting factors. Factors XI and RCF were essentially absent, and factors II, VII, and XII were lower than human. Fibrinogen (I) and factors IX, X, XIII, and

TABLE 13.2
Inhibition of Thrombin Time in Pooled Bat Plasma

Mixture	Thromb T (sec)
0.1 ml Human plasma	
(a) +0.1 ml Human plasma	14.5
•(b) +0.1 ml Bat plasma	45.1
(c) +0.1 ml Bat plasma, $BaCl_2$-treated	21.2
(d) +0.1 ml Human plasma, $BaCl_2$-treated	19.9
(e) +0.1 ml Bat plasma, 56°C, 5 min	26.0
(f) +0.1 ml Human plasma, 56°C, 5 min	20.0
(g) +0.1 ml Protamine (0.1 mg/ml)	11.3
0.1 ml Bat plasma	
(h) +0.1 ml Bat plasma	120+
(i) +0.1 ml Protamine (0.1 mg/ml)	17.5

TABLE 13.3
Coagulation Factors in the Fruit Bat

Factor	Bat No. 1♂	2♂	3♀	4♀	5♀	*N*	Mean	SD	Normal human range
I (mg/dl)	425	375	260	525	265	5	370	112	150–450
II (U/ml)	0.36	0.48	0.35	0.46	0.65	5	0.46	0.1	0.70–1.30
V (U/ml)	5.75	4.00	5.50	4.40	5.90	5	5.11	0.9	0.65–1.45
VII (U/ml)	0.18	0.21	0.21	0.38	0.41	5	0.28	0.1	0.50–1.30
X (U/ml)	1.18	1.63	1.15	1.24	1.98	5	1.44	0.4	0.75–1.25
VIII (U/ml)	2.0+	1.65	1.40	1.78	1.66	5	1.70	0.2	0.75–1.40
IX (U/ml)	0.98	0.65	0.34	0.70	1.40	5	0.81	0.4	0.65–1.70
XI (U/ml)	[a]	[a]	[a]	[a]	[a]	5	[a]	—	0.75–1.40
XII (U/ml)	0.20	0.17	0.09	0.25	0.36	5	0.21	0.1	0.50–1.45
ATIII (U/ml)	—	0.55	—	0.95	1.08	3	0.86	0.3	0.80–1.20
RCF (U/ml)	[a]	[a]	—	[a]	[a]	4	[a]	—	0.50–1.50
XIII (R)	16	32	4	16	4	5	14	11	4–16

[a]Less than 0.01 U/ml.

ATIII were within the human range. Factor V was about 5 times greater than human, and factor VIII was slightly above human.

Bat plasma showed no ristocetin cofactor (RCF) activity; when it was mixed with fixed or fresh human platelets, no aggregation occurred with the addition of ristocetin (9 mg/ml). Human plasma caused aggregation in 15.4 sec. Doubling the concentration of ristocetin did not induce aggregation with bat plasma, but did cause some precipitation.

3.3. Fibrinolytic Enzyme System

Table 13.4 shows that the bat fibrinolytic enzyme system was activated by both streptokinase (SK) (250 U/ml) and urokinase (UK) (500 U/ml). Not shown is the finding that bat plasma inhibited the fibrinolytic action of trypsin to the same extent as did human plasma.

3.4. Serum Proteins

Table 13.5 shows the distribution of serum proteins as determined by electrophoresis on cellulose acetate. Differences from human were not striking. Electrophoresis on the Panagel system gave a different picture. Albumin and other negatively charged proteins migrated more rapidly than human. There were a number of bands in the α_2-globulin area that did not have familiar human counterparts.

TABLE 13.4
Fibrinolytic Activity with Streptokinase or Urokinase in the Fruit Bat

	^{131}I Fibrin lysed (%)		
	Plasma		
	Bat No.		
Reagent	4♀	5♀	Human
Buffer	0%	0%	0%
SK (250 U/ml)	61%	55%	74%
UK (500 U/ml)	36%	75%	83%

3.5. Miscellaneous Tests

Bat plasma gave a positive staphylococcal clumping reaction in dilutions up to 1:2048.

3.6. Cellular Elements

Table 13.6 shows that the HCT and HGB fell within the human range, but the RBC count was much higher. Thus, the MCV and MCH were low. RBCs looked small, the diameters measuring 5.0–6.2 μm. On SEM (Fig. 13.1), the

TABLE 13.5
Serum Proteins by Cellulose Acetate Electrophoresis in the Fruit Bat

	Bat No.								Normal human
Protein	1♂	2♂	3♀	4♀	5♀	*N*	Mean	SD	range
TP (g/dl)	8.0	6.0	8.4	8.3	6.4	5	7.4	1.1	6.3–7.9
Albumin (g/dl)	3.8	3.5	3.7	3.8	2.7	5	3.5	0.5	3.2–4.4
α_1-Globulin (g/dl)	0.4	0.4	1.1	0.9	0.6	5	0.7	0.3	0.2–0.4
α_2-Globulin (g/dl)	1.8	1.1	1.8	1.0	1.0	5	1.3	0.4	0.4–1.0
β-Globulin (g/dl)	0.5	0.4	0.8	1.1	0.4	5	0.6	0.3	0.5–1.0
γ-Globulin (g/dl)	1.5	0.6	1.0	1.5	1.7	5	1.3	0.5	0.6–1.8

TABLE 13.6
Cellular Elements of the Blood in the Fruit Bat

Element	Bat No.								Normal human range
	1♂	2♂	3♀	4♀	5♀	N	Mean	SD	
HCT (%)	33	46	38	32	34	5	37	5.7	37–52
HGB (g/dl)	9.7	16.7	13.0	11.8	12.2	5	12.7	2.6	12–18
RBC ($\times 10^6/mm^3$)	7.1	9.99	7.99	6.98	7.98	5	8.0	1.2	4.2–6.2
MCV (fl)	46	46	48	46	43	5	46	1.8	79–97
MCH (pg)	14	17	16	17	15	5	16	1.3	27–31
MCHC (g/dl)	29	36	34	37	36	5	34	3.2	32–36
WBC ($\times 10^3/mm^3$)	7.3	6.0	27.0	26.2	16.9	5	16.7	10	5–10
Neutr (%)	38	—	16	23	—	3	25	11	55–75
Lymph (%)	57	—	76	47	—	3	60	15	20–40
Mono (%)	5	—	8	30	—	3	14	14	2–10
Eos (%)	0	—	0	0	—	3	0	0	0–3
Bas (%)	0	—	0	0	—	3	0	0	0–3
Plat C ($\times 10^3/mm^3$)	875	875	810	742	795	5	819	57	150–450

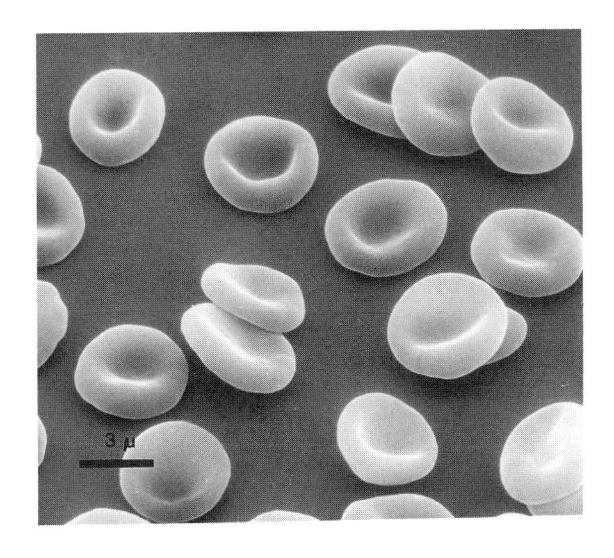

FIGURE 13.1 SEM of bat erythrocytes.

TABLE 13.7
Platelet Aggregation and Diameter in the Fruit Bat

Parameter	Bat No. 2♂	Bat No. 4♀	Bat No. 5♀	Human: Controls No. 4 and 5 (mean)	Human: Normal range
Aggregation (%)					
ADP (10 μM)	100	100	100	95	70–100
Collagen (0.19 mg/ml)	95	100	100	100	80–100
Risto (0.9 mg/ml)	0	0	0	100	80–100
Platelet diameter (μ)	<2	<2	<2	3	2–3

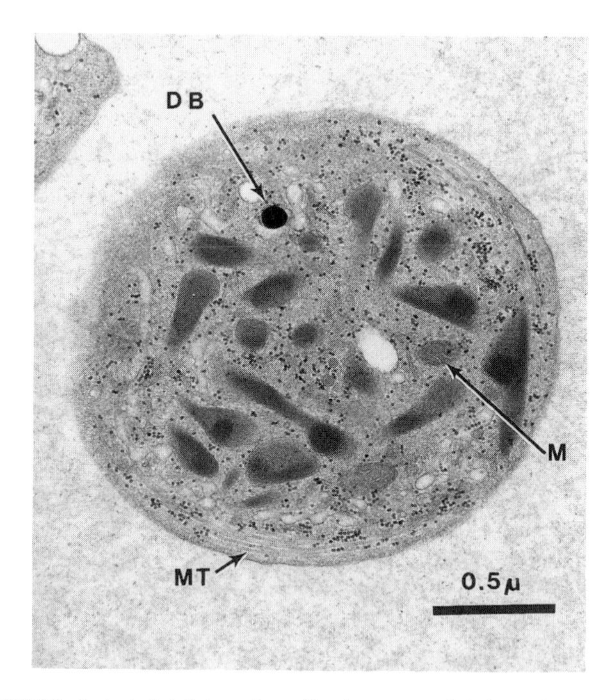

FIGURE 13.2 TEM of a bat platelet sectioned in the equatorial plane. Internal structures: (DB) dense body; (M) mitochondrion; (MT) microtubules.

erythrocytes appeared to have the typical mammalian doughnut shape. The WBCs showed varied counts and differentials in the bats. No eosinophils or basophils were seen. The platelets were small and numerous.

3.7. Platelet Aggregation

Bat platelets (Table 13.7) aggregated completely with ADP or collagen. No aggregation was observed with ristocetin.

3.8. Platelet Ultrastructure

Figure 13.2 presents a TEM of a bat platelet sectioned in the equatorial plane. A clear-cut band of circumferential microtubules was just below the surface of the platelet. Mitochondria, dense bodies, an open canalicular system, and scattered glycogen particles could be seen. Most of the alpha granules were elongated rather than round.

4. SUMMARY

About 15 ml of blood was obtained by cardiac puncture from each of five Asian fruit bats (*Pteropus giganteus*). Bat blood clotted more slowly than human and showed prolonged APTTs and long thrombin times. These long APTTs appeared to be due to very low or absent factor XI activity. A heparinlike inhibitor was found that caused the long thrombin times. The fibrinolytic system was activated by both SK and UK. Staphylococcal clumping titer was 1:2048. Erythrocytes and platelets were moderately small and numerous. Bat platelets aggregated with ADP and collagen, but not with ristocetin. Ultrastructure was not remarkable.

REFERENCES

Krutzch, P. H., and Wimsatt, W. A., 1963, Some normal values of the peripheral blood in the vampire bat, *J. Mammal.* **44:**556.

Lewis, J. H., 1977, Comparative hematology: Studies on chiroptera, *Pteropus giganteus*, *Comp. Biochem. Physiol. A* **58:**103.

Smith, D. E., Lewis, Y. S., and Svihla, G., 1954a, Blood clotting time and tissue mast cell number of the bat (*Myotis lucifugus*) in different physiological states, *Proc. Soc. Exp. Biol. Med.* **86:**473.

Smith, D. E., Lewis, Y. S., and Svihla, G., 1954b, Prolongation of clotting time in the dormant bat (*Myotis lucifugus*), *Experientia* **10:**218.

SUGGESTED READINGS

Cartwright, T., 1974, The plasminogen activator of vampire bat saliva, *Blood* **43**:317.

Cartwright, T., and Hawkey, C., 1969, Activation of the blood fibrinolytic mechanism in birds by saliva of the vampire bat (*Diaemus youngi*), *J. Physiol. (London)* **201**:45.

Hawkey, C., 1966, Plasminogen activator in saliva of the vampire bat, *Desmodus rotundus, Nature (London)* **211**:434.

Manwell, C., and Kerst, K. V., 1966, Possibilities of biochemical taxonomy of bats using hemoglobin, lactate dehydrogenase, esterases and other proteins, *Comp. Biochem. Physiol.* **17**:741.

Rasweiler, J. J., IV, and De Bonilla, H., 1972. Laboratory maintenance methods for some nectarivorous and frugivorous phyllostomatid bats, *Lab. Anim. Sci.* **22**:658.

CHAPTER 14

THE PRIMATES

1. INTRODUCTION

The mammalian order Primates is divided into two suborders: Prosimii, which includes tree shrews, lemurs, and bush babies, and Anthropoidea, which has three superfamilies: (1) New World monkeys and marmosets, (2) Old World monkeys and baboons, and (3) apes, orangutans, gorillas, chimpanzees, and man. Phylogenetically, man is moderately close to baboons and rhesus monkeys, but still closer to chimpanzees. These primates have been favorite models for investigators. Baboons have served in important studies of induced thrombosis and its treatment (Harker and Hanson, 1979). Very recently, baboons have been the source of livers for baboon-to-human liver xenografts (Starzl *et al.*, 1993). Previously published data (Lewis, 1977) concerning rhesus monkeys are reprinted here with the kind permission of the editor of *Comparative Biochemistry and Physiology.*

2. SOURCE

Baboon blood was obtained from three sources. Baboons No. 1 and 2 were of the species *Papio anubis;* their blood was collected at the Primate Research Center, University of Pittsburgh. Baboons No. 3–6 were *P. cynocephalus;* blood samples were obtained for us by Dr. Peter C. Johnson, Department of Plastic Surgery, University of Pittsburgh. Baboons No. 7–9 were also *P. cynocephalus;* blood samples were received through the Transplant Center. All baboons were adult males weighing 22–35 kg. Chimpanzee blood was acquired from 7 animals at the National Institutes of Health, Bethesda, Maryland, through the help of Dr. Sayah Nedjar. The chimpanzees (*Pan troglodytes*) were West African in origin, weighed about 100 lb (45 kg), and were 3 1/2–4 feet (110–120 cm) tall. Numbers 1, 2, and 5 were female. A total of 14 adult (>4-yr-old) monkeys were studied at the Primate Research Center, University of Pittsburgh. These included 8 females (Nos. 4–11) weighing 3.9–6.1 kg and 6 males (Nos. 1–3 and 12–14) weighing 8–9 kg.

All primates were tranquilized, and blood samples were collected from the antecubital or femoral veins. These primates were all involved in other investigations, and as far as possible, our samples were taken before others or after a quiescent period.

3. TESTS CONDUCTED

3.1. Bleeding Times

Bleeding times on 22 baboons (Peter Johnson, personal communication) varied from 150 to 618 sec with a mean ± SD of 313 ± 139 and a median of 254 sec. Bleeding times were not done on the chimpanzees. Simplate bleeding times without blood pressure cuff on four monkey forearms were 2, 2 ½, 2 ½, and 3 min (mean: 2 ½ min).

3.2. General Coagulation Tests and Factors

As shown in Table 14.1, primate clotting times in glass and siliconized tubes were short, and the clots were retracted partially or completely at 4 hr. At 2 hr, clots from two chimpanzees were lysed, and there was high RBC fallout in the others. SPTs on the baboons and monkeys were similar to human. Prothrombin times with Simplastin and RVV and APTTs were within the human range in

TABLE 14.1
General Coagulation Tests in Primates

	Baboon			Chimpanzee[a]			Monkey			Normal human range
	N	Mean	SD	*N*	Mean	SD	*N*	Mean	SD	
Clot T (min)										
Glass	7	6	0.9	5	3	1.0	10	5	1.3	6–12
Silicone	7	21	3.6	5	3	1.1	10	11	1.9	20–59
Clot Retr	7	4–4	—	5	2–4	—	10	3–4	—	3–4+
Clot Lys	7	0	0	5	2/5	0	10	0	0	0
SPT (sec)										
Human	7	22.6	3.6	—	—	—	10	31.4	12	20+
Homolog	—	—	—	—	—	—	4	37.9	6.4	—
PT (sec)										
Simplastin	9	12.1	0.5	5	10.6	0.4	10	11.5	1.4	10–13
RVV	6	17.6	1.7	5	15.9	1.4	10	21.6	1.1	13–18
APTT (sec)	9	33.1	4.1	5	22.0	1.8	10	35.5	2.9	24–34
Recal T (sec)	8	280	28	5	73	13	10	111	19	90–180
Lys	8	0	0	5	0	0	10	0	0	0
Lys MCA	8	0	0	5	0	0	10	0	0	0
Thromb T (sec)										
Bov Thr	9	24.5	2.5	5	24.5	2.4	6	22.1	3.7	11–18
CHH	6	25.7	5.8	5	18.1	0.6	—	—	—	11–18
Atroxin	9	14.5	1.1	5	13.2	0.6	—	—	—	11–18

TABLE 14.2
Thromboplastin Generation Test in Primates

Generating mixture			Substrate clotting time (sec)[a]		
Human	Baboon	Monkey (mean)	Human	Baboon	Monkey
H (mean of 3)	—	—	**9.9**	9.8	10.2
—	B (1)	—	10.4	**11.1**	—
—	—	M (4)	10.7	—	**11.3**

[a]Homologous systems are in **boldface.**

baboons and chimpanzees. For the monkeys, prothrombin times with Simplastin were also within the human range, but with RVV, the times were slightly longer. APTTs were also slightly longer. The Recal Ts were moderately long for baboons, short for chimpanzees, and within the human range for monkeys. The Recal clots did not lyse alone or in 1% MCA. Thrombin times with bovine thrombin and *Crotalus horridus horridus* venom (CHH) were slightly above the human range. *Bothrops atrox* venom (Atroxin) times in baboons and chimpanzees resembled human.

TGTs (Table 14.2) were done on one baboon and four monkeys. The substrate clotting times were close to those of human mixtures.

Coagulation factor levels are shown in Table 14.3. The baboon factor pattern was particularly striking. Factor VII levels were very high, factor XII moderately high, and factors IX and XI low compared to humans. Chimpanzee factors were within the human range except for XI and XII, which were moderately high. Baboon and chimpanzee plasminogens assayed low, perhaps due to specificity of this assay for human plasminogen. Monkey factors were remarkably similar to human factors, except for moderately high VII and high XIII.

3.3. Baboon-to-Human Liver Xenograft

The striking differences between human and baboon coagulation tests allowed the recognition of baboon liver function in a xenografted human recipient. Table 14.4 shows the coagulation factor changes during 19 postoperative days. The donee was a 35-year-old male with severe liver disease due to hepatitis B who survived only 20 days posttransplant. His preoperative coagulation pattern (in the "−2" column) was typical of that seen in severe liver disease. All factors except VIII were low to moderately low. After transplant, the pattern changed to reflect baboon levels: Factors **VII** and **XII** rose above 1.0 U/ml (normal human);

TABLE 14.3
Coagulation Factors in Primates

Factor	Baboon			Chimpanzee			Monkey			Normal human range
	N	Mean	SD	*N*	Mean	SD	*N*	Mean	SD	
I (mg/dl)	9	196	40	5	288	58	10	305	107	150–450
II (U/ml)	9	0.93	0.16	5	1.12	0.1	10	1.14	0.1	0.70–1.30
V (U/ml)	9	1.09	0.27	5	1.06	0.08	10	1.05	0.1	0.65–1.45
VII (U/ml)	9	3.63	0.8	5	1.32	0.13	6	1.85	0.3	0.50–1.30
X (U/ml)	9	0.90	0.16	5	1.07	0.07	10	0.80	0.2	0.75–1.25
VIII (U/ml)	9	0.78	0.17	5	1.32	0.31	10	1.09	0.2	0.75–1.40
IX (U/ml)	9	0.44	0.14	5	1.44	0.24	9	1.01	0.2	0.65–1.70
XI (U/ml)	9	0.40	0.06	5	2.24	0.31	5	1.00	0.4	0.75–1.40
XII (U/ml)	9	2.80	0.6	5	2.86	0.64	5	0.93	0.4	0.50–1.45
Flet (U/ml)	9	0.93	0.33	5	1.17	0.15	—	—	—	0.50–1.50
Fitz (U/ml)	9	0.61	0.31	5	1.37	0.19	—	—	—	0.50–1.50
ATIII (U/ml)	3	1.17	0.18	5	1.26	0.02	6	0.79	0.2	0.80–1.20
Prot C I (U/ml)	3	0.89	0.05	5	0.76	0.17	—	—	—	0.60–1.40
Prot C A (U/ml)	3	1.62	0.19	5	0.77	0.12	—	—	—	0.60–1.40
Prot S I (U/ml)	3	0.44	0.05	5	0.94	0.19	—	—	—	0.60–1.40
Plgn (U/ml)	3	0.09	0.01	5	0.38	0.03	—	—	—	0.80–1.20
AntiPl (U/ml)	3	1.17	0.04	5	1.00	0.08	—	—	—	0.80–1.20
RCF (U/ml)	3	2.68	1.8	5	0.87	0.34	—	—	—	0.50–1.50
XIII (R)	6	18.7	10.9	—	—	—	10	34.4	26.4	8–16

TABLE 14.4
Coagulation Factors following Baboon to Human Liver Xenograft No. 2

Factor	Man (postoperative day) −2	3	11	19	Baboon[a]	Normal human range
I (mg/dl)	80	70	92	96	196	150–450
II (U/ml)	0.29	0.25	0.26	0.32	0.93	0.70–1.30
V (U/ml)	0.17	0.41	0.41	0.92	1.09	0.65–1.45
VII (U/ml)	**0.11**	**0.31**	**1.41**	**1.54**	**3.63**	0.50–1.30
X (U/ml)	0.44	0.28	0.53	0.70	0.90	0.75–1.25
VIII (U/ml)	2.20	1.68	1.01	2.45	0.78	0.75–1.40
IX (U/ml)	*0.33*	*0.27*	*0.28*	*0.47*	*0.44*	0.65–1.70
XI (U/ml)	*0.18*	*0.22*	*0.13*	*0.33*	*0.40*	0.75–1.40
XII (U/ml)	**0.49**	**0.56**	**0.80**	**1.33**	**2.80**	0.50–1.45

[a]The case is described in Starzl *et al.*, 1993.

factors *IX* and *XI* stayed at low baboonlike levels. Unfortunately, the disease persisted.

3.4. Fibrinolytic Enzyme System

Table 14.5 shows the results of a radial diffusion protease assay in Rhesus monkeys. Neither human nor monkey plasmas showed any lytic activity alone. With streptokinase and urokinase the activities of human and monkey plasma

TABLE 14.5
Radial Diffusion Protease Assay in the Rhesus Monkey

	Diameter of cleared zones (mm)			
		Activator		
Plasma	Saline	SK (50 U/ml)	UK (100 U/ml)	Inhibition of trypsin
Human	0	9.0	10.0	9.7
Monkey				
No. 9	0	9.6	6.0	7.9
No. 10	0	5.8	13.0	7.7
No. 11	0	6.2	6.0	7.7
No. 12	0	9.9	12.0	9.0
No. 13	0	6.0	5.6	7.8
No. 14	0	9.7	12.0	7.8
Buffer	0	0	0	19.0

were variable but similar. Both human and monkey plasma decreased the activity of trypsin by more than 50%.

3.5. Miscellaneous Tests

Baboon plasma gave a staphylococcal clumping titer of 1:512. The ethanol and protamine gel tests and the Wellco and D-Dimer tests were negative on serum. The euglobulin lysis times were more than 3 hr. Monkey plasma staphylococcal clumping titer was 1:1024.

3.6. Cellular Elements

Table 14.6 lists the mean values for elements of the peripheral blood found in baboons, chimpanzees, and monkeys. The HCT and HGB of the baboon are slightly below the normal human range while those of the chimpanzees and monkey fall within this range. The calculated indices are low for MCV in monkey and MCH in all three. MCHCs were within the human range. WBCs are normal or slightly elevated. In baboons and monkeys the lymphocytes are somewhat more frequent than neutrophils, a status that is frequent in very young humans. Platelet counts were within the human range in baboons and chimpanzees and slightly above human in monkey. Platelet size did not appear remarkable.

TABLE 14.6
Cellular Elements of the Peripheral Blood in Primates

	Baboon			Chimpanzee			Monkey			Normal human
Element	*N*	Mean	SD	*N*	Mean	SD	*N*	Mean	SD	range
HCT (%)	7	36	3.8	5	39	2.6	13	44	3.3	37–52
HGB (g/dl)	7	11.4	1.2	5	12.6	0.9	13	13.9	1.1	12–18
RBC (× 10^6/mm^3)	7	4.5	0.5	5	4.8	0.3	13	5.7	0.6	4.2–6.2
MCV (fl)	7	80	2.5	5	82	3.6	13	78	5.7	79–97
MCH (pg)	7	26	0.5	5	26	0.9	13	25	1.9	27–31
MCHC (g/dl)	7	32	1.3	5	33	1.3	13	32	0.9	32–36
WBC (× 10^3/mm^3)	7	7.9	2.6	5	13.3	4.8	13	10.7	3.8	5–10
Neutr (%)	7	44	13	5	61	17	13	41	12	55–75
Lymph (%)	7	46	13	5	35	17	13	53	12	20–40
Mono (%)	7	7	3.9	5	2	1.5	13	1	1.3	2–10
Eos (%)	7	2	1.0	5	1	0.9	13	4	2.1	0–3
Bas (%)	7	1	1.0	5	0	0	13	0	0	0–3
Plat C (× 10^3/mm^3)	7	299	60	5	406	76	9	510	181	150–450
MVP (μM^3)	7	6.3	0.6	5	8.5	0.7	—	—	—	5.6–10.4

3.7. Biochemical Tests

Biochemical tests were done on five baboons and five monkeys (Table 14.7). Baboons showed levels above humans in P, chloride, Na, Alk Phos, and LDH. The values for cholesterol, uric acid, creatinine, and CO_2 were lower than human values. Monkeys had high levels of total protein, Ca, BUN, Na, Alk Phos, CPK, LDH, and SGOT. Low levels were found in uric acid and CO_2.

3.8. Platelet Activities

Baboon platelets aggregated with collagen, ristocetin (Risto), CHH, and pig plasma (Table 14.8). With ADP, aggregation was biphasic; that is, the platelets aggregated and then disaggregated shortly thereafter. Chimpanzee platelet reactions were similar to baboon. Three of the five animals showed biphasic curves with ADP at 20, 10, and 5 μM. All five curves were biphasic with 2.5 μM ADP. Figure 14.1 shows aggregation with ADP at four concentrations. Only the highest—40 μM—underwent only minimal disaggregation. Figure 14.2 shows

TABLE 14.7
Biochemical Tests in Primates

Substance	Baboon			Monkey			Normal human range
	N	Mean[a]	SD	*N*	Mean[a]	SD	
TP (g/dl)	5	6.7	0.5	5	**8.9**	0.7	6.0–8.0
Albumin (g/dl)	5	4.4	0.2	5	4.2	0.3	3.0–5.5
Ca (mg/dl)	5	9.7	0.3	5	**11.5**	0.1	8.5–10.5
P (mg/dl)	5	**5.8**	1.0	5	**4.9**	0.9	2.5–4.5
Cholesterol (mg/dl)	5	*127*	17	5	147	25	150–200
Glucose (mg/dl)	5	79	9.0	5	94	22	65–110
Uric acid (mg/dl)	3	*0.1*	0	5	*0.3*	0.1	2.5–8.0
Creatinine (mg/dl)	5	*0.5*	0.2	5	1.5	0.2	0.7–1.4
T bilirubin (mg/dl)	5	0.2	0.1	5	0.3	0.1	0.2–1.4
BUN (mg/dl)	5	17	4.0	5	**27**	3.1	10–20
Chloride (meq/liter)	5	**110**	2.4	5	105	1.6	95–105
CO_2 (meq/liter)	5	*22*	1.6	5	*19*	0.8	24–32
K (meq/liter)	5	3.7	0.4	5	44	0.6	3.5–5.0
Na (meq/liter)	5	**150**	2.8	5	**150**	3.0	135–145
Alk Phos (IU/liter)	4	**598**	247	5	**203**	57	30–85
CPK (IU/liter)	—	—	—	5	**306**	212	0–110
LDH (IU/liter)	5	**333**	112	5	**477**	91	60–200
SGOT (IU/liter)	5	29	15	5	**71**	7.9	0–41
SGPT (IU/liter)	5	43	21	—	—	—	10–50

[a]Values **higher** than human are in **boldface**, values *lower* than human in *italics*.

TABLE 14.8
Platelet Activities in Primates

Activity	Baboon			Chimpanzee			Monkey			Human	
	N	Mean	SD	*N*	Mean	SD	*N*	Mean	SD	Controls (mean)	Normal range
Plat Glass Ret Ind (%)	—	—	—	—	—	—	5	63	11	42	>20
Aggregation (%)											
ADP (40 μM)	5	60[a]	1.7	—	—	—	—	—	—	86	70–100
(30 μM)	5	55[a]	2.0	—	—	—	—	—	—	83	70–100
(20 μM)	7	57[a]	13	5	48[a]	12	—	—	—	86	70–100
(10 μM)	7	44[a]	14	5	45[a]	13	5	98	4.5	90	70–100
(5 μM)	2	22[a]	2.0	5	40[a]	16	5	76	20	77	50–100
(2.5 μM)	2	5[a]	0	5	26[a]	15	—	—	—	44	40–100
Collagen (0.19 mg/ml)	6	89	14	5	61	30	5	95	7.1	88	80–100
Risto (0.9 mg/ml)	6	97	9.1	5	90	3.0	5	100	0	91	80–100
Risto ½ (0.45 mg/ml)	6	3	2.1	5	8	4.0	5	5	3.0	5	<18
Bov Thr (0.4 U/ml)	—	—	—	5	22	10	—	—	—	Clot	60–100
(0.25 U/ml)	—	—	—	5	8	8.0	—	—	—	48[a]	60–100
CHH (0.05 mg/ml)	6	59	15	5	24[a]	17	—	—	—	83	60–100
Arach A (0.5 mg/ml)	6	67	11	5	3	4.0	—	—	—	83	60–100
Pig Pl (1:10)	6	66	32	5	51	6.0	—	—	—	95	60–100
Hu Pl (1:10)	—	—	—	5	12	3.1	—	—	—	—	<10

[a]At least half of curves were biphasic, showing aggregation followed by disaggregation.

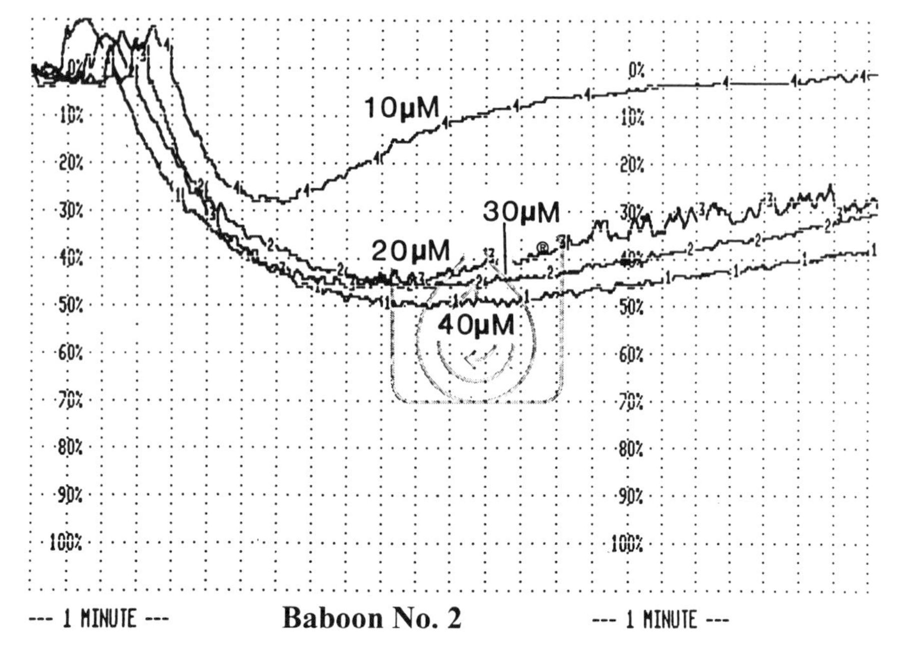

FIGURE 14.1 Aggregation and disaggregation with four strengths of ADP.

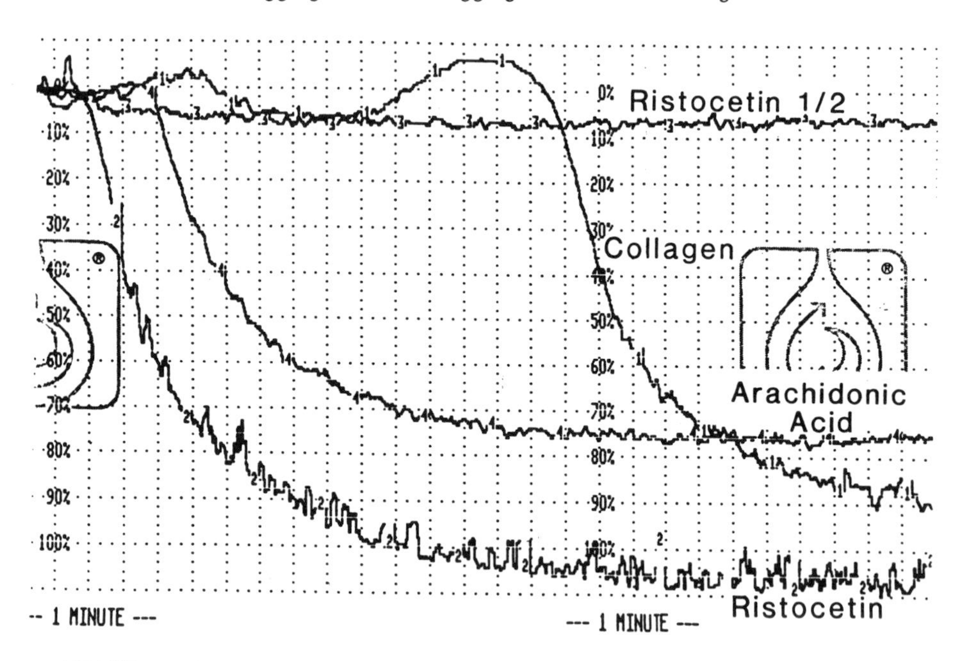

FIGURE 14.2 Aggregation of baboon No. 2 platelets with bovine soluble collagen (0.5 mg/ml), ristocetin (0.9 mg/ml), ristocetin ½ (0.45 mg/ml), and arachidonic acid (0.5 mg/ml).

aggregation of baboon No. 2 platelets with collagen, Arach A, and Risto. These curves resembled human platelet-rich plasma (PRP). Three of the five chimpanzees aggregated with collagen in the human pattern—a lag period of 2–3 min, then a short decrease in light transmission (size change) followed by increasing light transmission (aggregation). Two of the chimpanzees did not react to collagen. The mean for collagen aggregation was a low 45%. All Risto curves followed the human pattern. Arach A did not aggregate these chimpanzee platelets, perhaps because of the tranquilizer given to the chimpanzees before the blood samples were obtained. Pig plasma clearly caused aggregation in all chimpanzees, but it was somewhat less than in human or baboon PRP.

3.7. Platelet Ultrastructure

TEMs of baboon (Fig. 14.3), chimpanzee (Fig. 14.4), and monkey platelets (Figs. 14.5 and 14.6) show that most of the internal structures are similar to those seen in man. Monkey TEMs show large clumps of glycogen and structures of unknown character circumscribed with fibrous strands.

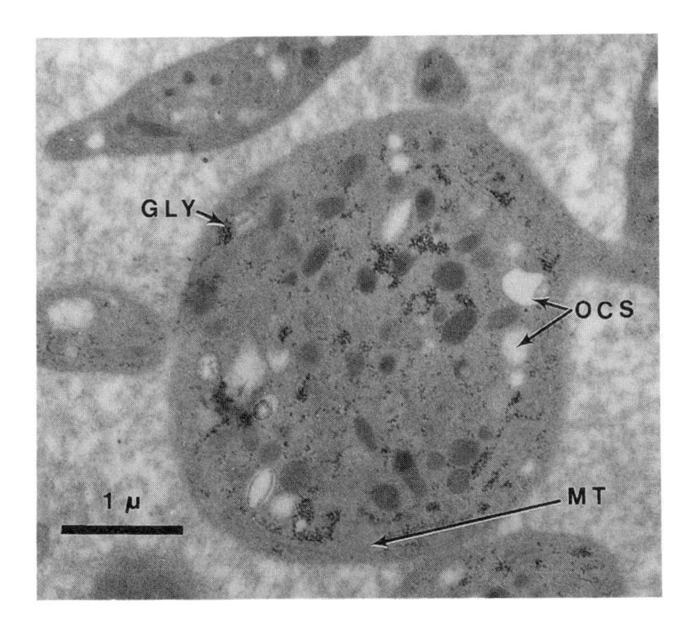

FIGURE 14.3. TEM of a baboon platelet. Internal structures: (OCS) open canalicular system; (MT) microtubules.

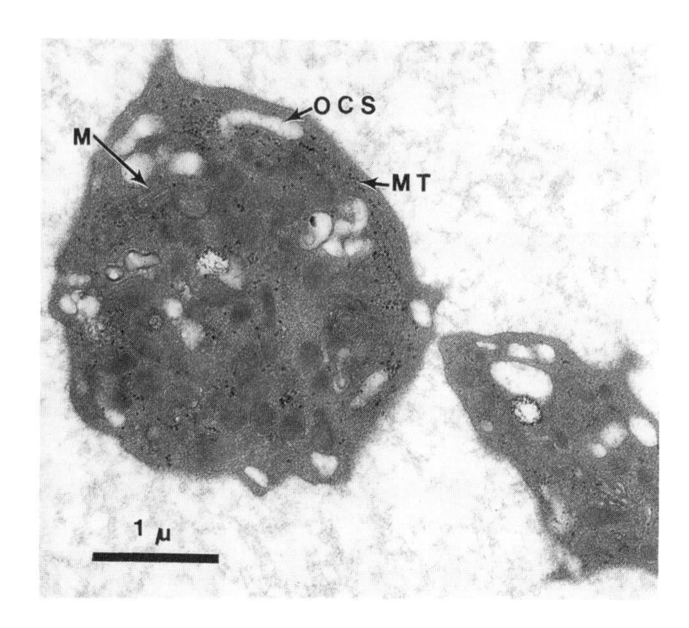

FIGURE 14.4. TEM of a chimpanzee platelet. Internal structures: (M) mitochondrion; (OCS) open canalicular system; (MT) microtubulons.

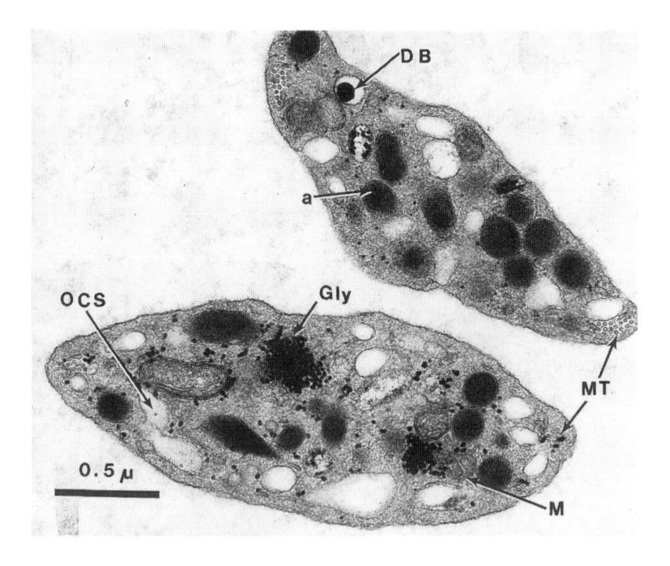

FIGURE 14.5. TEM of a monkey platelet.

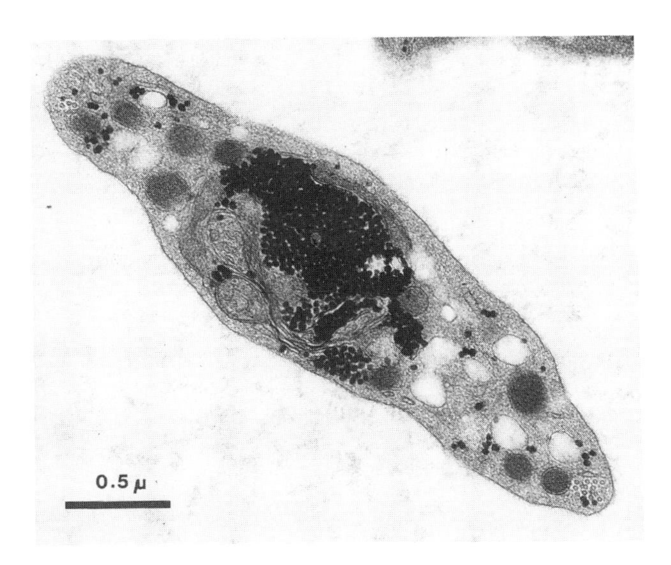

FIGURE 14.6. TEM of monkey platelets. Internal structures: (DB) dense bodies; (a) alpha granules; (Gly) glycogen particles; (MT) microtubules; (OCS) open canalicular system; (M) mitochondria.

4. SUMMARY

The primates are closely related, but do differ considerably in coagulation factor levels. The differences between human and baboon coagulation factor levels allowed the detection of baboon liver function after transplant into a human. Baboon, chimpanzee, and monkey biochemical tests and cellular elements resembled human. The platelet activities were quite different in that biphasic curves were common and arachidonic acid did not aggregate chimpanzee platelets. The latter result may have been due to tranquilizer or ketamine treatment.

REFERENCES

Hampton, J. W., and Matthews, C., 1966, Similarities between baboon and human blood clotting, *J. Appl. Physiol.* **21:**1713.

Harker, L. A., and Hanson, S. R., 1979, Experimental arterial thromboembolism in baboons, *J. Clin. Invest.* **64:**559.

Lewis, J. H., 1977, Comparative hematology: Rhesus monkeys (*Macaca mulatta*), *Comp. Biochem. Physiol. A* **56:**379.

Starzl, T. E., Fung, J., and Tzakis, A., 1993, Baboon to human liver transplantation, *Lancet* **341:**65.

SUGGESTED READINGS

Abildgaard, C. F., Harrison, J., and Johnson, C. A., 1971, Comparative study of blood coagulation in non-human primates, *J. Appl. Physiol.* **30**:400.

Addonizio, V. P., Jr., Edmunds, L. H., Jr., and Colman, R. W., 1978, The function of monkey (*M. mulatta*) platelets compared to platelets of pig, sheep and man, *J. Lab. Clin. Med.* **91**:989.

Birndorf, N. I., Pearson, J. D., and Wredman, A., 1971, The coagulation system of monkeys: A comparison of coagulation factors and tests between cynomolgus monkeys (*Macaca irus*) and humans, *Comp. Biochem. Physiol. A* **38**:157.

Clevenger, A. B., Marsh, W. L., and Peery, T. M., 1971, Clinical laboratory studies on the gorilla, chimpanzee, and orangutan, *Am. J. Clin. Pathol.* **55**:479.

Cooper, R. G., and Sorenson, M. W., 1969, Blood coagulation in the tree shrew, *Tupaia chinensis*, *Lab. Anim. Care* **19**:513.

De LaPena, A., Matthijssen, C., and Goldzieher, J., 1970, Normal values for blood constituents of the baboon (*Papio* species), *Lab. Anim. Care* **20**:251.

Foy, H., Kondi, A., and Mbaya, V., 1965, Haematologic and biochemical indices in the East African baboon, *Blood* **26**:682.

Hall, B. E., 1929, The morphology of the cellular elements of the blood of the monkey, *Macaca rhesus*, *Folia Haematol.* **38**:30.

Hampton, J. W., and Matthews, C., 1966, Similarities between baboon and human blood clotting, *J. Appl. Physiol.* **21**:1713.

Hodson, H. H., Lee, B. D., Wisecup, W. G., and Fineg, J., 1967, Baseline blood values of the chimpanzee. I. The relationship of age and sex and haematological values, *Folia Primatol.* **7**:1.

Loeb, W. F., and Mackey, B., 1973, A comparative study of platelet aggregation in primates, *J. Med. Primatol.* **2**:195.

Matthews, C., McKee, P., Kramer, J., and Hampton, J. W., 1968, Functional and analytical characteristics of baboon plasminogen, *Proc. Int. Congr. Primatol.* **3**:113.

McKee, P. A., Lemmon, W. B., and Hampton, J. W., 1971, Streptokinase and urokinase activation of human, chimpanzee and baboon plasminogen, *Thromb. Diath. Haemorrh.* **26**:512.

Melville, G. S., Whitcomb, W. H., and Martinez, R. S., 1967, Haematology of the *Macaca mulatta* monkey, *Lab. Anim. Care* **17**:189.

Picard, J., Heremans, J., and Vandebroek, G., 1962, Serum proteins found in primates. I. Proteins of man, *Macaccus irus*, and *Lemur mongoz*, *Vox Sang.* **7**:190.

Ponder, E., Yeager, J. F., and Charipper, H. A. 1928. Haematology of the primates, *Zoologica* **11**:9.

Rao, G. N., Yasukawa, J. J., and Shipley, E. G., 1970, Plasma clotting of adult female rhesus monkeys (*Macaca mulatta*), *Lab. Anim. Care* **20**:88.

Robinson, F. R., and Ziegler, R. F., 1968, Clinical laboratory data derived from 102 *Macaca mulatta*, *Lab. Anim. Care* **18**:50.

Rollins, J. B., Hobbs, C. H., Spertzel, R. O., and McConnell, S., 1970, Hematologic studies of the rhesus monkey (*Macaca mulatta*), *Lab. Anim. Care* **20**:681.

Seaman, A. J., and Malinow, M. R., 1968, Blood clotting in non-human primates, *Lab. Anim. Care* **18**:80.

Shukers, C. F., Langston, W. C., and Day, P. L., 1938, The normal blood picture of the young rhesus monkey, *Folia Haematol.* **60**:416.

Stanley, R. E., and Cramer, M. B., 1968, Hematologic values of the monkey (*Macaca mulatta*), *Am. J. Vet. Res.* **29**:1041.

Suarez, R. M., Dias Rivera, R. S., and Hernandes Morales, F., 1942, Haematological studies in normal rhesus monkeys (*Macaca mulatta*), *P. R. J. Public Health Trop. Med.* **18**:212.

Switzer, J. W., 1967, Bone marrow composition in the adult rhesus monkey (*Macaca mulatta*), *J. Am. Vet Med. Assoc.* **151:**823.

Todd, M. E., McDevitt, E., and Goldsmith, E. I., 1972, Blood-clotting mechanisms of nonhuman primates: Choice of the baboon model to simulate man, *J. Med. Primatol.* **1:**132.

Vogin, E. E., and Oser, F., 1971, Comparative blood values in several species of nonhuman primates, *Lab. Anim. Sci.* **21:**937.

CHAPTER 15

THE NINE-BANDED ARMADILLO

1. INTRODUCTION

Order Edentata (Xenarthra), the "toothless" mammals, consists of only three families: armadillos, sloths, and anteaters. Edentata flourished in the Lower Tertiary Age (30 MYA), but only these three genera have survived. Until recently, there were a number of species with and without teeth that were confined to South America. In the last century, they have spread to the warm areas of Central America, Mexico, and the southeastern United States.

The nine-banded Texas armadillo (*Dasypus novemcinctus*) is a primitive, cat-sized (4–6 kg) omnivorous animal that is chiefly nocturnal. It is almost unique among mammals in having a protective armor composed of small, jointed, bony plates. It can curl up into a ball if attacked. The armadillo is neither endangered by nor endangering to mankind and therefore may flourish again. Its meat is not desirable and is eaten by man only out of necessity. Despite the belief of some, the shell has no magic power, only souvenir value. Armadillos lack endearing qualities and are not widely sought as pets. They are good swimmers and expert burrowers, and they live in swampy, often nonarable land. Because they consume large numbers of insect larvae, they may have economic value for insect control. Their reproductive pattern is unique in that the fertilized females produce identical quadruplets about once a year. Armadillos served as research models in early studies of leprosy.

Various biological, morphological, and genetic studies were reviewed in a monograph by Talmage and Buchanan (1954). Coagulation studies performed in this laboratory (Lewis and Doyle, 1964) on four armadillos are reprinted here with the kind permission of the editor of *Comparative Biochemistry and Physiology.*

2. SOURCE

A total of six armadillos were shipped to our laboratories, four in 1964 and two in 1975, and maintained on diets of rabbit chow, lettuce, carrots, and water ad lib. Their average weight was 5.5 kg. The animals were under light Nembutal (intraperitoneal) anesthesia when their blood was obtained by cardiac puncture.

3. TESTS CONDUCTED

3.1. Bleeding Times

The skin bleeding times performed on the underbelly of the first four were short: mean: 3 min; range: 2–4 min.

3.2. General Coagulation Tests and Factors

Table 15.1 shows the results of general coagulation tests. The prothrombin test with Simplastin was slightly longer and with armadillo brain, human brain, and RVV slightly shorter than with human plasma. The APTT was also shorter, but the thrombin time with either bovine thrombin or Atroxin was much longer than human.

Table 15.2 illustrates the results of the TGT employing all armadillo or all human materials. The substrate clotting times were longer with the armadillo generating mixtures. When armadillo platelets or adsorbed plasmas were substituted for human materials, they were equally effective, but armadillo serum allowed longer clotting times, suggesting that one or more of the serum factors (VII, IX, X, XI, XII) were less active than human factors.

Table 15.3 shows coagulation factor assay results. Fibrinogen (factor I) and factors V, VIII, IX, XI, XII, and XIII were higher than human; ATIII and II were about the same; VII and X were slightly below the lower edge of the human range.

3.3. Tissue Factor

Armadillo tissues were washed free of blood, ground, and suspended, 1 g/2ml in imidazole-buffered saline (I/S). The suspension was allowed to sediment for 5 min, and the relatively clear top was tested as shown in Table 15.4. Lung and brain were high in tissue thromboplastin (tissue factor) activity when tested with armadillo plasma.

3.4. Fibrinolytic Enzyme System

Neither human nor armadillo plasma lysed ^{131}Ifibrin in the presence of buffer. When 100 U/ml or 1000 U/ml streptokinase (SK) was added, human plasma lysed 40–60% of the fibrin, but armadillo only 2–11%. With 440 U/ml urokinase (UK), human plasma lysed 90% and armadillo 52%. These results are shown in Table 15.5.

TABLE 15.1
General Coagulation Tests in the Nine-Banded Armadillo

Test	Armadillo No.									Human		
										Controls		
										(4)	(2)	Normal
	1	2	3	4	5	6	*N*	Mean	SD	(means)		range
Clot T (min)												
Glass	8	5	6	8	5	6	6	6	1.3	12	8	6–12
Silicone	13	55	68	13	—	15	5	33	27	48	28	20–59
Clot Retr	2/2	2/4	2/2	4/4	4	4/4	6	2–4	—	4/4	4/4	3–4+
Clot Lys	0	0	0	0	0	0	6	0	0	0	0	0
SPT (sec)												
Human	42.4	39.8	43.4	63.0	48.8	66.0	6	50.6	11	37.4	43.2	20+
Armadillo	44.4	45.4	41.4	42.8	—	—	4	43.5	18	24.8	—	—
PT (sec)												
Simplastin	15.3	14.2	15.8	16.0	17.5	17.0	6	15.9	1.2	12.1	12.4	10–13
Arm Br	—	—	—	—	10.4	10.8	2	10.5	0.1	—	28.4	—
Hu Br	9.1	8.8	9.0	9.0	9.1	9.0	6	9.0	0.1	12.2	12.2	11–15
RVV	11.0	11.2	13.1	11.8	8.8	8.8	6	10.8	1.7	15.0	17.4	13–18
APTT (sec)	—	—	—	—	24.4	34.1	2	29.2	6.9	45.4	48.2	45–55
Recal T (sec)	215	275	275	275	120	90	6	200	78	120	120	90–180
Lys	0	0	0	0	+	0	6	1/6	—	0	0	0
Lys MCA	0	0	0	0	0	0	6	0	0	0	0	0
Thromb T (sec)												
Bov Thr	66	86	64	68	60	60	6	67	9.7	15.0	15.4	11–18
Atroxin	—	—	—	—	58	60	2	59	1.4	—	16.1	11–18

TABLE 15.2
Thromboplastin Generation Test in the Nine-Banded Armadillo

Generating mixture		Substrate Clot T (sec)[a]	
Human	Armadillo		
Means		Human	Armadillo
H (2)	—	**9.9**	10.7
—	Arm (2)	13.3	**14.6**

[a]Homologous systems are in **boldface.**

3.5. Serum Proteins

α_1-Globulin and β-globulin were slightly higher than in human blood. Otherwise, the protein distribution was very similar, as shown in Table 15.6.

3.6. Biochemical Tests

Biochemical test results are shown in Table 15.7. The albumin by this method was much lower than by electropheresis. Cholesterol, uric acid, chloride, CO_2, and Na were each somewhat lower than values found in humans. On the other hand, most of the liver function test values were higher than those of humans.

3.7. Miscellaneous Tests

Armadillo plasma (No. 5) in dilutions up to 1:1024 clumped the standard killed staphylococcal culture preparation. The euglobulin lysis time was greater than 2 hr.

3.8. Cellular Elements

Table 15.8 shows that the erythrocyte values were comparable to human values. The RBCs were somewhat more numerous and slightly smaller. The mean RBC diameter was 6.2 μM. The leukocytes had much the same distribution and appearance as their human counterparts. The platelets varied in size from very small to very large. The counts were within the human range.

TABLE 15.3
Coagulation Factors in the Nine-Banded Armadillo

Factor	Armadillo No. 1	2	3	4	5	6	*N*	Mean	SD	Normal human range
I (mg/dl)	630	730	750	750	318	470	6	608	179	150–450
II (U/ml)	0.58	0.43	0.45	0.46	2.0	0.68	6	0.77	0.6	0.70–1.30
V (U/ml)	2.0+	2.0+	2.0+	2.0+	3.7	4.2	6	2.0+	—	0.65–1.45
VII (U/ml)	—	—	—	—	0.51	0.64	2	0.58	0.1	0.50–1.30
X (U/ml)	—	—	—	—	0.54	0.52	2	0.53	0.1	0.75–1.25
VIII (U/ml)	2.0+	2.0+	2.0+	2.0+	3.8	3.6	6	2.0+	—	0.75–1.40
IX (U/ml)	2.0+	2.0+	2.0+	2.0+	5.0	4.5	6	2.0+	—	0.65–1.70
XI (U/ml)	—	—	—	—	2.0	3.0	2	2.0+	—	0.75–1.40
XII (U/ml)	2.0+	2.0+	2.0+	2.0+	2.0+	2.0+	6	2.0+	—	0.50–1.45
ATIII (U/ml)	0.65	0.70	—	—	1.10	0.95	4	0.90	0.2	0.80–1.20
XIII (R)	16	16	16	32	32	32	6	24	8.8	4–16

TABLE 15.4
The Effects of Armadillo Tissues on Clotting Time[a]

Agent	Clotting time (sec)	
	Armadillo	Human
Buffer (I/S)	51.4	103.0
Armadillo tissue suspension		
Lung	10.0	30.4
Brain	13.4	23.8
Heart	17.2	28.8
Testis	20.4	40.4
Skin	23.4	31.8
Liver	28.8	29.2
Spleen	30.2	40.4
Human brain	9.0	12.2

[a]Test system: 0.1 ml armadillo tissue suspension or other agent + 0.1 ml armadillo or human plasma + 0.1 ml 0.025 M $CaCl_2$.

3.9. Platelet Aggregation

Armadillo platelet aggregation with ADP and collagen did not differ from human, as shown in Table 15.9.

3.10. Platelet Ultrastructure

TEMs (Figs. 15.1 and 15.2) show greatly enlarged armadillo platelets. Strikingly large inclusion bodies surrounded by fibrillar material were seen in

TABLE 15.5
Fibrinolytic Activity with Streptokinase or Urokinase in the Nine-Banded Armadillo

Reagent	[^{131}I]Fibrin lysed (%)	
	Plasma	
	Armadillo	Human
Buffer	0.41%	0.68%
SK (100 U/ml)	2.3%	62.4%
(1000 U/ml)	11.3%	39.4%
UK (110 U/ml)	14.7%	41.7%
(440 U/ml)	52.0%	90.5%

TABLE 15.6
Serum Proteins by Paper Electrophoresis in the Nine-Banded Armadillo

Protein	Armadillo No. Pool (1–4)	5	6	*N*	Mean	SD	Normal human range
Total protein (g/dl)	8.7	6.5	6.8	6	8.0	1.1	6.3–7.9
Albumin (g/dl)	3.5	2.5	2.5	6	3.2	0.5	3.2–4.4
α_1-Globulin (g/dl)	0.7	0.8	0.9	6	0.8	0.1	0.2–0.4
α_2-Globulin (g/dl)	0.8	0.8	0.8	6	0.8	0.0	0.4–1.0
β-Globulin (g/dl)	1.4	0.9	0.9	6	1.2	0.3	0.5–1.0
γ-Globulin (g/dl)	1.2	1.5	1.9	6	1.4	0.3	0.6–1.8

TABLE 15.7
Biochemical Tests in the Nine-Banded Armadillo

Substance	Armadillo No.[a] Pool (1–4)	5	6	*N*	Mean	SD	Normal human range
TP (g/dl)	**8.7**	6.3	7.8	6	**8.2**	1.0	6.0–8.0
Albumin (g/dl)	*0.8*	*0.7*	*0.5*	6	*0.7*	0.1	3.5–5.0
Ca (mg/dl)	10.2	**12.1**	**10.8**	6	**10.6**	0.7	8.5–10.5
P (mg/dl)	4.2	**9.7**	3.8	6	**5.1**	2.3	2.5–4.5
Cholesterol (mg/dl)	*128*	*147*	*93*	6	*125*	17	150–200
Glucose (mg/dl)	**172**	**329**	**141**	6	**193**	67	65–110
Uric acid (mg/dl)	*0.5*	*0.4*	*0.3*	6	*0.5*	0.1	2.5–8.0
Creatinine (mg/dl)	1.1	1.2	0.7	6	1.1	0.2	0.7–1.4
T bilirubin (mg/dl)	0.2	0.0	0.1	6	0.2	0.1	0.1–1.4
BUN (mg/dl)	16	17	19	6	17	1.2	10–20
Chloride (meq/liter)	104	—	*92*	5	*102*	5.4	95–105
CO_2 (meq/liter)	*15*	—	*17*	5	*15*	0.9	24–32
K (meq/liter)	**5.1**	—	**5.8**	5	**5.2**	0.3	3.5–5.0
Na (meq/liter)	*124*	—	*130*	5	*125*	2.7	135–145
Alk Phos (IU/liter)	60	56	67	6	61	3.6	30–85
SGOT (IU/liter)	**58**	—	**90**	5	**64**	14	0–41
SGPT (IU/liter)	22	11	9	6	18	6.2	10–50

[a]Values **higher** than human are in **boldface**, values *lower* than human in *italics*.

TABLE 15.8
Cellular Elements of the Blood in in the Nine-Banded Armadillo

Element	Armadillo No. 1	2	3	4	5	6	N	Mean	SD	Normal human range
HCT (%)	41	40	41	49	41	42	6	42	3.3	37–52
HGB (g/dl)	12.9	14.1	13.7	14.5	13.2	13.6	6	13.7	0.6	12–18
RBC ($\times$ 10^6/mm^3)	5.1	5.1	4.8	6.9	6.9	5.8	6	5.8	0.9	4.2–6.2
MCV (fl)	80	78	85	71	59	72	6	75	9.0	79–97
MCH (pg)	25	28	29	21	19	23	6	24	3.9	27–31
MCHC (g/dl)	31	35	33	30	32	32	6	32	1.7	32–36
WBC ($\times$ 10^3/mm^3)	10.2	4.9	6.6	9.1	10.9	8.4	6	8.3	2.3	5–10
Neutr (%)	60	55	63	64	71	65	6	63	5.3	55–75
Lymph (%)	20	26	18	11	26	25	6	21	5.9	20–40
Mono (%)	7	10	7	6	2	5	6	6	2.6	2–10
Eos (%)	5	3	6	16	3	3	6	6	5.1	0–3
Bas (%)	8	6	6	3	0	2	6	4	3.0	0–3
Plat C ($\times$ 10^3/mm^3)	266	322	246	432	555	318	6	357	117	150–450

many platelets. Circumferential microtubules as well as alpha granules and glycogen particles were easily visible. An open canalicular system and dense bodies were minimal or difficult to observe.

4. SUMMARY

Blood was collected by cardiac puncture from six armadillos (*Dasypus novemcinctus*). The blood clotted effectively, but the clots retracted poorly. Most

TABLE 15.9
Platelet Aggregation in the Nine-Banded Armadillo

Activity	Armadillo No. 1	5	6	N	Mean	SD	Human Control	Human Normal range
Aggregation (%)	100	100	100	3	100	0	100	70–100
ADP (10 μM)								
Collagen								
Bov (.19 mg/ml)	100	100	100	3	100	0	75	80–100
Armadillo (.19 mg/ml)	—	100	68	2	84	23	100	—

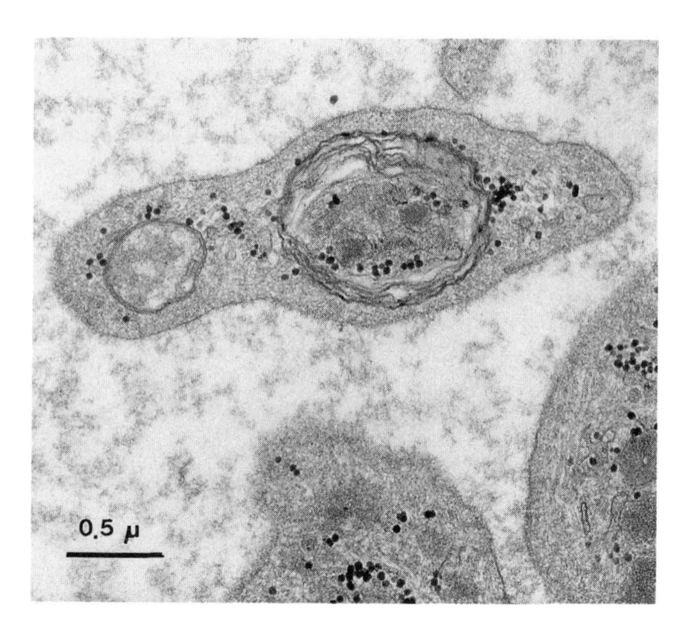

FIGURE 15.1. TEM of an armadillo platelet showing two large inclusions banded by circles of fibrillar strands and containing microtubules (MT), glycogen particles (GLY), and alpha granules (a GRAN).

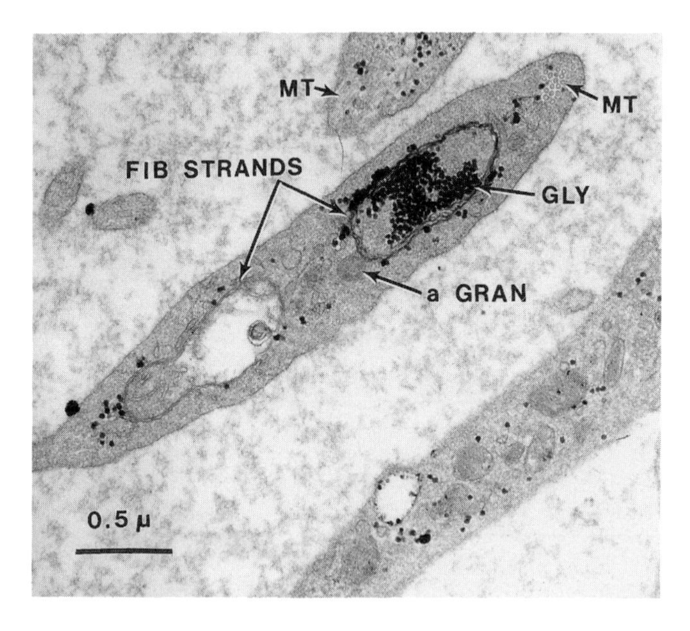

FIGURE 15.2. TEM of another armadillo platelet, also showing two inclusions banded by circles of fibrillar strands. One inclusion is almost empty; the other contains a large group of glycogen particles.

of the coagulation factors showed greater activity (or higher concentration) than their human counterparts. Armadillo tissues, especially lung and brain, were far more thromboplastic to armadillo plasma than to human plasma. Spontaneous fibrinolytic activity was seen only in the recalcification time of armadillo No. 5. SK and UK in high concentrations activated armadillo plasminogen. Serum proteins and their electrophoretic patterns were similar to human. Biochemical tests showed high levels of liver enzymes. Cellular elements differed only slightly from human. Erythrocytes were small and numerous; leukocyte values fell within the human range. Platelets were variable in size, numerous, and aggregated with ADP and collagen. Platelet appearance on electron microscopy was distinctive in that about one in six platelets contained numerous multiwalled rings that appeared to circumscribe normal platelet elements.

REFERENCES

Lewis, J. H., and Doyle, A. P., 1964, Coagulation, protein and cellular studies on armadillo blood, *Comp. Biochem. Physiol.* **12:**61.

Talmage, R. V., and Buchanan, G. D., 1954, The Armadillo (*Dasypus novemcinctus*): A Review of Its Natural History, Ecology, Anatomy and Reproductive Physiology, *Rice Inst. Pamph.* **1954**(No. 41).

SUGGESTED READINGS

Anderson, J., and Benirscheke, K., 1966, The armadillo, *Dasypus novemcinctus,* in experimental biology, *Lab. Anim. Care* **16:**202.

Binford, C. H., Issar, S. L., Storrs, E. E., and Walsh, G. P., 1974, Leprosy in the nine-banded armadillo (*Dasypus novemcinctus*): Summary of post-mortem observations made on seven animals, *Int. J. Lepr.* **42:**123.

Burchfield, H. P., Storrs, E. E., Walsh, G. P., and Vidrine, M. F., 1976, Improved caging for nine-banded armadillos, *Lab. Anim. Sci.* **26:**234.

D'Addamio, G. H., Walsh, G. P., Harris, L., Borne, R., and Derouen, M. S., 1978, Hematologic parameters for wild and captive nine-banded armadillos (*Dasypus novemcinctus*), *Lab. Anim. Sci.* **28:**607.

Dhinsa, D. S., Hoversland, A. S., and Metcalfe, J., 1971, Comparative studies of the respiratory functions of mammalian blood. VII. Armadillo (*Dasypus novemcinctus*), *Respir. Physiol.* **13:**198.

Ebaugh, F. G., and Benson, M. A., 1964. Armadillo hemoglobin characteristics and red cell survival, *J. Cell. Comp. Physiol.* **64:**183.

Giacometti, L., Berntzen, A. K., and Bliss, M. L., 1972, Hematologic parameters of the nine-banded armadillo (*Dasypus novemcinctus*), *Comp. Biochem. Physiol. A* **43:**213.

Negassi, K., Closs, O., and Harboe, M., 1979, Cross-reactions between serum proteins and water soluble liver tissue antigens of the nine-banded armadillo (*Dasypus novemcinctus* Linn.) and man, *Clin. Exp. Immunol.* **38:**135.

Newman, H. H., and Patterson, J. T., 1909, A case of normal identical quadruplets in the armadillo and its bearing on the problem of identical twins and sex determination, *Biol. Bull.* **17:**181.

Purtilo, D. T., Walsh, G. P., Storrs, E. E., and Gannon, C., 1975, The immune system of the nine-banded armadillo (*Dasypus novemcinctus* Linn.), *Anat. Rec.* **181:**725.

Storrs, E. E., 1971, The nine-banded armadillo: A model for leprosy and biomedical research, *Int. J. Lepr.* **39:**703.

Storrs, E. E., and Greer, W. E., 1973, Maintenance and husbandry of armadillo colonies, *Lab. Anim. Sci.* **23:**823.

Strozier, L. M., Blair, C. B., Jr., and Evans, B. H., 1971, Armadillos: Basic profiles. II. Serum proteins, *Lab. Anim. Sci.* **21:**602.

CHAPTER 16

RABBITS

1. INTRODUCTION

The small, cat-sized animals of order Lagomorpha are among the most common mammals found throughout the world. They differentiated in the mid-Eocene Age, developing into at least 150 species. The domestic rabbit, a descendant of *Oryctolagus cuniculus,* was introduced by man into New Zealand and Australia, where it rapidly became the most prevalent wild mammal. Rabbits have claws on their toes, sharp incisors and grinding molars, and jaws with mainly lateral motion. They are totally herbivorous, eating grasses, leaves, stems, bark, and other vegetable matter. Rabbits and hares have long ears and acute hearing, enabling them to hear danger approaching. They have long hindlegs that enable them to leap and run to escape danger (e.g., carnivores, birds of prey). In the wild, they use their droppings, "peas," to mark their territory. Females reach maturity at 6 months or earlier and may bear two litters of 6–8 per year. Rabbits are used by man as sources of fur and food and as pets. They are also specially bred for research purposes.

Although rabbits are frequently the subjects of experimental research, there are few references that describe normal rabbit hemostatic factors, particularly as compared to human. Blood cell counts and platelet ultrastructure and aggregation studies are listed in the Suggested Readings and in veterinary hematology textbooks.

2. SOURCE

New Zealand outbred white (albino) rabbits of both sexes weighing about 3 kg were obtained from a local reliable rabbitry in two lots of six each. Blood was obtained by cardiac puncture from lightly anesthetized animals.

3. TESTS CONDUCTED

3.1. Bleeding Times

Bleeding times were performed on the posterior ear surface, in areas free from visible vessels, by the Simplate method (rabbits No. 1–6) or by simply measuring the bleeding time from a nick made with scissors on the ear margin (rabbits No. 7–12). The Simplate method gave the following times: 4 $^1/_2$, 4, 3 $^1/_2$,

3 1/2, 4 and 4 min (mean ± SD: 3.87 ± 0.41 min). The nick gave: 2, 4, 3 1/2, 2 1/2, 3 1/2, and 2 min (mean ± SD: 2.91 ± 0.85 min).

3.2. General Coagulation Tests and Factors

Table 16.1 shows that rabbit blood clotted rapidly in glass and somewhat more slowly in siliconized tubes. The clots retracted well and did not lyse. The prothrombin times with Simplastin were fast. On the first three animals, the tests were done on an instrument that could not read a clot end point shorter than 8.9 sec. The second three were done manually and were considerably faster than on human plasma. The rabbit brain thromboplastin made in our laboratory was not as active as the commercially prepared Simplastin. The first three rabbits were tested with an old preparation and the second three with a new acetone-dehydrated brain thromboplastin. With these preparations, rabbit and human clotting were almost the same. With RVV, the prothrombin times on human and rabbit plasma were similar. APTTs also fell within the normal human range. The Recal T of rabbit plasma was somewhat faster than human. The clots did not lyse alone or in 1% MCA. Thrombin times with bovine thrombin were slower and with Atroxin much slower on rabbit than on human plasma.

Rabbit thromboplastin generation mixtures (Table 16.2) clotted human and rabbit substrate in about the same times. The human generating mixture was more potent than that of the rabbits.

Factor assays (Table 16.3) showed extremely high levels of factor V. Rabbit plasma was diluted 100-fold and still showed activity similar to human. Factors I, IX, XII, and ATIII fell within the normal human range. Factors II, VII, VIII, X, XI, and XIII were above this range, but not extraordinarily high. On the other hand, Fletcher and Fitzgerald factors assayed in the very low range.

3.3. Fibrinolytic Enzyme System

Rabbit plasminogen activated poorly with streptokinase, but almost completely with urokinase and staphylokinase. Antiplasmin was present in rabbit plasma in about the same concentration as in human plasma.

3.4. Serum Proteins

Rabbit serum proteins compared to human showed that the α_1-globulin was slightly elevated and α_2-globulin slightly depressed, as shown in Table 16.4.

TABLE 16.1
General Coagulation Tests in Rabbits

	Rabbit No.									Human	
Test	1♀	2♂	3♂	4♀	5♀	6♀	N	Mean	SD	Controls (6) (mean)	Normal range
Clot T (min)											
Glass	3	2	6	4	5	6	6	4	1.7	—	6–12
Silicone	4	8	29	27	25	27	6	20	11	—	20–59
Clot Retr	4/4	4/4	3/3	4/3	4/3	4/4	6	3–4	—	—	3–4+
Clot Lys	0	0	0	0	0	0	6	0	0	—	0
PT (sec)											
Simplastin	8.9	8.9	8.9	6.0	6.0	6.2	6	7.5	1.5	12.1	10–13
Rab Br	18.9	18.2	18.2	9.1	11.1	12.1	6	14.6	4.3	15.8	—
RVV	15.3	17.2	19.0	10.7	11.3	12.6	6	14.3	3.3	15.4	13–18
APTT (sec)	31.0	30.9	37.6	28.9	29.0	39.5	6	32.8	4.5	33.9	24–34
Recal T (sec)	—	—	54	60	56	54	4	56	2.8	133	90–180
Lys	—	—	0	0	0	0	4	0	0	0	0
Lys MCA	—	—	0	0	0	0	4	0	0	0	0
Thromb T (sec)											
Bov Thr	30.5	30.5	19.6	19.5	19.6	18.1	6	22.9	5.9	16.2	11–18
Atroxin	60+	60+	—	60+	60+	60+	5	60+	0	15.3	11–18

TABLE 16.2
Thromboplastin Generation Test in Rabbits

Generating mixture		Substrate Clot T (sec)[a]	
Human	Rabbit (mean)	Human	Rabbit
H	—	**9.9**	9.0
—	R (3)	12.5	**10.8**

[a]Homologous systems are in **boldface.**

3.5. Biochemical Tests

Biochemical test results on six rabbits are shown in Table 16.5. Although the total protein was within the human range, the albumin was very low compared to that measured by the electrophoretic method or by this biochemical method in humans. This discrepancy was not uncommon in animals. Other differences between rabbit and human were not striking.

3.6. Miscellaneous Tests

Staphylococcal clumping was positive to a dilution of 1:2048 of rabbit plasma. Euglobulin lysis times were greater than 2 hours. Ethanol and protamine gel tests were 1–3+ on all rabbit plasma. D-Dimer was positive on rabbits No. 1–4 but negative on 5 and 6.

3.7. Cellular Elements

Table 16.6 shows that the rabbit RBC parameters were not very different from human. RBCs were slightly smaller and had slightly less HGB. In these rabbits, WBC differential counts showed that the lymphocytes predominated. Platelets were numerous and slightly smaller than human.

3.8. Platelet Activities

Up to 95% of rabbit platelets were retained on glass beads, as shown in Table 16.7. Rabbit platelets aggregated with ADP, collagen, arachidonic acid,

TABLE 16.3
Coagulation Factors in Rabbits

Factor	Rabbit No.									
	1♀	2♂	3♂	4♀	5♀	6♀	*N*	Mean	SD	Normal human range
I (mg/dl)	200	210	170	185	240	170	6	196	27	150–450
II (U/ml)	2.30	230	2.30	3.00	2.70	2.00	6	2.40	0.3	0.70–1.30
V (U/ml)	125	105	105	100+	100+	100+	6	100+	—	0.65–1.45
VII (U/ml)	3.70	2.70	2.10	3.30	4.04	1.96	6	3.00	0.8	0.50–1.30
X (U/ml)	2.10	1.72	1.20	2.50	2.00	1.42	6	1.80	0.5	0.75–1.25
VIII (U/ml)	3.70	1.85	3.25	3.95	3.75	2.65	6	3.20	0.8	0.75–1.40
IX (U/ml)	1.70	0.80	0.50	1.65	1.80	1.00	6	1.20	0.5	0.65–1.70
XI (U/ml)	3.70	2.50	2.45	2.05	2.50	1.55	6	2.50	0.7	0.75–1.40
XII (U/ml)	0.54	0.70	0.78	1.05	0.65	0.90	6	0.80	0.2	0.50–1.45
Flet (U/ml)	0.03	0.04	0.01	0.04	0.05	0.03	6	0.03	0.01	0.50–1.50
Fitz (U/ml)	0.03	0.05	0.05	0.08	0.09	0.02	6	0.05	0.03	0.50–1.50
ATIII (U/ml)	0.90	0.96	0.82	0.19	0.75	0.65	6	0.80	0.1	0.80–1.20
XIII (R)	16	32	32	32	16	32	6	27	8.3	4–16

TABLE 16.4
Serum Proteins by Cellulose Electrophoresis in Rabbits

	Rabbit No.													
Protein	2♂	3♂	4♀	5♀	6♀	7	8	9	10	11	*N*	Mean	SD	Normal human range
Total protein (g/dl)	6.9	6.9	7.2	7.0	7.3	6.9	7.1	6.7	7.1	6.9	10	7.0	0.2	6.3–7.9
Albumin (g/dl)	4.9	4.5	4.2	4.9	4.9	4.1	3.7	3.6	4.1	3.4	10	4.2	0.6	3.2–4.4
α_1-Globulin (g/dl)	0.5	0.4	1.0	0.5	0.5	1.3	1.2	1.0	1.2	1.3	10	0.9	0.4	0.2–0.4
α_2-Globulin (g/dl)	0.3	0.4	0.4	0.4	0.4	0.4	0.1	0.2	0.2	0.3	10	0.3	0.1	0.4–1.0
β-Globulin (g/dl)	0.9	0.9	1.0	0.4	0.8	0.6	1.3	1.2	1.2	1.0	10	0.9	0.3	0.5–1.0
γ-Globulin (g/dl)	0.3	0.7	0.4	0.5	0.6	0.6	0.8	0.7	0.4	0.9	10	0.6	0.2	0.6–1.8

TABLE 16.5
Biochemical Tests in Rabbits

Substance	Rabbit No.[a]									
	7	8	9	10	11	12	*N*	Mean[a]	SD	Normal human range
TP (g/dl)	6.4	6.8	6.1	6.9	6.9	6.4	6	6.6	0.4	6.0–8.0
Albumin (g/dl)	*0.9*	*0.8*	*0.8*	*0.9*	*0.9*	*0.8*	6	*0.9*	0.0	3.5–5.0
Ca (mg/dl)	**14.3**	**13.2**	**14.2**	**15.3**	**14.2**	**16.0**	6	**14.6**	0.9	8.5–10.5
P (mg/dl)	**4.7**	**4.9**	3.2	3.7	3.7	—	5	4.0	0.7	2.5–4.5
Cholesterol (mg/dl)	*52*	*50*	*59*	*52*	*54*	*59*	6	*54*	3.9	150–200
Glucose (mg/dl)	**146**	**146**	**175**	**151**	**158**	**147**	6	**153**	11	65–110
Uric acid (mg/dl)	*0.4*	*0.3*	*0.4*	*0.2*	*0.3*	*0.3*	6	*0.3*	0.0	2.5–8.0
Creatinine (mg/dl)	**1.5**	**1.5**	**1.6**	1.4	**1.9**	**1.8**	6	**1.6**	0.2	0.7–1.4
T bilirubin (mg/dl)	>1	>1	>1	>1	>1	>1	6	>1	0.0	0.1–1.4
BUN (mg/dl)	**22**	**24**	19	**23**	**24**	**26**	6	**23**	2.4	10–20
Chloride (meq/liter)	104	**106**	**108**	**112**	102	103	6	105	3.7	95–105
CO_2 (meq/liter)	*22*	*20*	*21*	*19*	*16*	*17*	6	*19*	2.3	24–32
K (meq/liter)	**5.0**	3.9	**5.2**	**6.2**	**5.3**	**6.6**	6	**5.4**	0.9	3.5–5.0
Na (meq/liter)	145	**148**	145	**146**	**147**	**147**	6	**146**	1.4	135–145
Alk Phos (IU/liter)	31	41	*23*	34	34	38	6	34	6.2	30–85
LDH (IU/liter)	144	83	**320**	180	144	120	6	165	82	60–200
SGOT (IU/liter)	36	35	**78**	**50**	26	28	6	**42**	19	0–41
SGPT (IU/liter)	43	36	**107**	48	30	32	6	49	29	10–50

[a]Values **higher** than human are in **boldface,** values *lower* than human in *italics*.

TABLE 16.6
Cellular Elements of the Peripheral Blood in Rabbits

Element	Rabbit No.													
	1	2	3	5	6	7	12	13	14	16	*N*	Mean	SD	Normal human range
HCT (%)	39	38	35	40	40	39	40	36	40	40	10	39	1.8	37–52
HGB (g/dl)	12.4	12.4	11.3	13.7	13.4	13.4	13.1	12.7	12.9	13.1	10	12.8	0.7	12–18
RBC ($\times 10^6/mm^3$)	5.0	5.0	4.9	5.7	6.2	6.0	5.6	4.8	5.7	5.6	10	5.5	0.5	4.2–6.2
MCV (fl)	78	76	71	70	65	65	71	75	70	71	10	71	4.3	79–97
MCH (pg)	25	25	23	24	22	22	23	26	23	23	10	24	1.3	27–31
MCHC (g/dl)	32	33	32	34	34	34	33	35	32	33	10	33	1.0	32–36
WBC ($\times 10^3/mm^3$)	7.4	7.7	5.1	7.2	6.9	7.0	8.5	9.7	9.4	5.5	10	7.3	1.5	5–10
Neutr (%)	39	42[a]	40[a]	11	21	28	14	21	6	17	10	24	13	55–75
Lymph (%)	57	58	60	86	79	72	85	77	82	80	10	74	11	20–40
Mono (%)	2	0	0	3	0	0	1	0	0	2	10	1	1.1	2–10
Eos (%)	2	0	0	0	0	0	0	0	12	0	10	1	3.7	0–3
Plat C ($\times 10^3/mm^3$)	317	274	220	580	281	416	394	580	704	348	10	411	159	150–450
MPV (μM^3)	5.6	5.9	5.2	—	4.7	6.0	6.2	5.8	6.8	8.6	9	6.1	1.1	5.6–10.4

[a]Includes bands (premature cells).

TABLE 16.7
Platelet Activities in Rabbits

Activity	Rabbit No. 1♀	2♂	3♂	6♀	N	Mean	SD	Normal human range
Plat Glass Ret Ind (%)	90	93	82	95	4	90	5.7	>20
Aggregation (%)								
ADP (20 μM)	54	15	66	51	4	47	22	70–100
(10 μM)	42	22	65	46	4	44	18	70–100
(5 μM)	42	16	59	35	4	38	18	50–100
(2.5 μM)	32	14	34	29	4	27	9.1	40–100
Collagen (0.19 mg/ml)	52	31	53	70	4	52	15	80–100
Risto (0.9 mg/ml)	8	12	7	1	4	7	0.05	80–100
Risto ½ (0.45 mg/ml)	10	6	3	8	4	7	0.03	<18
Arach A (0.5 mg/ml)	73	38	77	65	4	63	17	60–100
Bov Thr (1.0 U/ml)	100	84	100	Clot	3	97	12	Clots
(0.5 U/ml)	9	8	74	55	4	36	33	60–100
Bov Fib (0.1%)	6	3	5	4	3	5	0.01	60–100
CHH (0.05 mg/ml)	70	39	18	—	4	42	26	60–100
Pig Pl (1:10)	10	5	2	5	4	5	0.01	60–100

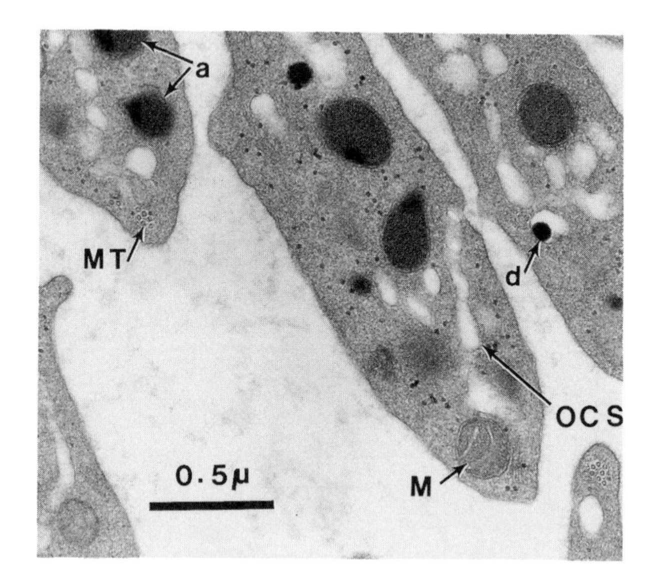

FIGURE 16.1 TEM of rabbit platelets. Internal structures: (a) alpha granules (rather large); (d) dense body in a vacuole; (M) mitochondrion; (MT) cross-sectioned microtubules; (OCS) open canalicular system.

thrombin, and CHH, but to a lesser extent than did human platelets. Rabbit platelets did *not* aggregate with ristocetin (Risto), bovine fibrinogen (Bov Fib), or pig plasma (Pig Pl).

3.9. Platelet Ultrastructure

Figure 16.1 is a high-magnification TEM of rabbit platelets. All the usual platelet ultrastructural components could be identified and are identified in the figure caption.

4. SUMMARY

Rabbit coagulation was characterized by short prothrombin times, long thrombin and Atroxin times, and very high levels of factor V. Platelets aggregated poorly with Risto, Bov Fib, or Pig Pl. Ultrastructure was not remarkable. RBCs were moderately small (MCV: 71 fl). WBC counts were within the human range. Lymphocytes predominated.

SUGGESTED READINGS

Baumgartner, H. R., and Born, G. V. R., 1968, 5-Hydroxytryptamine in rabbit platelets and their aggregation, *J. Physiol.* **194:**92P.

Botzaris, A., 1938, Contribution a étude de l'hematocytologie des cobayes normaux, *Folia Haematol.* **60:**222.

Bushnell, L. D., and Bangs, E. F., 1926, A study of the variation in number of blood cells of normal rabbits, *J. Infect. Dis.* **29:**291.

Daver, J., Colinot, M., and Bourse, R., 1968, Étude de la coagulation du sang du cobaye après catheterisme veineux et artériel sous anaesthésie générale, *Pathol. Biol.* **16:**633.

Hovig, T., 1962, The ultrastructure of rabbit blood platelet aggregates, *Thromb. Diath. Haemorrh.* **8:**455.

Hovig, T., 1963, Release of a platelet aggregating substance (adenosine diphosphate) from rabbit blood platelets induced by saline "extract" of tendons, *Thromb. Diath. Haemorrh.* **9:**264.

Hovig, T., 1965, The effect of various enzymes on the ultrastructure, aggregation, and clot retraction ability of rabbit blood platelets, *Thromb. Diath. Haemorrh.* **13:**84.

Jackson, J. W., and Stovall, W. D., 1930, Normal blood count of the rabbit, *J. Lab. Clin. Med.* **16:**82.

Johnson, A. J., and Tillett, W. S., 1952, The lysis in rabbits of intravascular blood clots by the streptococcal fibrinolytic system (streptokinase), *J. Exp. Med.* **95:**499.

Korninger, C., Stassen, J. M., and Collen, D., 1981, Turnover of human extrinsic (tissue-type) plasminogen activator in rabbits, *Thromb. Diath. Haemorrh.* **46:**658.

Lollar, P., and Owen, W. G., 1980, Clearance of thrombin from circulation in rabbits by high-affinity binding sites on endothelium, *J. Clin. Invest.* **66:**1222.

McFarlane, A. S., Irons, L., Koj, A., and Regoeczi, E., 1965, The measurement of synthesis rates of albumin and fibrinogen in rabbits, *Biochem. J.* **95:**536.

Millar, G. J., Jaques, L. B., and Henriet, M., 1964, The prothrombin time response of rabbits to dicumarol, *Arch. Int. Pharmacodyn. Ther.* **150:**197.

Pearce, L., and Casey, A. E., 1930, Blood counts in normal rabbits, *J. Exp. Med.* **51:**83.

Reimers, H. J., Buchanan, M. R., and Mustard, J. F., 1973, Survival of washed rabbit platelets in vivo, *Proc. Soc. Exp. Biol. Med.* **142:**1222.

Rick, M., Wampler, D. E., and Hoyer, L. W., 1977, Rabbit factor VIII: Identification of size heterogeneity, *Blood* **49:**209.

Rosahn, P. D., Pearce, L., and Hu, C., 1934, Comparison of the hemocytological constitution of male and female rabbits, *J. Exp. Med.* **60:**687.

Soria, J., Soria, C., Vairel, E., Prost, R. J., and Warneson, G., 1973, Experimental study of rabbit fibrinogen degradation products: Influence of plasminogen level for the interpretation of serum concentration of fibrinogen degradation products in consumption coagulopathies, *Thromb. Diath. Haemorrh.* **29:**115.

Sutherland, D. A., Minton, P., and Lanz, H., 1959, The life span of the rabbit erythrocyte, *Acta Haematol.* **21:**36.

Weinstein, R. E., Rickles, F. R., and Walker, F. J., 1990, Purification and preliminary characterization of rabbit vitamin K–dependent coagulation proteins, *Thromb. Res.* **59:**759.

Weintraub, A. H., and Karpatkin, S., 1974, Heterogeneity of rabbit platelets. II. Use of the megathrombocyte to demonstrate a thrombopoietic stimulus, *J. Lab. Clin. Med.* **83:**896.

CHAPTER 17

RODENTS

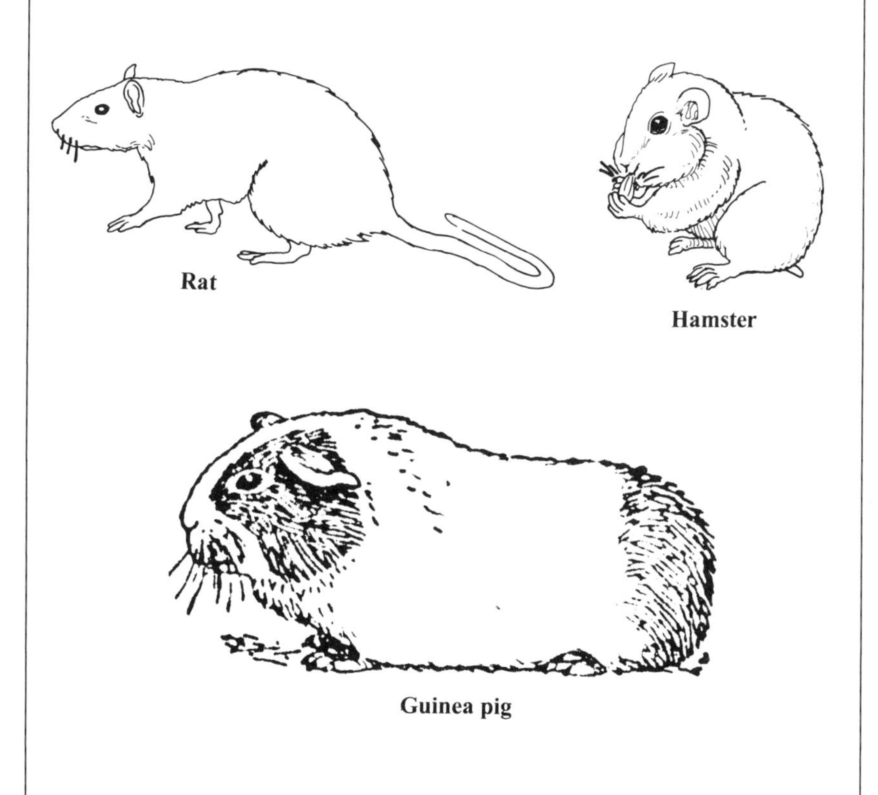

Rat

Hamster

Guinea pig

1. INTRODUCTION

Rodentia is an extremely successful order of small, furry, gnawing animals found worldwide. Some are entirely herbivorous, others mostly insectivorous. All are characterized by high fertility and reproductive capacity. There are some endearing wild rodents, including squirrels, chipmunks, woodchucks, gophers, beavers, muskrats, and guinea pigs, and some not so well liked, such as rats, mice, and porcupines. Rats and mice cause billions of dollars worth of food and property damage and may carry various diseases such as plague and typhus. Because of their small size and high fertility rate, rats, hamsters, and guinea pigs are bred to achieve various special strains for biological research.

The animals studied in this report were all inbred research strains obtained from reliable suppliers. Data from previous publications concerning rats (Lewis *et al.*, 1985) and guinea pigs (Lewis, 1992) are reprinted here with the kind permission of the editor of *Comparative Biochemistry and Physiology.*

2. SOURCE

Guinea pigs (*Cavia porcellus*) were outbred from the original strain (Hartley) by the Charles River Breeding Laboratory. Golden Syrian hamsters (*Cricetus auratus*) were obtained from local suppliers. Two strains of rats (*Rattus rattus*) were studied, the Sprague-Dawley (S-D) rat and the Wistar (Lewis) rat. They were both obtained from commercial suppliers. The guinea pigs weighed 300–500 g, the S-D rats 200–220 g, the Wistar rats approximately 340 g, and the hamsters 100–160 g.

All animals were lightly anesthetized. Blood from guinea pigs was obtained by external cardiac puncture. In hamsters and rats, the abdomen was quickly opened and blood was obtained from the abdominal aorta or vena cava with a No. 19 needle or butterfly.

3. TESTS CONDUCTED

3.1. Bleeding Times

On guinea pigs and rats, hair was removed with the depilatory cream Nair. Bleeding times on the guinea pigs were done by Dr. Peter C. Johnson (personal communication, 1991) using the Simplate method and a neonatal blood pressure cuff inflated to 40 mm Hg. Twenty tests gave a mean of 4 min on the foreleg, 4 1/2 min on the hindleg, and 3 1/2 min on the paw. For guinea pigs, the overall mean was 4 ± 1 1/2 min. On S-D rats, eight bleeding times were done on hindlegs with a mean of 3 1/2 ± 1 min.

3.2. General Coagulation Tests and Factors

The clotting times in glass or siliconized tubes shown in Table 17.1 were fast for these small mammals. The clots retracted well and did not lyse. SPTs for guinea pigs were long on both substrates. There was a marked lack of conformity in the prothrombin times with Simplastin. Those for guinea pigs were very long, for hamsters and S-D rats short, and for Wistar rats just longer than humans. With RVV, they were much closer together. APTTs were shorter than human in most of the animals. Thrombin times with bovine thrombin, CHH, and Atroxin were longer than human in all.

Table 17.2 shows the results of the TGT in guinea pigs and in both strains of rats. All results were within the human range except those of the Wistar rats, the generating mixtures of which clotted rat and human substrates in about 13.0 sec.

Table 17.3 shows the results of coagulation factor assays. There are striking differences from species to species. The guinea pig shows factors I, IX, Fitzgerald, ATIII, ProtC, and Antipl within the human range. Factors V, VIII, XI, XII, Fletcher, and XIII were high and II, VII, X, plgn, and RCF low compared to human. Hamsters gave quite different results. Factors VIII, IX, XI, XII, Fletcher, and ATIII fell within the normal human range. Factors V, VII, and X assayed in the high range, and factor, I, II, Fitzgerald, ProtC, plgn, and antipl were in the low or very low range. In the Wistar rat, factors I, IX, XII, Fletcher, ATIII, and antipl fell within the normal human range; factors V, VII, and VIII were high; and II, X, XI, Fitzgerald, ProtC, ProtS, and plgn were low or very low. The S-D rat assayed in the normal human range for factors I, II, V, XI, Fletcher, XIII; above the human range for VII, VIII, XII, and ATIII; and low in IX, X, and RCF.

The long thrombin time observed in the Wistar rats was explored in the tests shown in Table 17.4. The inhibitor was not lessened by $BaCl_2$ or heat treatment of rat plasma. Thus, it was defined as a non-heparin-like heat-stable inhibitor.

3.3. Hamster-to-Rat Liver Xenograft

The marked differences in clotting tests between hamster and rat (Wistar) allowed detection of the functioning of the new rat liver in the hamster host. The greatest differences between the two species were in factor X and protein C

TABLE 17.1
General Coagulation Tests in Rodentia

Test	Guinea pig			Hamster			Rat[a] Wistar			Rat[a] S-D			Human	
	N	Mean	SD	N	Mean	SD	N	Mean	SD	N	Mean	SD	Controls (4) (mean)	Normal range
Clot T (min)														
Glass	11	3	0.7	7	5	0.8	6	4	0.8	6	3	0.3	6	6–12
Silicone	11	7	2.6	7	8	1.6	6	8	0.8	6	9	0.7	26	20–59
Clot Retr	10	3–4	—	7	4–4	—	6	3–4	—	6	3–4	—	4–4	3–4+
Clot Lys	10	0	0	7	0	0	6	0	0	6	0	0	0	0
SPT (sec)														
Human	11	50	11	—	—	—	—	—	—	—	—	—	31.2	20+
Homolog	11	113	38	—	—	—	—	—	—	6	15.9	1.7	[b]	—
PT (sec)														
Simplastin	11	26.0	2.5	21	9.0	0.8	15	15.3	2.1	6	10.1	0.1	11.8	10–13
Homolog Br	11	19.7	1.6	16	19.3	2.0	11	28.1	6.2	6	18.5	2.1	[b]	—
RVV	11	14.4	1.8	7	15.2	4.2	6	10.6	1.3	6	13.8	1.7	14.9	13–18
APTT (sec)	11	28.7	3.8	9	22.2	2.1	6	22.6	3.5	6	18.8	1.8	33.3	24–34
Recal T (sec)	10	50	6.2	7	97	14	6	180	24	6	85	5.8	132	90–180
Lys	10	0	0	7	0	0	6	0	0	6	0	0	0	0
Lys MCA	10	0	0	7	0	0	6	0	0	6	0	0	0	0
Thromb T (sec)														
Bov Thr	11	24.5	6.8	7	[b]	—	6	[b]	—	6	66.0	14.5	16.4	11–18
CHH	6	26.2	2.9	7	[b]	—	6	[b]	—	—	—	—	15.2	11–18
Atroxin	6	[b]	—	7	39.6	13.3	6	58.5	19	6	33.0	6.5	14.8	11–18

[a]The rat brain SPT on human plasma was more than 120 sec; on hamster plasma, 33.7 sec.
[b]More than 120 sec.

TABLE 17.2
Thromboplastin Generation Test in Rodentia

Generating mixture[a]				Substrate clotting time (sec)[a,b]			
		Rat				Rat	
H	GP	W	S-D	H	GP	W	S-D
H (4)	—	—	—	**9.3**	11.1	11.8	14.6
—	GP (2)	—	—	11.2	**11.7**	—	—
—	—	W (3)	—	13.3	—	**13.6**	—
—	—	—	SD (2)	8.6	—	—	**9.8**

[a]Plasmas: (H) human; (GP) guinea pig; rat: (W) Wistar; (S-D) Sprague-Dawley.
[b]Homologous systems are in **boldface.**

bioassays. Results following transplant of hamster liver to rat are shown in Table 17.5. The first column shows the mean value for 15 normal rats, and the last column shows the mean value for 14 normal hamsters. Between these are the results from the transplanted rats on POD 8, 40, and 100 ($\pm$). On each occasion, samples from two rats were obtained and the rats sacrificed. Note that by 8 days after the transplant the PTs were shorter, and factors X and protein C had risen. By POD 40, factor X seemed to be in overproduction, and protein C had reached "hamster" levels. By POD 100, the coagulation tests on the transplanted rats resembled the mean tests obtained in the 14 normal hamsters.

3.4. Fibrinolytic Enzyme System

Clots prepared with guinea pig or rat plasma and streptokinase (SK) (250 U/ml) or urokinase (UK) (500 U/ml) and $CaCl_2$ lysed very slowly (about 240 min) compared to human plasma clots, which lysed in 2 min with UK and 7 1/2 min with SK. Guinea pig, hamster, and S-D and Wistar rat euglobulin lysis times were all longer than 4 hr.

3.5. Serum Proteins

Table 17.6 shows total protein (TP) and electrophoretic distribution of serum proteins in guinea pigs and Wistar rats. The TP and albumin of both were below and the α_1 globulin was above the human range. Guinea pig α_2 globulin was also above this range.

TABLE 17.3
Coagulation Factors in Rodentia

Factor	Guinea pig			Hamster			Rat: Wistar			Rat: Sprague-Dawley			Normal human range
	N	Mean	SD	N	Mean	SD	N	Mean	SD	N	Mean	SD	
I (mg/ml)	6	262	60	11	141	22	6	164	16	7	268	22	150–450
II (U/ml)	6	0.42	0.1	9	0.68	0.2	6	0.45	0.09	7	0.70	0.2	0.70–1.30
V (U/ml)	5	6.10	0.7	9	5.30	2.0	6	2.82	0.9	5	1.42	0.2	0.65–1.45
VII (U/ml)	6	0.09	0.03	9	9.45	3.3	6	3.91	0.9	7	1.63	0.5	0.50–1.30
X (U/ml)	6	0.12	0.02	17	3.43	1.1	15	0.30	0.07	7	0.36	0.07	0.75–1.25
VIII (U/ml)	6	4.50	1.3	9	1.03	0.3	6	1.74	0.7	7	4.00	0.02	0.75–1.40
IX (U/ml)	6	0.97	0.1	9	0.78	0.2	6	0.77	0.2	7	0.41	0.08	0.65–1.70
XI (U/ml)	6	1.71	0.6	9	0.79	0.1	6	0.63	0.2	7	1.28	0.3	0.75–1.40
XII (U/ml)	6	7.00	0.7	9	1.14	0.2	6	1.28	0.2	7	1.77	0.3	0.50–1.45
Flet (U/ml)	6	1.74	0.5	9	1.33	0.4	6	1.26	0.3	7	1.44	0.2	0.50–1.50
Fitz (U/ml)	6	1.00	0.1	9	0.22	0.04	6	0.30	0.08	—	—	—	0.50–1.50
ATIII (U/ml)	6	1.25	0.1	3	0.86	0.02	6	1.00	0.07	7	2.38	0.4	0.80–1.20
Prot C (U/ml)	3	0.72	0.2	14	0.51	0.06	15	[a]	—	—	—	—	0.60–1.40
Prot S (U/ml)	—	—	—	—	—	—	6	0.04	0.01	—	—	—	0.60–1.40
Plgn (U/ml)	5	0.02	0.01	3	0.15	0.02	6	0.13	0.05	—	—	—	0.80–1.20
Antipl (U/ml)	5	0.97	0.17	3	0.72	0.3	6	1.18	0.1	—	—	—	0.80–1.20
RCF (U/ml)	6	[a]	0	—	—	—	—	—	—	7	0.49	0.2	0.50–1.50
XIII (R)	6	59	13	—	—	—	—	—	—	7	9	3.6	4–16

[a]Less than 0.01 U/ml.

TABLE 17.4
Inhibition of Thrombin Time in Pooled Wistar Rat Plasma

Mixture	Thromb T (sec)
0.1 ml Human plasma	
(a) + 0.1 ml Normal human plasma	15.0
(b) + 0.1 ml Rat plasma	21.8
(c) + 0.1 ml Human plasma, $BaCl_2$-treated	15.2
(d) + 0.1 ml Rat plasma, $BaCl_2$-treated	21.0
(e) + 0.1 ml Human plasma, 56°C, 5 min	16.4
(f) + 0.1 ml Rat plasma, 56°C, 5 min	23.0
0.1 ml Rat plasma	
(h) + 0.1 ml Rat plasma	45.5

3.6. Miscellaneous Tests

Guinea pig plasma did *not* aggregate staphylococci at any dilution. Hamster and both rats in dilutions up to 1:1024 clumped the standard staphylococcal preparation.

3.7. Cellular Elements

The RBC and WBC parameters are shown in Table 17.7. Guinea pig RBCs were somewhat larger than those of hamster or rat, both of which fell below the

TABLE 17.5
Hamster-to-Rat Liver Xenograft

Test	Wistar rats[a] (15)	Rats (postoperative day)[b]			Hamsters[a] (14)
		8	40	100+	
Prothromb T (sec)	15.3	9.4	8.2	9.9	9.0
Hamster Br (sec)	42.7	20.1	17.7	18.4	19.3
Factor X (U/ml)	0.30	0.90	6.10	2.10	3.43
Protein C (U/ml)	<0.01	0.14	0.61	0.60	0.51

[a]Preoperative data. [b]Means of two rats each postoperative day.

TABLE 17.6
Serum Proteins by Paper Electrophoresis in Rodentia

Protein	Guinea pig			Rat (Wistar) (pool 1, 2, 4)	Normal human range
	N	Mean	SD		
TP (g/dl)	5	5.8	0.6	5.9	6.3–7.9
Albumin (g/dl)	5	2.1	0.4	2.1	3.2–4.4
α_1-Globulin (g/dl)	5	0.5	0.1	1.5	0.2–0.4
α_2-Globulin (g/dl)	5	1.2	0.2	0.5	0.4–1.0
β-Globulin (g/dl)	5	0.9	0.1	0.8	0.5–1.0
γ-Globulin (g/dl)	5	1.1	0.2	1.0	0.6–1.8

normal human range. Lymphocytes were more numerous than neutrophils in these rodents. Platelets were plentiful and appeared to be slightly smaller than human platelets.

3.8. Platelet Activities

Table 17.8 shows that platelet glass retention indices were within the normal human range in guinea pigs and Wistar rats. Aggregation with ADP fell within the human range. Guinea pig platelets aggregated moderately with ristocetin (Risto), arachidonic acid (Arach A), and pig plasma (Pig Pl), and very slightly with collagen and thrombin. Rat platelets aggregated moderately with Arach A, but hardly at all with collagen, Risto, thrombin, CHH, or bovine (Bov) or Pig Pl.

3.9. Platelet Ultrastructure

Figures 17.1, 17.2, and 17.3 are TEMs of guinea pig, Wistar rat, and mouse platelets. Circumferential microtubules,, alpha granules, dense bodies, glycogen particles, and an open canalicular system were similar to those found in human preparations.

4. SUMMARY

Blood was obtained in relatively small amounts and with some difficulty from these small mammals. It clotted rapidly, retracted well, and did not lyse. Prothrombin times with Simplastin, homologous brain, and RVV differed a good bit from rodent to rodent. In guinea pigs, the prothrombin times were the same as

TABLE 17.7
Cellular Elements of the Blood in Rodentia

Element	Guinea pig			Hamster			Rat: Wistar			Rat: Sprague-Dawley			Normal human range
	N	Mean	SD	*N*	Mean	SD	*N*	Mean	SD	*N*	Mean	SD	
HCT (%)	6	41	2.9	6	42	1.9	4	34	0.8	12	36	1.5	37–52
HGB (g/dl)	6	13.5	0.7	6	15.2	0.6	4	13.0	3.1	12	13.7	1.0	12–18
RBC (× 10^6/mm^3)	6	4.8	0.4	6	7.1	0.2	4	6.6	0.5	12	6.2	0.5	4.2–6.2
MCV (fl)	6	82	0.7	6	59	1.0	4	51	2.8	12	58	3.5	79–97
MCH (pg)	6	27	0.4	6	21	0.8	4	20	1.4	12	22	1.9	27–31
MCHC (g/dl)	6	33	0.8	6	36	0.8	4	39	0.8	12	38	2.2	32–36
WBC (× 10^3/mm^3)	9	11.2	4.1	6	4.7	0.8	4	4.3	0.9	12	5.9	1.9	5–10
Neut (%)	9	47	6.7	6	24	9	1	18	—	12	13	8.4	55–75
Lymph (%)	9	48	6.3	6	74	9	1	76	—	12	86	8.7	20–40
Mono (%)	5	2	1.4	6	2	1	1	4	—	12	0.5	0.05	2–10
Eos (%)	8	3	3.2	6	0	0	1	0.2	—	12	0.5	0.08	0–3
Bas (%)	3	1	0.8	6	0	0	1	0	—	12	0	0	0–3
Plat C (× 10^3/mm^3)	7	434	42	6	368	71	4	656	185	12	770	193	150–450
MVP (μM^3)	5	4.8	0.5	6	6.2	0.8	4	6.9	0.4	2	4.7	0.4	5.6–10.4

TABLE 17.8
Platelet Activities in Rodentia

Activity	Guinea pig			Human control	Wistar rat		Human	
	N	Mean	SD		N	Mean	Control	Normal range
Plat Glass Ret Ind (%)	3	59	138	42	8	66	3	>20
Aggregation (%)								
ADP (20 μM)	7	71	6.8	84	10	78	80–100	70–100
(10 μM)	3	69	4.0	92	10	68	80–100	70–100
(5 μM)	3	68	3.0	91	—	—	80–100	50–100
Collagen								
Bov (0.19 mg/ml)	5	8	7.7	86	8	7	80–100	80–100
Rat (0.2 mg/ml)	—	—	—	—	4	0	—	—
Ristocetin (0.9 mg/ml)	7	34	37	90	4	0	80–100	80–100
Risto ½ (0.45 mg/ml)	6	5	2.9	6	[a]	0	0	<18
Arach A (0.5 mg/ml)	7	59	16	95	[a]	43	80–100	60–100
Bov Thr (0.1 U/ml)	—	—	—	—	6	10	77	60–100
(0.2 U/ml)	4	4	5.2	33	[a]	3	89	60–100
(0.4 U/ml)	—	—	—	—	[a]	2	97	60–100
(0.7 U/ml)	3	3	2.1	68	—	—	—	60–100
CHH (0.05 mg/ml)	—	—	—	—	[a]	6	100	60–100
Bov Pl (1:10)	—	—	—	—	[a]	6	100	60–100
Pig Pl (1:10)	3	68	2.9	87	6	0	100	60–100

[a] Pool.

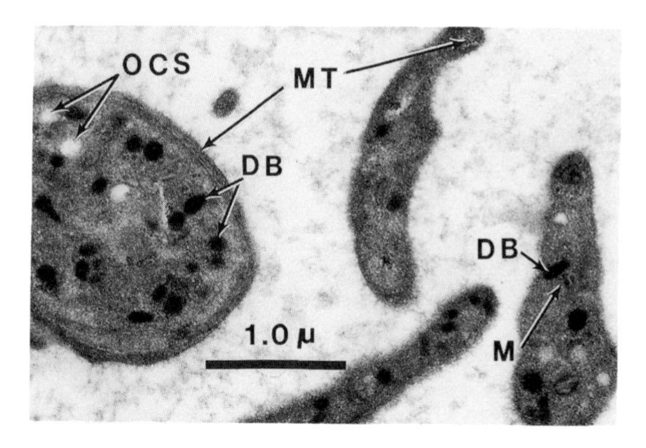

FIGURE 17.1 TEM of guinea pig platelets. Internal structures here and in Figs. 17.2 and 17.3: (a-gran) alpha granules; (DB) dense bodies; (GLY) glycogen particles; (M) mitochondrion; (MT) microtubules; (OCS) open canalicular system.

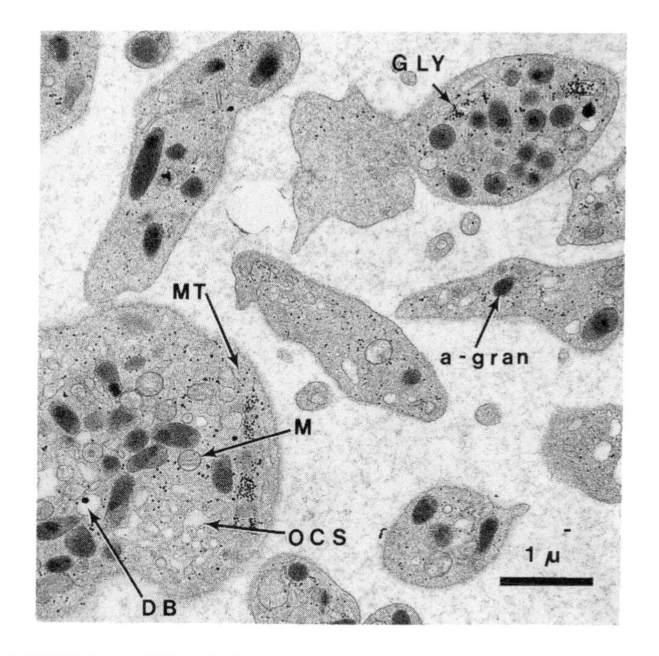

FIGURE 17.2. TEM of Wistar rat platelets. Key in Fig. 17.1 caption.

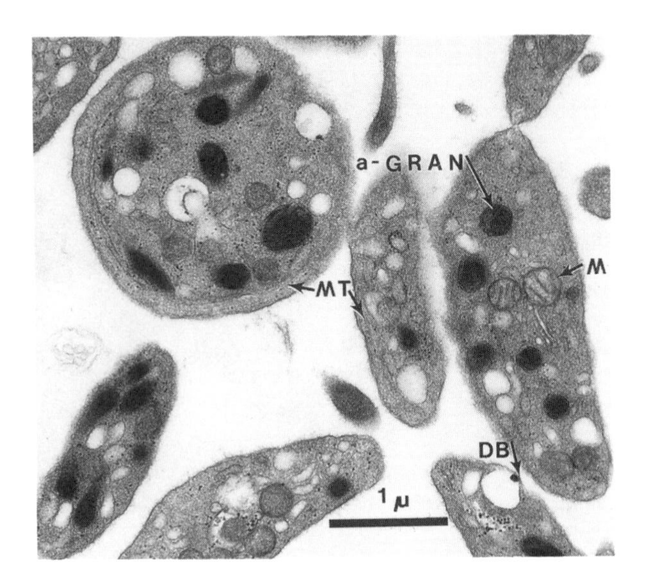

FIGURE 17.3. TEM of mouse platelets. Key in Fig. 17.1 caption.

or longer than APTTs. This result reflects the very low levels of factor VII found in these animals. Hamsters and rats differed strikingly in factor X and protein C levels. In hamster-to-rat liver xenografts, the very low levels of these two factors rose with the functioning of the new liver.

Rodent platelets showed ultrastructures similar to human platelets, but their responses to aggregating agents were quite different. Guinea pigs and rats aggregated well with ADP, but poorly (<50%) or not at all with the other agents.

REFERENCES

Lewis, J. H., 1992, Comparative hematology studies on guinea pigs (*Cavia porcellus*), *Comp. Biochem. Physiol. A* **102:**507.

Lewis, J. H., Van Thiel, D. H., Hasiba, U., Spero, J. A., and Gavaler, J., 1985, Comparative hematology: Studies on *Rodentia* (rats), *Comp. Biochem. Physiol.* **82:**211.

SUGGESTED READINGS

Rodents

Fritz, T. E., Tolle, D. V., and Flynn, R. J., 1968, Hemorrhagic diathesis in laboratory rodents, *Proc. Soc. Exp. Biol. Med.* **128:**228.

Guinea Pig

Archambault, A., and Tremblay, J., 1965, Étude du mecanisme de la coagulation chez le cobaye. I. La thromboplastino-formation, *Rev. Can. Biol.* **24:**2.
Archambault, A., and Tremblay, J., 1965, Étude du mecanisme de la coagulation chez le cobaye. II. La thrombino-formation et la fibrino-formation, *Rev. Can. Biol.* **24:**2.
Archambault, A., and Tremblay, J., 1965, Study of coagulation in the guinea pig. I. Thromboplastin formation, *Rev. Can. Biol.* **24:**101.
Archambault, A., and Tremblay, J., 1965, Mechanism of coagulation in the guinea pig. II. Thrombin formation and fibrin formation, *Rev. Can. Biol.* **24:**111.
Bilbey, D. L. J., and Nicol, T., 1955, Normal blood picture of the guinea pig, *Nature (London)* **176:**1218.
Chenkin, T., and Weiner, M., 1965, A comparison of some human clotting abnormalities with the clotting pattern of normal guinea pigs, *Exp. Med. Surg.* **23:**398.
Constantine, J. W., 1966, Aggregation of guinea pig platelets by adenosine diphosphate, *Nature (London)* **210:**162.
Hill, B. F., 1972, The guinea pig: Management and research use, *Charles River Digest,* Vol. 11, p. 3, Charles River Breeding Laboratories, Wilmington, MA.
Hurt, J. P., and Krigman, M., 1970, Selected procoagulants in the guinea pig, *Am. J. Physiol.* **218:**(3)**:**832.
Johnson, P. C., 1991, Guinea pig bleeding times, Personal communication.
King, E. S., and Lucas, M., 1941, A study of the blood cells of normal guinea pigs, *J. Lab. Clin. Med.* **26:**1364.
Latallo, Z., Niewiarowski, S., and Copley, A. L., 1958, Fibrinolytic system of guinea pig serum, *Am. J. Physiol.* **196:**779.

Hamster

Amin, A., and Shamloo, K. D., 1963, Distribution of the serum proteins of Syrian hamster as revealed by starch-gel electrophoresis, *Nature (London)* **198:**485.
Denyes, A., and Carter, J. D., 1961, Clotting-time of cold-exposed and hibernating hamsters, *Nature (London)* **190:**450.
Haydon, G. B., and Taylor, D. A., 1965, Microtubules in hamster platelets, *J. Cell Biol.* **26:**637.
Lyman, C. P., Weiss, L. P., O'Brien, R. C., and Barbeau, A. A., 1957, The effect of hibernation on the replacement of blood in the golden hamster, *J. Exp. Zool.* **136:**471.
Maupin, B., 1967, Les plaquettes sanguines des mammiferes, *Biol. Med.* **56:**143.
Ottis, K., and Tauber, O. E., 1952, Blood platelet counts of the golden hamster (*Cricetus auratus*), *Blood* **7:**948.
Stewart, M. O., Florio, L., and Mugrage, E. R., 1944, Hematological findings in the golden hamster (*Cricetus auratus*), *J. Exp. Med.* **80:**189.
Svihla, A., Bowman, H., and Pearson, R., 1952, Prolongation of blood clotting time in the dormant hamster, *Science* **114:**272.
Valdivia, L. A., Lewis, J. H., Celli, S., Bontempo, F. A., Fung, J. J., Demetris, A. J., and Starzl, T. E., 1993, Hamster coagulation and serum proteins in rat recipients of hamster xenografts, *Transplantation* **56:**489.

Rat

Behnke, O., 1967, Electron microscopic observations on the membrane systems of the rat blood platelet, *Anat. Rec.* **158:**121.

Burns, K. F., Timmons, E. H., and Poiley, S. M., 1971, Serum chemistry and hematological values for axenic (germfree) and environmentally associated inbred rats, *Lab. Anim. Sci.* **21:**415.

Cameron, D. G., and Watson, G. M., 1949, The blood counts of adult albino rats, *Blood* **4:**816.

Cotchin, E., and Roe, F. J. C., Eds., 1967, *Pathology of Laboratory Rats and Mice,* Blackwell, Oxford.

Creskoff, A. J., Fitz-Hugh, T., Jr., and Farris, E. J., 1942, Hematology of the rat—methods and standards, in: *The Rat in Laboratory Investigation,* 2nd ed. (E. J. Ferris, ed.), pp. 406–420, J. B. Lippincott, Philadelphia.

DeClerck, F., Goossens, J., and Renneman, R., 1976, Effects of anti-inflammatory, anticoagulant and vasoactive compounds on tail bleeding time, whole blood coagulation time and platelet retention by glass beads in rats, *Thromb. Res.* **8:**179.

DeGabriele, G., and Penington, D. G., 1967, Physiology of the regulation of platelet production, *Br. J. Haematol.* **13:**202.

DeGaetano, G., Donati, M. B., Poggi, A., Innocenti, I. R., and Roncaglione, M. C., 1975, A peculiar haemostatic mechanism in the rat, *Thromb. Res.* **6:**201.

DeGaetano, G., Donati, M. B., Reyers-Degla, I., Innocenti, M., and Roncaglione, M. C., 1975, Defective ristocetin and bovine factor VIII–induced platelet aggregation in rats, *Experientia* **31:**500.

Dejana, E., Callioni, A., Quintana, A., and deGaetano, G., 1979, Bleeding time in laboratory animals. II. A comparison of different assay conditions in rats, *Thromb. Res.* **15:**191.

DeLong, E., Uhley, H. N., and Friedman, M., 1959, Change in blood clotting time of rats exposed to a particular form of stress, *Am. J. Physiol.* **196:**429.

Duvolon, S., 1947, Contribution a l'étude hematologique du rat blanc normal, *Donn. Morphol. Numer. Sang* **18:**205.

Ebbe, S., Stohlman, F., Jr., Donovan, J., and Howard, D., 1966, Platelet survival in the rat as measured with tritium-labeled diisopropylflourophosphate, *J. Lab. Clin. Med.* **68:**233.

Evans, G. O., and Flynn, R. M., 1990, Further observations on in-vitro aggregation of rat platelets with different collagens, *Thromb. Res.* **57:**301.

Farris, E. J., and Griffith, J. Q. (eds.), *The Rat in Laboratory Investigation,* J. B. Lippincott, Philadelphia.

Gallimore, M. J., Tyler, H. M., and Shaw, J. T. B., 1971, The measurement of fibrinolysis in the rat, *Thromb. Diath. Haemorrh.* **26:**295.

Godwin, K. O., Fraser, F. J., and Ibbotson, R. N., 1964, Haematological observations on healthy (SPF) rats, *Br. J. Exp. Pathol.* **45:**514.

Gogjiam, M. A., Barry, K., and Stein, B. M., 1981, Measurement of fibrinolytic activity after epsilon-aminocaproic acid administration in rats, *Lab. Anim. Sci.* **31:**710.

Gruner, O. P. N., and Endressen, G. K. M., 1971, Platelet count and platelet stickiness in rats, *Thromb. Diath. Haemmorh.* **26:**389.

Gunn, F. E., and Vaughan, S. L., 1930, Bone marrow reactions. II. The blood count in the albino rat: Blood platelets, *Anat. Rec.* **45:**59.

Hilgard, P., 1972, Comparative studies of the fibrinolytic system in rats and humans, *Haemostasis* **1:**101.

Hogstorp, H., Jakobsson, H., and Carlin, G., 1981, Studies on α_2-antiplasmin in the rat, *Thromb. Res.* **21:**247.

Hohage, R., 1971, Methodical studies on platelet aggregation in the rat, *Thromb. Diath. Haemorrh.* **26:**393.

Kozma, C. K., Weisbroth, S. H., Stratmen, S. L., and Conejeros, M., 1969, Normal biological values for Long-Evans rats (Part II), *Lab. Anim. Care* **19:**746.

Li, L., and Olson, R. E., 1967, Purification and properties of rat prothrombin, *J. Biol. Chem.* **242:**5611.

Losonczy, G., and Harsing, L., 1982, Simultaneous activation of coagulation and fibrinolysis after severe renal ischemia in rats, *Nephron* **32:**180.

MacGregor, L., Morazain, R., and Renaud, S., 1980, A comparison of the effect of dietary short and long chain saturated fatty acids on platelet functions, platelet phospholipids, and blood coagulation in rats, *Lab. Invest.* **43:**438.

Mogenson, G. J., and Jaques, L. B., 1957, The effect of psychological stress procedure on the prothrombin time of rats, *Blood* **12:**649.

Morland, J., Holm, T., and Pyrdz, H., 1976, The influence of ethanol feeding on the activities of factors II, VII, VIII, IX, and X in rats, *Thromb. Haemost.* **36:**532.

Nadelhaft, I., and Lamy, F., 1972, Fibrinogen turnover in rats as a function of age, *J. Lab. Clin. Med.* **79:**724.

Pyorala, K., 1965, Determinants of the clotting factor response to warfarin in the rat, *Ann. Med. Exp. Biol. Fenn.* **43:**1.

Quimby, F. H., Saxon, P. A., and Goff, L. G., 1948, Total white cell counts of peripheral and heart blood of the rat, *Science* **107:**447.

Raymond, S. L., and Dodds, W. J., 1975, Characterization of the fawn-hooded rat as a model for hemostatic studies, *Thromb. Diath. Haemorrh.* **33:**361.

Reich, C., and Dunning, W. F., 1943, Studies on the morphology of the peripheral blood of rats, *Cancer Res.* **3:**248.

Renaud, S., and Lecompte, F., 1970, Hypercoagulability induced by hyperlipemia in rat, rabbit and man, *Circ. Res.* **27:**1003.

Renaud, S., Kuba, K., Goulet, C., Lemire, Y., and Allard, C., 1970, Relationship between fatty-acid composition of platelets and platelet aggregation in rat and man, *Circ. Res.* **26:**553.

Sinakos, Z., and Caen, J. P., 1966, Le comportement des plaquettes chez le rat, *Thromb. Diath. Haemorrh.* **21:**163.

Soulban, G., and Labrecque, G., 1989, Circadian rhythms of blood clotting time and coagulation factors II, VII, IX and X in rats, *Life Sci.* **45:**2485.

Surodeikina, L. N., 1965, Protein composition of blood serum of intact rats, *Lab. Delo* **4:**230.

Suttie, J. W., 1967, Control of prothrombin and factor VII biosynthesis by vitamin K_1, *Arch. Biochem. Biophys.* **118:**166.

Thwelis, E. W., and Meyer, O. O., 1942, The blood count of normal white rats, *Anat. Rec.* **82:**115.

Vainieri, H., and Wingard, L. B., 1977, Effect of warfarin on the kinetics of the vitamin K–dependent clotting factors in rats, *J. Pharmacol. Exp. Ther.* **201:**507.

Vondruska, J. F., and Greco, R. A., 1973, Certain hematologic and blood chemical values in Charles River CD albino rats, *Bull. Am. Soc. Vet. Clin. Pathol.* **22:**3.

Waaler, B. A., Gustafsson, B. E., Hauge, A., Nilsson, D., and Amundsen, E., 1964, Plasma levels of various blood clotting factors in germ free rats, *Proc. Soc. Exp. Biol. Med.* **117:**444.

Wingard, L. B., Jr., and Levy, G., 1973, Kinetics of anti-coagulant effect of dicumarol in rats, *J. Pharmacol. Exp. Ther.* **184:**253.

Zoucas, E., Bergqvist, D., Goransson, G., and Bengmark, S., 1982, Effect of acute ethanol intoxication on primary haemostasis, coagulation factors and fibrinolytic activity, *Eur. Surg. Res.* **14:**33.

CHAPTER 18

THE BOTTLE-NOSED PORPOISE

1. INTRODUCTION

The order Cetacea comprises more than 40 species of marine mammals that live their long lives in the waters of the great oceans or, much less frequently, in rivers (the Amazon) or estuaries. All adult Cetacea are large, and one, the great blue whale, is the largest animal that ever lived—larger than the largest dinosaur. The great blue whale is almost as long as a football field and weighs up to 150 tons (136 metric tons). It is presumed that Cetacea developed from primitive land carnivores, but their origin and return to the sea lack fossil definition. In the Eocene Age, Cetacea were already abundant. They have large torpedolike bodies with forelimbs serving as flippers, no hindlimbs, a dorsal fin, and notched, flattened tails called "flukes" that are powerful aids to swimming and diving. Over time, their nostrils have moved to the top of the head as single or double blowholes. When whales exhale, a cloud of water drops and vapor is blown out. These blow clouds form a unique pattern that differs from species to species. Ordinarily, Cetacea breathe every 2–3 min, but they can dive to great depths for food and stay under water for up to 40 min without breathing. The porpoises and about half the whales have teeth and are carnivorous, eating only flesh, mostly fish or squid. The other whales have no teeth but rows of baleen (whalebone) that strain krill, the plankton and larval mixture that constitutes their principal food. The gestation period in Cetacea is almost a year. The young are then nursed for another year. Their life expectancy is long, 25–40 years, so that each female may produce 10–15 offspring in a lifetime.

The bottle-nosed porpoise (*Tursiops truncatus*) is the most common species in the Atlantic Ocean. It is also the species most commonly found in aquariums or marinelands. These intelligent mammals are easily trained to perform acrobatic feats. Man has been fascinated by porpoises since early times. Their ability to play in small herds, to leap, and to somersault has been recognized for centuries. Many an enchanting tale has described their intelligence and their devotion to and friendship with humans, including their rescue of drowning individuals. Despite our interest in and attraction to these animals, humans and their commercial fishing activities destroy thousands every year.

Harboe and Schrumpf (1953) recognized the large size of whale erythrocytes. Medway and Geraci (1964) also gave data concerning blood counts and erythrocyte parameters. Coagulation studies showing lack of factor XII activity in Cetacea were published by Lewis *et al.* (1969) and Robinson *et al.* (1969). Those from Lewis et al. (1969) are reprinted here with the kind permission of the editor of *Comparative Biochemistry and Physiology.* An additional eight animals were studied later.

2. SOURCE

The first porpoise was studied at the Wometco Seaquarium in Miami, Florida, through the courtesy of Messrs. Burton Clark and Warren Zeiller. The next three were studied at the Aquarama Theatre of the Sea in Philadelphia, Pennsylvania, through the courtesy of William Medway, D.V.M., of the University of Pennsylvania School of Veterinary Medicine. Because a deficiency of factor XII was found in these four animals, it was decided to study eight more. This study was done at the Wometco Sea Aquarium in Miami, Florida, through the courtesy of Jesse White, D.V.M. In order to take the blood, each animal was confined to a small tank and the water level was lowered until the porpoise rested on a foam mattress. Blood samples were then obtained from the center vein of the tail fluke or the vertebral artery. The porpoises weighed an average of 173 kg. There were seven females and five males. Their sex is noted in the tables.

3. TESTS CONDUCTED

3.1. Bleeding Times

A short cut was made on the first porpoise in the flipper area. Bleeding was limited to less than 2 min. Bleeding times were not done on the other animals.

3.2. General Coagulation Tests and Factors

Table 18.1 shows the results of general coagulation tests in the 12 porpoises. All clotting times in glass and ten in siliconized tubes were much longer than human times. Clotting in siliconized tubes was usually over an hour. Clots retracted and did not lyse. SPTs on either human or porpoise substrate were short. Simplastin and RVV times were similar to human times, as were thrombin times. APTTs and Recal Ts were long. TGTs in four porpoises (Table 18.2) showed that no plasma thromboplastin activity was generated. When human barium-treated plasma (BaPl), was substituted for porpoise Ba Pl, the defect was partially corrected.

Coagulation factors are shown in Table 18.3. The pattern was strikingly different from that of humans: Factor XII was essentially absent in all animals, no anti-XII could be found in porpoise plasma, and factor VIII was very high. The other coagulation factor levels fell within the human range.

TABLE 18.1
General Coagulation Tests in the Bottle-Nosed Porpoise

	Porpoise No.															Human	
Test	1♂	2♂	3♀	4♀	5♂	6♀	7♀	8♀	9♀	10♂	11♂	12♀	N	Mean	SD	Controls (12) (mean)	Normal range
Clot T (min)																	
Glass	38	39	26	30	45	14	50	33	57	37	42	36	12	37	11	7	6–12
Silicone	150	150	90	65	120	16	65	45	150	360	240	150	12	133	94	25	20–59
Clot Retr	4/4	3/4	3/4	4/4	3/4	3/4	3/4	3/4	2/4	3/4	4/4	4/4	12	3–4	—	4/4	3–4+
Clot Lys	0	0	0	0	0	0	0	0	0	0	0	0	12	0	0	0	0
SPT (sec)																	
Human	14.8	—	15.8	18.4	19.2	19.8	19.0	17.0	18.2	18.0	18.2	19.7	11	18.0	1.6	44.1	20+
Porpoise	13.8	—	13.2	15.4	15.2	15.7	15.4	14.5	14.2	15.8	15.0	14.5	11	14.8	0.8	23.5	—
PT (sec)																	
Simplastin	13.2	12.7	12.4	12.4	12.5	14.7	14.2	14.0	14.5	12.0	12.0	13.0	12	13.1	1.0	12.6	10–13
Hu Br	20.2	21.7	20.8	22.8	—	—	—	—	—	—	—	—	4	21.4	1.1	11.4	11–15
RVV	28.8	27.8	27.8	—	28.0	27.2	23.5	21.0	28.8	24.0	21.2	23.5	11	25.6	3.0	19.0	13–18
APTT (sec)	420	400	375	400	120	360	360	280	300	300	300	325	12	328	80	42.1	45–55
Recal T (sec)	1800	1800	1800	1800	1800	1800	1800	1800	1800	1800	1800	1800	12	1800	—	160	90–180
Lys	0	0	0	0	0	0	0	0	0	0	0	0	12	0	0	0	0
Lys MCA	0	0	0	0	0	0	0	0	0	0	0	0	12	0	0	0	0
Thromb T (sec)																	
Bov Thr	25.2	27.8	26.8	28.2	14.8	12.0	12.5	12.6	15.5	14.0	12.0	12.5	12	17.9	7.0	11.2	11–18

TABLE 18.2
Thromboplastin Generation Test in the Bottle-Nosed Porpoise

Generating mixture		Substrate Clot T (sec)[a]	
Human	Porpoise	Human	Porpoise
Means			
H (2)	—	**9.3**	10.6
—	P (4)	40+	**40+**
P8 serum + plat + H Ba Pl		13.8	14.4

[a]Homologous systems are in **boldface.**

3.3. Fibrinolytic Enzyme System

Table 18.4 shows that fibrinolytic activity approximately equal to human was generated in the plasmas of six porpoises by activation with streptokinase, urokinase, and 2% staphylokinase. Antiplasmin was also present in porpoise plasma.

3.4. Serum Proteins

Table 18.5 shows serum proteins (determined by paper electropheresis) in the 12 porpoises. The total protein content and electrophoretic distribution were striking in their likeness to the human pattern. Albumin was slightly lower and α_2- and γ-globulins slightly higher than human.

3.5. Miscellaneous Tests

Porpoise plasma (from Nos. 2, 3, and 4) clumped the standard staphylococcal preparation in dilutions up to 1:2048.

3.6. Cellular Elements

The cellular elements of the blood are shown in Table 18.6. HCTs and HGBs fell within the human range, but RBC counts were low. Thus, the RBCs were large with high HGB levels. The WBC counts were similar to human counts, and the differential counts showed high numbers of eosinophils. Platelet counts were slightly lower than human counts.

TABLE 18.3
Coagulation Factors in the Bottle-Nosed Porpoise

Factor	Porpoise No.												
	1–4 (mean)	5♂	6♀	7♀	8♀	9♀	10♂	11♂	12♀	*N*	Mean	SD	Normal human range
I (mg/dl)	418	320	409	478	478	410	270	385	410	12	403	58	150–450
II (U/ml)	1.17	1.48	1.28	1.33	1.43	1.19	1.33	1.28	1.43	12	1.29	0.10	0.70–1.30
V (U/ml)	1.63	0.83	0.54	1.00	0.85	0.93	0.71	0.58	0.63	12	1.05	0.45	0.65–1.45
VII (U/ml)	—	1.11	0.81	1.07	0.81	0.74	1.18	0.97	0.92	8	0.95	0.17	0.50–1.30
X (U/ml)	—	1.40	1.40	1.10	1.20	1.06	1.40	1.02	1.40	8	1.25	0.14	0.75–1.25
VIII (U/ml)	4.6	9.00	10+	10+	9.00	10+	9.00	10+	10+	12	10+	—	0.75–1.40
IX (U/ml)	1.30	0.95	1.0	1.10	1.20	0.95	1.25	1.25	1.05	12	1.12	0.10	0.65–1.70
XI (U/ml)	0.95	0.80	0.8	1.20	1.40	1.0	1.35	1.10	1.10	12	1.08	0.20	0.75–1.40
XII (U/ml)	[a]	[a]	[a]	[a]	[a]	[a]	[a]	[a]	[a]	12	[a]	0	0.50–1.45
XIII (R)	32	32	16	16	16	32	8	8	16	12	23	10.1	4–16

[a]Less than 0.01 U/ml.

TABLE 18.4
Fibrinolytic Activity with Streptokinase or Urokinase in the Bottle-Nosed Porpoise

	[^{125}I]Fibrin lysed (%)												
	Porpoise No. and sex											Human	
Reagent	2♂	3♀	4♀	5♂	6♀	7♀	8♀	9♀	*N*	Mean	SD	Control	Normal range
Plasmin	0	0	0	0	0	0	0	0	8	0	—	0	0
Antiplasmin	65	75	55	71	36	26	70	78	8	59	19	40	35–55
SK (250 U/ml)	53	100	—	49	28	43	17	—	6	48	29	67	50–90
UK (50 U/ml)	29	36	—	16	25	15	14	—	6	23	9	26	20–60
Staph K (2%)	65	73	—	61	51	63	40	—	6	59	12	25	20–60

TABLE 18.5
Serum Proteins by Paper Electrophoresis in the Bottle-Nosed Porpoise

Protein	Porpoise No.												N	Mean	SD	Normal human range
	1♂	2♂	3♀	4♀	5♂	6♀	7♀	8♀	9♀	10♂	11♂	12♀				
TP (g/dl)	8.8	7.8	7.0	6.0	8.1	7.4	7.5	8.0	7.9	7.4	7.7	7.9	12	7.6	0.7	6.3–7.9
Albumin (g/dl)	3.2	2.8	3.0	2.4	2.8	2.8	2.9	3.5	2.9	3.1	3.2	3.9	12	3.0	0.4	3.2–4.4
α_1-Globulin (g/dl)	0.4	0.6	0.4	0.3	0.6	0.6	0.6	0.7	0.8	0.6	0.5	0.6	12	0.6	0.1	0.2–0.4
α_2-Globulin (g/dl)	2.0	1.8	1.2	1.2	1.8	1.4	1.7	1.5	1.4	1.3	1.2	1.4	12	1.5	0.3	0.4–1.0
β-Globulin (g/dl)	0.8	0.6	0.2	0.7	0.6	0.6	0.7	0.7	0.6	0.7	0.5	0.6	12	0.6	0.1	0.5–1.0
γ-Globulin (g/dl)	2.4	2.0	2.2	1.4	2.3	2.0	1.6	1.6	2.2	1.7	2.2	1.4	12	1.9	0.4	0.6–1.8

TABLE 18.6
Cellular Elements of the Peripheral Blood in the Bottle-Nosed Porpoise

Element	Porpoise No.															Normal human range
	1♂	2♂	3♀	4♀	5♂	6♀	7♀	8♀	9♀	10♂	11♂	12♀	*N*	Mean	SD	
HCT (%)	48	50	49	50	47	36	40	38	40	45	39	39	12	43	5.2	37–52
HGB (g/dl)	17.2	19.0	18.3	17.8	16.8	13.4	15.0	14.6	15.0	16.8	14.8	15.0	12	16.1	1.7	12–18
RBC (× 10^6/mm^3)	3.1	4.0	4.0	3.6	3.7	3.3	3.7	3.3	3.7	3.9	3.3	3.6	12	3.6	0.3	4.2–6.2
MCV (fl)	155	125	123	139	127	109	108	115	108	115	118	108	12	121	14	79–97
MCH (pg)	56	48	46	49	45	41	41	44	41	43	45	42	12	45	4.4	27–31
MCHC (g/dl)	36	38	37	36	36	37	38	38	38	37	38	39	12	37	1.1	32–36
WBC (× 10^3/mm^3)	5.9	9.8	8.4	9.1	10.0	25.0	10.0	9.3	9.5	10.0	8.3	10.0	12	10.4	4.7	5–10
Neutr (%)	55	75	65	63	62	87	84	60	70	78	61	88	12	71	12	55–75
Lymph (%)	32	18	29	28	25	9	6	28	5	13	4	8	12	17	11	20–40
Mono (%)	0	0	0	2	0	0	1	1	4	4	2	0	12	0.8	0.5	2–10
Eos (%)	13	4	5	7	13	4	7	11	20	5	33	4	12	11	8.6	0–3
Bas (%)	0	3	1	0	0	0	2	0	1	0	0	0	12	0.6	0.6	0–3
Plat C (× 10^3/mm^3)	120	130	134	128	155	160	120	143	175	120	130	115	12	136	19	150–450

TABLE 18.7
Platelet Activities in the Bottle-Nosed Porpoise

Activity	Porpoise No.						N	Mean	SD	Normal human range
	5♂	6♀	7♀	8♀	10♂	12♀				
Plat Glass Ret Ind (%)	45	17	60	23	19	19	6	31	18	>20
Aggregation (%)										
ADP (20 μM)	100	90	100	100	80	90	6	93	8.2	70–100
(10 μM)	80	90	90	90	80	70	6	83	8.2	70–100
Collagen (0.19 mg/ml)	15	20	10	75	25	70	6	36	29	80–100

3.7. Platelet Activities

Table 18.7 shows that half the porpoises had platelet glass bead retention indices below 20%. The human range was above 20%. Aggregation of porpoise platelets was over 80% at two levels of ADP. With collagen, aggregation was much less.

3.8. Platelet Ultrastructure

Figure 18.1 is a TEM of porpoise platelets. Most of the platelets are sectioned end-to-end. The open canalicular system, peripheral bundles of microtubules, mitochondria, large alpha granules, small dense bodies in vacuoles, and occasional glycogen particles can be seen.

4. SUMMARY

Porpoises showed a consistent species lack of factor XII. General coagulation tests supported the presence of a deficiency. The long clotting time, APTT,

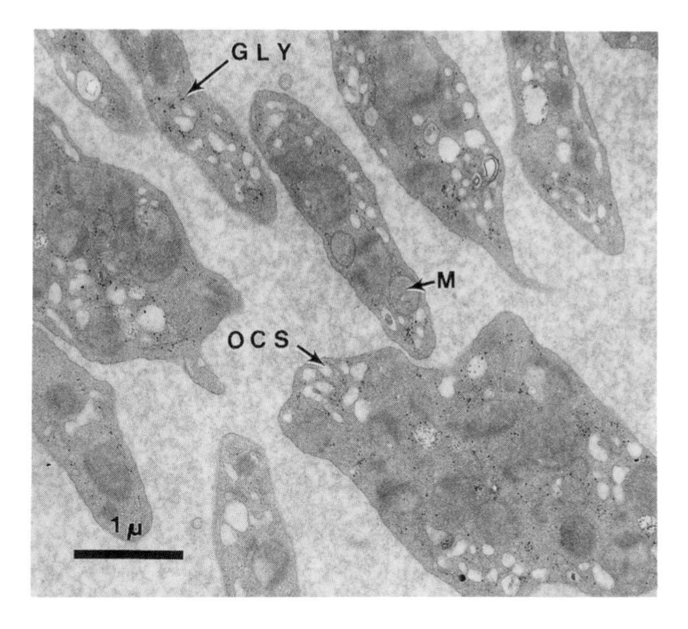

FIGURE 18.1 TEM of porpoise platelets. Internal structures: (GLY) glycogen particles; (M) mitochondria; (OCS) open canalicular system.

and TGT and the short SPT were all suggestive of a real defect rather than a problem with species specificity. About the same time these findings were published, Robinson *et al.* (1969) also reported the lack of factor XII in the blood of three *Tursiops truncatus* and three killer whales (*Orcinus orca*). This finding certainly suggests that the lack of factor XII is a true species characteristic that does not affect the hemostatic mechanism or the life expectancy of this order of mammals. RBCs are large (MCV = 121 fl).

REFERENCES

Harboe, A., and Schrumpf, A., 1953, The red blood cell diameter in blue whale and humpback whale, *Acta Haematol.* **9:**54.

Lewis, J. H., Bayer, W. L., and Szeto, I. L. F., 1969, Short communication: Coagulation factor XII deficiency in the porpoise, *Tursiops truncatus, Comp. Biochem. Physiol.* **31:**667.

Medway, W., and Geraci, J. R., 1964, Hematology of the bottlenose dolphin (*Tursiops truncatus*), *Am. J. Physiol.* **207:**1367.

Robinson, A. J., Kropatkin, M., and Aggeler, P. M., 1969, Hageman factor (factor XII) deficiency in marine mammals, *Science* **166:**1420.

SUGGESTED READINGS

Anderson, S., 1966, The physiological range of formed elements in the blood of the harbour porpoise, *Phocaena phocaena, Nord. Veterinaermed.* **18:**51.

Baluda, M. C., Kulu, D. D., and Sparkes, R. S., 1972, Cetacean hemoglobins: Electrophoretic findings in nine species, *Comp. Biochem. Physiol. B* **41:**647.

DeMonte, T., and Pilleri, G., 1968, Haemoglobin of *Delphinus delphis L.* and plasma protein fractions in some species of the family Delphinidae, determined by microelectrophoresis on cellulose acetate gel, *Blut* **17:**25.

Geraci, J. R., and Medway, W., 1973, Simulated field blood studies in the bottle-nosed dolphin, *Tursiops truncatus, J. Wildl. Dis.* **9:**29.

Harkness, D. R., and Grayson, V., 1969, Erythrocyte metabolism in the bottlenosed dolphin, *Tursiops truncatus, Comp. Biochem. Physiol.* **28:**1289.

Horvath, S. M., Chiodi, H., Ridgeway, S. H., and Azar, S., 1968, Respiratory and electrophoretic characteristics of hemoglobin of porpoises and sea lion, *Comp. Biochem. Physiol.* **24:**1027.

Matsuno, M., and Hashimoto, K., 1964, Assay of anticoagulant unit of W-heparin newly isolated from whale's lung and intestine and its toxicity, *Tohoku J. Exp. Med.* **83:**143.

Medway, W., 1974, Some coagulation factors in plasma from a California gray whale, *Eschrichtius robustus, MFR Paper 1050, Mar. Fisheries Rev.* **36**(4).

Medway, W., and Geraci, J. R., 1965, Blood chemistry of the bottlenose dolphin (*Tursiops truncatus*), *Am. J. Physiol.* **209:**169.

Medway, W., and Moldovan, F., 1966, Blood studies on the North Atlantic pilot (pothead) whale, *Globicephala melaena, Physiol. Zool.* **39:**110.

Medway, W., Schryver, H. F., and Bell, B., 1966, Clinical jaundice in a dolphin, *J. Am. Vet. Med. Assoc.* **149:**891.

Morimoto, Y., Takata, M., and Sudzuki, M., 1921, Untersuchungen über Cetacea, *Tohuko J. Exp. Med.* **1921:**2.

Myhre, B. A., Simpson, J. G., and Ridgeway, S. H., 1971, Blood groups in the Atlantic bottlenosed porpoise (*Tursiops truncatus*), *Proc. Soc. Exp. Biol. Med.* **137**:404.

Ratnoff, O. D., Saito, H., and Arneson, U., 1976, Studies on the blood clotting and fibrinolytic systems in the plasma from a Sei (baleen) whale, *Proc. Soc. Exp. Biol. Med.* **152**:503.

Ridgeway, S. H., and Johnston, D. G., 1966, Blood oxygen and ecology of porpoises of three genera, *Science* **151**:456.

Ridgeway, S. H., Simpson, J. G., Patton, G. S., and Gilmartin, W. G., 1970, Hematological findings in certain small Cetaceans, *J. Am. Vet. Med. Assoc.* **157**:566.

Saito, H., Poon, M. C., Goldsmith, G. H., Ratnoff, O. D., and Arnason, U., 1976, Studies on the blood clotting and fibrinolytic system in the plasma from a Sei (baleen) whale, *Proc. Soc. Exp. Biol. Med.* **152**:503.

Travis, J. C., and Sanders, B. G., 1972, Whale immunoglobulins. I. Light chain types, *Comp. Biochem. Physiol. B* **43**:627.

Travis, J. C., and Sanders, B. G., 1972, Whale immunoglobulins. II. Heavy chain structure, *Comp. Biochem. Physiol. B* **43**:637.

Yosizawa, Z., 1964, A new type of heparin, "W-heparin," isolated from whale organs, *Biochem. Biophys. Res. Commun.* **16**:336.

CHAPTER 19

THE CARNIVORES

Coon hound

Cat

Raccoon

1. INTRODUCTION

The mammalian order Carnivora is named for the dietary habits of the great majority of its 250-plus species that are meat or flesh eaters. There are an exceptional few carnivores that do not eat flesh. For example, the giant panda feeds on bamboo, and some other bears are primarily herbivorous.

With the exception of the Australian area, these predators are found worldwide and include seven families. Examples from four of these families and seven species within them (listed on page 8) have been studied. Cats and dogs have been domesticated for millennia as pets or for jobs requiring special training such as sheep-herding. The pelts of many carnivores are valuable as fur, and the animals have been hunted or farmed for their skins. A few, like the ferret, have been especially raised for laboratory studies.

Animals belonging to the order Carnivora are quick-witted and quick-footed and equipped with sharp claws and teeth for catching, killing, and devouring their victims. Carnivores bear living offspring, sometimes singly or as twins, more often in litters of six to eight, which they nurse from multiple mammary glands and nurture for months or years until they reach an age sufficient to survive alone.

Common carnivores have been studied hematologically with some quite variable conclusions. In general, stress, diet, age, and other factors may affect the results. The figures reported here may not necessarily be transferrable to other animal studies. Special attention to controls is needed in studies involving carnivores that are genetically dissimiliar. Former studies on cats (Lewis, 1981) and dogs (Lewis *et al.*, 1983) are included here with the kind permission of the editor of *Comparative Biochemistry and Physiology.* Another group of dogs was studied in 1991.

One cat in an early study on many species (Didisheim *et al.*, 1959) and another in the more recent cat study (Lewis, 1981) were found to be deficient in factor XII. Results on the factor XII-deficient cats studied here are excluded from the tables and presented in Section 3.3. Another cat with factor XII deficiency was described by Green and White (1977). This deficiency appears to occur in about 5% of miscellaneous house cats of mixed breed.

2. SOURCE

The cats (*Felis catus*) were mixed breeds weighing 6.8–9 kg that were subjects in other research projects. They were housed at the University of Pittsburgh Medical School Animal Facility in care of Paul Bramson, D.D.S. Our blood samples were obtained at least 1 month after any other blood sampling. The dogs (*Canis familiaris*) were large coon hounds (9–18 kg) purchased as animal models in the Resuscitation Research Laboratory under the direction of Peter Safar, M.D. Our blood samples taken through the jugular vein were part of the first taken in these animals, which received ketamine anesthesia. The ferrets (*Mustela putorius furo*) (5–8 kg) were being kept for reproduction studies by Dr. Kathleen Ryan at the Magee Womens Hospital and were housed in the animal facility there. The foxes (*Vulpes fulva* and *V. urocyon*) (10 and 18 kg) and raccoons (*Procyon lotor*) (about 15 kg) were wild animals, trapped as nuisances, and bled in a local laboratory. The minks (*Mustela vison*) (6 kg) were raised on a local fur farm.

3. TESTS CONDUCTED

3.1. Bleeding Times

Bleeding times were done on the Nair-depilated forearms of six cats (Nos. 9–14) without tourniquet and employing a Simplate. The times from cut to cessation of bleeding were 3, 3 1/2, 3 1/2, 3, 4, and 4 min (mean: 3 1/2 min). For dogs, the bleeding times were done with a Simplate on the relatively hairless inguinal area with times of 2, 2 1/2, 2, 2, 4, and 6 min (mean: 3.1 min). Bleeding times were not done on other carnivores.

3.2. General Coagulation Tests and Factors

Clotting times of Carnivora blood (Table 19.1) in glass and siliconized tubes were short, and the clots retracted well and did not lyse. SPTs were normal for all except mink, which was shorter than human on human and mink substrates. Prothrombin times with Simplastin were all short. They were often so short (<7.0 sec) that it was difficult to obtain accurate results when using semiautomated instruments. Prothrombin times with homologous brain or RVV were also shorter than human. APTTs were short and, except in the raccoon, much shorter than on human plasma. Thrombin, CHH, and Atroxin times fell into the human range except those for the ferret. These times were considerably longer than human and suggest that ferrets have a dysfibrinogenemia or a non-heparin-like inhibitor.

TGTs (Table 19.2) using generating mixtures that each contained only homologous products gave very short substrate clotting times on human or homologous substrates.

TABLE 19.1
General Coagulation Tests in Carnivora

Element	Cat			Dog			Ferret			Fox			Mink			Raccoon			Normal human range
	N	Mean	SD	*N*	Mean	SD	*N*	Mean	SD	*N*	Mean	SD	*N*	Mean	SD	*N*	Mean	SD	
Clot T (min)																			
Glass	8	6	1.4	6	4	0.7	6	2	0.5	2	6	1.3	3	4	1.0	3	4	0.6	6–12
Silicone	8	30	10	6	16	4.9	6	3	0.9	2	25	2.5	3	8	1.5	3	14	4.4	20–59
Clot Retr	8	3–4	—	6	3–4	—	6	1–3	—	2	3–4	—	3	5–5	—	3	2–3	—	3–4+
Clot Lys	8	0	0	6	0	0	6	0	0	2	0	0	3	0	0	3	0	0	0
SPT (sec)																			
Human	8	58.5	11	6	[a]	0	—	—	—	2	68.2	20	3	17.2	1.7	3	45.3	17	20+
Homolog	8	35.0	4.6	6	23.2	2.4	—	—	—	—	—	—	3	14.1	1.7	3	18.5	0.1	—
PT (sec)																			
Simplastin	8	8.2	0.7	6	7.6	1.0	6	10.3	0.1	2	10.2	0.3	3	6.9	0.6	3	9.6	0.6	10–13
Homolog Br	8	17.2	3.1	6	8.0	0.2	—	—	—	2	9.2	0.1	3	12.9	1.1	3	13.2	0.6	—
RVV	8	17.7	1.7	6	14.4	1.6	6	14.1	1.1	—	—	—	3	13.3	0.8	3	15.5	0.2	13–18
APTT (sec)	8	17.3	2.1	6	12.3	1.1	6	18.4	1.4	2	18.2	2.2	3	15.8	1.2	3	26.2	2.7	24–34
Recal T (sec)	8	81	8.4	6	99	9.4	6	66	6.0	2	104	10	3	67	4.2	3	87	5.8	90–180
Lys	8	0	0	6	0	0	6	0	0	2	0	0	3	0	0	3	0	0	0
Lys MCA	8	0	0	6	0	0	6	0	0	2	0	0	3	0	0	3	0	0	0
Thromb T (sec)																			
Bov Thr	8	15.8	1.2	6	16.0	1.3	6	28.8	8.7	2	17.6	1.8	3	14.1	0.3	3	16.9	1.4	11–18
CHH	—	—	—	6	14.9	0.9	5	[a]	—	—	—	—	—	—	—	—	—	—	11–18
Atroxin	8	21.2	1.2	6	13.6	0.7	6	49.0	12	—	—	—	—	—	—	—	—	—	11–18

[a]More than 120 sec.

TABLE 19.2
Thromboplastin Generation Test in Carnivora

Generating mixture						Substrate clotting time (sec)[a]					
Human	Cat	Dog		Mink	Raccoon						
Means			Fox	Mean		H	C	D	F	M	R
H (4)	—	—	—	—	—	**9.4**	14.9	9.0	7.2	6.8	8.8
—	C (2)	—	—	—	—	9.0	**12.8**	—	—	—	—
—	—	D (2)	—	—	—	8.5	—	**7.4**	—	—	—
—	—	—	F (1)	—	—	10.0	—	—	**7.6**	—	—
—	—	—	—	M (2)	—	10.6	—	—	—	**10.0**	—
—	—	—	—	—	R (2)	9.1	—	—	—	—	**9.1**

[a]Homologous systems are in **boldface**.

Coagulation factor levels (Table 19.3) were quite variable from species to species, but in general the factors were high. The only exceptions among the assayed factors were Fletcher in cat and dog and Fitzgerald, protein C, protein S, and plasminogen in dog. Very high levels of factor V (>4.0 U/ml) were found in cat, dog, raccoon and ferret; of factor VII in cat and dog; of factor VIII in cat and dog; and of factor XII in ferret. In fox, mink, and raccoon, dilution was insufficient in some tests, and results were expressed as 2.0+ U/ml.

3.3. XII-Deficient Cat

The factor XII (Hageman)–deficient cat (No. 2) in the first study (Didisheim *et al.*, 1959) was tested on three separate occasions and showed a level of less than 0.05 U/ml and a long TGT (40+ sec) corrected by addition of normal cat barium-treated plasma or serum. In the second report from this laboratory (Lewis, 1981), cat No. 3 showed a prolonged clotting time (glass = 27, silicone = 30 min) and APTT (57.4 sec). The factor XII assay was less than 0.05 U/ml.

3.4. Fibrinolytic Enzyme System

Table 19.4 shows the effects of streptokinase (SK), urokinase (UK), and staphylokinase (Staph K) on cat, dog, and mink plasminogens. Different assay systems were used, so the results are expressed relative to those obtained with human plasma. Cat, dog, and mink were all activated with SK. Dog required a

TABLE 19.3
Coagulation Factors in Carnivora

Factor	Cat			Dog			Ferret			Fox		Mink			Raccoon			Normal human range
	N	Mean	SD	*N*	Mean	SD	*N*	Mean	SD	*N*	Mean	*N*	Mean	SD	*N*	Mean	SD	
I (mg/ml)	8	477	76	6	215	50	5	188	37	2	688	3	155	44	3	410	119	150–450
II (U/ml)	8	1.33	0.3	6	1.43	0.3	5	0.81	0.2	2	1.50	3	2.30	21	3	2.04	0.3	0.70–1.30
V (U/ml)	8	11.9	3.9	6	6.02	0.7	5	5.41	1.4	2	1.90	3	2.85	—	3	3.47	0.8	0.65–1.45
VII (U/ml)	8	4.82	1.0	6	4.70	1.4	5	1.59	0.4	1	2.15	3	2.45	1.1	3	2.40	0.1	0.50–1.30
X (U/ml)	8	3.16	0.6	6	2.18	0.4	5	1.68	0.6	1	2.14	3	1.57	0.12	3	1.69	0.3	0.75–1.25
VIII (U/ml)	8	18.9+	—	6	4.49	1.0	5	2.50	0.5	2	2.0+	3	2.0+	—	3	2.0+	—	0.75–1.40
IX (U/ml)	8	1.84	0.3	6	1.12	0.4	5	1.14	0.3	2	2.0+	3	1.70	0.3	3	1.50	0	0.65–1.70
XI (U/ml)	8	2.84	0.4	6	2.79	0.6	5	0.98	0.3	1	1.50	3	1.70	0.3	3	1.50	0	0.75–1.40
XII (U/ml)	8	2.22	0.8	6	1.06	0.2	5	6.23	1.8	1	1.10	3	1.60	0.2	3	1.42	0.07	0.50–1.45
Flet (U/ml)	8	0.09	0.04	6	0.32	0.1	3	0.92	0.2	—	—	—	—	—	—	—	—	0.50–1.50
Fitz (U/ml)	—	—	—	6	0.30	0.05	3	1.50	0.2	—	—	—	—	—	—	—	—	0.50–1.50
ATIII (U/ml)	8	0.93	0.2	6	0.87	0.08	3	1.18	0.02	—	—	—	—	—	3	1.20	0.2	0.80–1.20
Prot C (U/ml)	—	—	—	6	0.15	0.04	3	0.77	0.03	—	—	—	—	—	—	—	—	0.60–1.40
Prot S (U/ml)	—	—	—	6	0.08	0.02	—	—	—	—	—	—	—	—	—	—	—	0.60–1.40
Plgn (U/ml)	—	—	—	6	0.25	0.06	3	0.44	0.1	—	—	—	—	—	—	—	—	0.80–1.20
Antipl (U/ml)	—	—	—	6	1.10	0.1	3	1.45	0.1	—	—	—	—	—	—	—	—	0.80–1.20
RCF (U/ml)	8	0.48	0.1	6	0.64	0.2	—	—	—	—	—	—	—	—	—	—	—	0.50–1.50
XIII (R)	8	20	10	4	19	10	4	16	0	1	8	3	43	19	3	16	0	4–16

TABLE 19.4
Activation of Plasminogen by Streptokinase, Urokinase, and Staphylokinase in Carnivora

Plasma	Buffer	SK		UK	Staph K
		100 U/ml	1000 U/ml	(500 U/ml)	(2%)
Human	0	++++	++++	++++	++++
Cat (mean of 4)	0	++++	++++	0	+
Dog (mean of 4)	0	++	++++	+	+++
Mink	0	+++	+++	+	++

higher concentration of SK. UK activated human fully, dog and mink slightly, and cat not at all. Staph K activated human fully, cat only slightly, and dog and mink to a greater extent than cat.

3.5. Serum Proteins

Table 19.5 shows paper electrophoresis results on cat, dog, and mink sera. Protein distributions for cat and dog did not differ greatly from human. All three mink patterns failed to differentiate α_1- and α_2-globulins.

3.6. Biochemical Tests

Biochemical test results on cat, dog, and raccoon are shown in Table 19.6. In cat and dog, the differences from the human range were not remarkable. Raccoon showed high cholesterol, blood glucose, and SGOT.

TABLE 19.5
Serum Proteins by Paper Electrophoresis in Carnivora

Protein	Cat			Dog			Mink			Normal human range
	N	Mean	SD	*N*	Mean	SD	*N*	Mean	SD	
TP (g/dl)	12	6.4	0.6	6	6.8	0.6	3	6.5	0.5	6.3–7.9
Albumin (g/dl)	12	2.3	0.4	6	2.8	0.2	3	3.4	0.5	3.2–4.4
α_1-Globulin (g/dl)	12	0.4	0.2	6	0.8	0.1	3[a]	0.9	.08	0.2–0.4
α_2-Globulin (g/dl)	12	1.2	0.4	6	0.7	0.2				0.4–1.0
β-Globulin (g/dl)	12	0.9	0.2	6	1.8	0.5	3	1.0	0.2	0.5–1.0
γ-Globulin (g/dl)	12	1.6	0.7	6	0.6	0.2	3	1.2	0.2	0.6–1.8

[a]Mink α_1- and α_2-globulins are combined into a single peak.

TABLE 19.6
Biochemical Tests in Carnivora

Substance	Cat			Dog			Raccoon			Normal human range
	N	Mean[a]	SD	*N*	Mean[a]	SD	*N*	Mean[a]	SD	
TP (g/dl)	—	—	—	6	6.6	0.9	1	6.4	—	6.0–8.0
Albumin (g/dl)	—	—	—	6	*3.3*	0.5	1	*1.8*	—	3.5–5.0
Ca (mg/dl)	4	9.1	0.5	6	10	0.9	2	9.7	0.4	8.5–10.5
P (mg/dl)	4	**6.8**	1.4	6	**7.0**	2.0	2	**4.6**	0.07	2.5–4.5
Cholesterol (mg/dl)	4	*82*	15	—	—	—	2	**231**	4.9	150–200
Glucose (mg/dl)	4	**11.4**	63	6	*60*	24	2	**124**	34	65–110
Uric acid (mg/dl)	4	*0.8*	0.2	6	*0.7*	0.3	2	4.3	0.07	2.5–8.50
Creatinine (mg/dl)	4	0.9	0.2	6	*0.6*	0.1	2	1.0	0	0.7–1.4
T bilirubin (mg/dl)	4	0.2	0.1	6	0.4	0.7	2	0.1	0	0.1–1.4
BUN (mg/dl)	4	**23**	4.7	3	**21**	3.6	2	10	2.8	10–20
Chloride (meq/liter)	4	**120**	31	6	**106**	5.9	2	**109**	2.1	95–105
CO_2 (meq/liter)	4	*17*	2.4	—	—	—	2	*18*	0	24–32
K (meq/liter)	4	**5.2**	0.5	6	4.8	0.6	2	**6.1**	0.4	3.5–5.0
Na (meq/liter)	4	**151**	0.5	6	141	0.6	2	145	7.1	135–145
Alk Phos (IU/liter)	4	*17*	7.9	—	—	—	2	74	21	30–85
SGOT (IU/liter)	4	**42**	14	—	—	—	2	**207**	9.9	0–41

[a]Values **higher** than human are in **boldface**, those *lower* then human in *italics*.

3.7. Miscellaneous Tests

Plasma from two cats, four dogs, one fox, and one raccoon clumped the standard staphyloccocal preparation up to the 12th dilution (1:2048). Plasmas, from three minks were tested to the 9th dilution and were positive.

3.8. Cellular Elements

Table 19.7 shows the species-to-species variations found in the limited numbers of carnivores studied. In general, the cells were smaller than human (MCV < 80) with less hemoglobin (MCH < 27). The single fox with low HCT may have bled during capture. The WBC counts were not greatly different from human. Neutrophils were more frequent than lymphocytes in cat, dog, fox, and mink. In ferret and raccoon, this ratio was reversed. Platelet counts were within the human range in cat, dog, ferret, and fox. In mink and raccoon, platelets were higher than human.

TABLE 19.7
Cellular Elements of the Peripheral Blood in Carnivora

Element	Cat			Dog			Ferret			Fox		Mink			Raccoon			Normal human range
	N	Mean	SD	*N*	Mean	SD	*N*	Mean	SD	*N*	Mean	*N*	Mean	SD	*N*	Mean	SD	
HCT (%)	14	30	4.1	6	43	3.5	5	43	3.9	1	27	3	54	0.6	3	37	27	37–52
HGB (g/dl)	14	10.4	1.4	6	14.9	1.1	5	15	1.2	1	11.6	3	20.3	0.3	3	13.5	0.4	12–18
RBC (× 10^6/mm^3)	14	5.7	0.9	6	6.3	0.5	5	6.3	0.5	1	3.2	3	7.5	0.5	3	7.6	0.5	4.2–6.2
MCV (fl)	14	55	11	6	69	3.9	5	69	3.6	1	84	3	72	1.5	3	49	4.6	79–97
MCH (pg)	14	19	4.0	6	24	1.0	5	23	1.1	1	36	3	27	1.0	3	18	1.2	27–31
MCHC (g/dl)	14	35	3.2	6	35	1.2	5	34	0	1	43	3	38	0.6	3	37	1.7	32–36
WBC (× 10^3/mm^3)	14	11.0	6.5	6	11.1	1.4	5	11	1.4	1	7.2	3	8.4	.48	3	10.8	3.0	5–10
Neut (%)	14	58	15	6	56	16	5	15	11	1	63	3	73	3.8	3	42	18	55–75
Lymph (%)	14	32	15	6	38	18	5	85	11	1	36	3	25	3.5	3	50	17	20–40
Mono (%)	14	1	1.0	6	4	2.1	5	7	4.0	1	1	3	1	0.6	3	4	2	2–10
Eos (%)	14	8	1.3	6	2	1.8	5	1	0.5	1	0	3	0	0	3	2	2	0–3
Bas (%)	14	1	1.3	6	1	0.5	5	1	1.4	1	0	3	1	0	3	2	2	0–3
Plat C (× 10^3/mm^3)	14	382	152	6	450	151	5	300	46	1	109	3	773	29	3	824	550	150–450
MPV (μM^3)	—	—	—	6	7.3	0.7	4	8.7	0.8	—	—	—	—	—	—	—	—	5.6–10.4

3.9. Platelet Activities

Blood volume available to study platelet activities was adequate in cat and dog, but limited in ferret, mink, and raccoon. Table 19.8 shows that the platelet glass retention index was within the human range for cat, dog, mink, and raccoon. Aggregation with ADP and collagen was within or slightly below the human range in all. Cat platelets reacted poorly or not at all to ristocetin (Risto), thrombin, bovine fibrinogen, and pig plasma (Pig Pl). They did aggregate with CHH. Dog platelets also reacted poorly to Risto and Pig Pl. They did aggregate with both thrombin and CHH. Ferret platelet aggregation was limited with Risto and Pig Pl.

3.10. Platelet Ultrastructure

Figure 19.1 is a TEM of cat platelets. One platelet, in the upper-left corner, was sectioned in the equatorial plane; others were sectioned end-to-end or are fragments. Microtubules were very evident; mitochondria, alpha granules, occasional dense bodies in vacuoles, and glycogen particles were seen. Figure 19.2 is a TEM of dog platelets, which were similar to cat. Ferret platelets (Fig. 19.3) showed an extensive open canalicular system. Raccoon platelets (Fig. 19.4) showed the same internal structures. The alpha granules were sometimes elongated.

4. SUMMARY

Blood from the Carnivora clotted quickly; the clots retracted well and did not lyse. In general, the prothrombin times, APTTs, Recal Ts, and TGTs were shorter than those of human plasma. Factors V, VII, X, VIII, and XI were high. Fibrinogen (factor I) and factors II, XIII, and ATIII were in the normal human range. Fletcher factor was low in cat, and Fletcher and Fitzgerald were low in dog but normal in ferrets.

Two cats in a total of 12 tested were deficient in factor XII. These were No. 2 in the 1959 study and No. 3 in the 1981 study.

Cat, dog, and mink plasminogens were activated with SK and partially with Staph K. UK at 500 U/ml activated cat, dog, and mink very poorly. Carnivores had small RBCs. The platelet counts were variable—within the human range in cats and dogs, high in ferrets, mink, and raccoon, and low in fox. In general, Carnivora platelets aggregated well with ADP and collagen, but poorly with Risto, arachidonic acid, and Pig Pl. The ultrastructure was not remarkable.

TABLE 19.8
Platelet Activities in Carnivora

Activity	Cat			Dog			Ferret			Mink			Raccoon			Normal human range
	n	Mean	SD	*n*	Mean	SD	*n*	Mean	SD	*n*	Mean	SD	*n*	Mean	SD	
Plat Glass Ret Ind (%)	6	52	12	6	52	15	—	—	—	3	50.3	1.5	3	72.3	2.5	>20
Aggregation (%)																
ADP (20 μM)	8	94	7.4	6	73	26	—	—	—	—	—	—	—	—	—	70–100
(10 μM)	8	81	12	5	53	37	5	67	9.9	3	67	7	3	100	0	70–100
(5 μM)	—	—	—	6	37	22	—	—	—	—	—	—	—	—	—	50–100
(2.5 μM)	—	—	—	6	14	10	—	—	—	—	—	—	—	—	—	40–100
Collagen (0.19 mg/dl)	8	95	4.6	6	89	7.6	5	74	1.6	3	69	6.3	3	100	0	80–100
Risto (1.8 mg/dl)	5	0	0	—	—	—	—	—	—	—	—	—	—	—	—	80–100
(0.9 mg/dl)	8	0	0	6	7	4.7	5	4	0.5	—	—	—	—	—	—	80–100
(0.45 mg/dl)	8	0	0	5	8	4.9	5	3	1.6	—	—	—	—	—	—	<18
Arach A (0.5 mg/ml)	—	—	—	6	22	23	—	—	—	—	—	—	—	—	—	60–100
Bov Thr (0.5 U/ml)	5	0	0	—	—	—	—	—	—	—	—	—	—	—	—	60–100
(0.4 U/ml)	—	—	—	6	88	11	—	—	—	3	39	0.6	—	—	—	60–100
(0.25 U/ml)	8	0	0	—	—	—	—	—	—	—	—	—	—	—	—	60–100
Bov Fib (0.1%)	3	0	0	—	—	—	—	—	—	—	—	—	—	—	—	60–100
CHH (0.05 mg/ml)	5	100	0	6	94	10	—	—	—	—	—	—	—	—	—	60–100
Pig Pl (1 : 10)	8	13	28	6	16	9.4	2	17	0	—	—	—	—	—	—	60–100

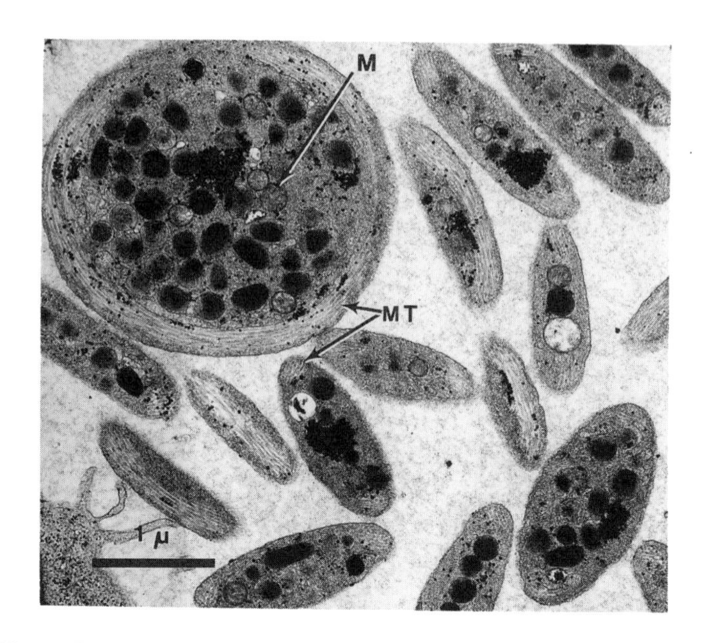

FIGURE 19.1. TEM of cat platelets. Internal structures here and in Figs. 19.2–19.4: (a-GRAN) alpha granules; (DB) dense bodies; (M) mitochondria; (MT) microtubules; (OCS) open canalicular system.

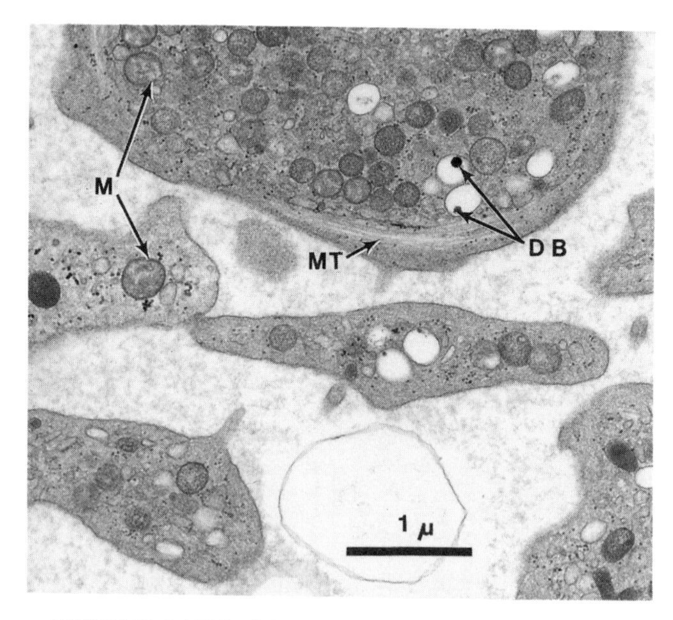

FIGURE 19.2. TEM of dog platelets. Key in Fig. 19.1 caption.

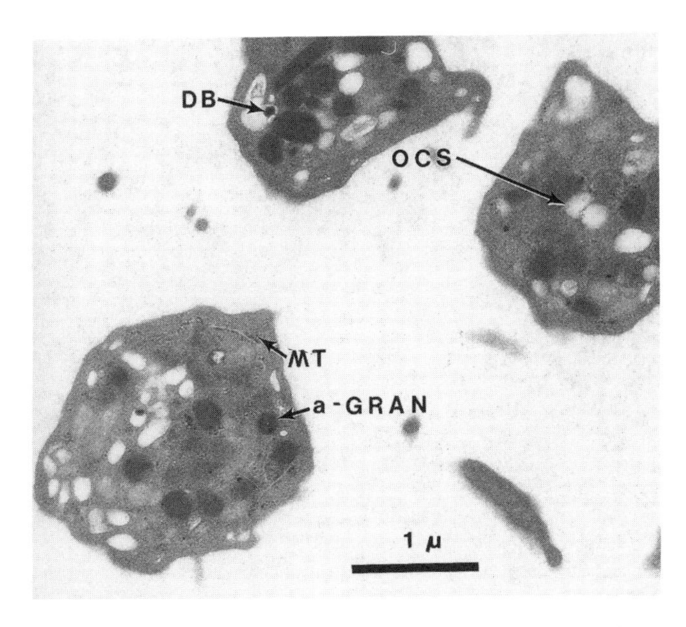

FIGURE 19.3. TEM of ferret platelets. Key in Fig. 19.1 caption.

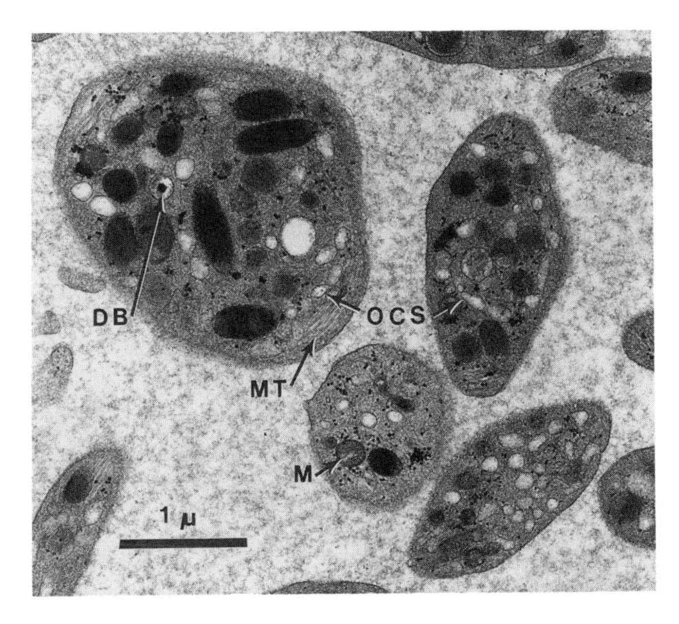

FIGURE 19.4. TEM of raccoon platelets. Key in Fig. 19.1 caption. Scale bar: 1 μm.

REFERENCES

Didisheim, P., Hattori, K., and Lewis, J. H., 1959, Hematological and coagulation studies in various animal species, *J. Lab. Clin. Med.* **53:**866.

Green, R. A., and White, F., 1977, Feline factor XII (Hageman) deficiency, *Am. J. Vet. Res.* **38:**893.

Lewis, J. H., 1981, Comparative hematology: Studies on cats including one with factor XII (Hageman) deficiency, *Comp. Biochem. Physiol.* **68:**355.

Lewis, J. H., Spero, J. A., and Hasiba, U., 1983, A hemophilic dog colony: Genetic and coagulation findings in hemophiliac and normal dogs, *Comp. Biochem. Physiol.* **75A:**147.

SUGGESTED READINGS

Cat

Ackart, R. J., Shaw, J. S., and Lawrence, J. S., 1940, Blood cell picture of normal cats, *Anat. Rec.* **76:**357.

Burdick, M. D., and Schaub, R. G., 1987, Human protein C induces anticoagulation and increased fibrinolytic activity in the cat, *Thromb. Res.* **45:**413.

Fleming, L. B., and Cliffton, E. E., 1965, In vivo activation of the fibrinolytic system of the cat, *J. Surg. Res.* **5:**153.

Gilmore, C. E., Gilmore, V. H., and Jones, T. C., 1964, Bone marrow and peripheral blood of cats: Technique and normal values, *Pathol. Vet.* **1:**18.

Hauser, P., 1963, Quantitätives und qualitätives Blutbild der gesunden Katze, *Schweiz. Arch. Tierheilkd.* **105:**438.

Jain, N. C., and Kono, C. S., 1972, Scanning electron microscopy of erythrocytes of dog, cat, cow, horse, sheep, and goat, *Res. Vet. Sci.* **13:**489.

Johnson, K. H., and Perman, V., 1968, Normal values for jugular blood in the cat, *Vet. Med. Small Anim. Clin.* **63:**851.

Kier, A. B., Bresnahan, J. F., White, F. J., and Wagner, J. E., 1980, The inheritance pattern of factor XII (Hageman) deficiency in domestic cats, *Can. J. Comp. Med.* **44:**309.

Kraft, H., 1957, Das morphologische Blutbild einer Junglöwen (*Felis leo*), *Blut* **3:**344.

Landsberg, J. W., 1940, The blood picture of normal cats, *Folia Haematol.* **64:**169.

O'Rourke, L., Feldman, B. F., and Ito, R. K., 1982, Coagulation, fibrinolysis, and kinin generation in adult cats, *Am. J. Vet. Res.* **43:**1478.

Penny, R. H. C., Carlisle, C. H., and Davidson, H. A., 1970, The blood and marrow picture of the cat, *Br. Vet. J.* **126:**459.

Riser, W. H., 1946, The normal blood count of the domestic cat, *North Am. Vet.* **27:**93.

Rohovsky, M. W., and Griesemer, R. A., 1969, Hematology of the germ free cat, *Lab. Anim. Care* **19:**60.

Sawitsky, A., and Meyer, L. M., 1947, The bone marrow of normal cats, *J. Lab. Clin. Med.* **32:**70.

Schryver, H. F., 1963, The bone marrow of the cat, *Am. J. Vet. Res.* **24:**1012.

Spink, R. R., Malvin, R. L., and Cohen, B. J., 1966, Determination of erythrocyte half life and blood volume in cats, *Am. J. Vet. Res.* **27:**1041.

Tschopp, T. B., 1969, Aggregation of cat platelets in vitro, *Thromb. Diath. Haemorrh.* **23:**601.

Velasco, M., Landaverde, M., Lifshitz, F., and Parra, A., 1971, Some constituents in normal cats, *J. Am. Vet. Med. Assoc.* **158:**763.

Weiser, M. G., and Kociba, G. J., 1984, Platelet concentration and platelet volume distribution in healthy cats, *Am. J. Vet. Res.* **45:**518.

Dog

Adelson, E., Kaufmann, R. M., Lear, A. A., Kirby, J. C., and Rheingold, J. J., 1963, Physiology of platelet destruction as revealed by tagging of cohorts. I. Studies of dogs, *J. Lab. Clin. Med.* **62:**385.

Afonsky, D., 1955, Blood picture in normal dogs, *Am. J. Physiol.* **180:**456.

Anderson, A. C., and Gee, W., 1958, Normal blood values in the beagle, *Vet. Med.* **53:**135.

Anderson, G. F., and Barnhart, M. I., 1964, Prothrombin synthesis in the dog, *Am. J. Physiol.* **206:**929.

Belleville, J., Thouverez, J. P., Mikaeloff, P., and J. Descotes, 1966, Étude des principaux tests de l'hemostase chez le chien normal, *Pathol. Biol.* **14:**41.

Benson, R. E., and Dodds, W. J., 1976, Immunologic characterization of canine factor VIII, *Blood* **48:**521.

Benson, R. E., and Dodds, W. J.;, 1976, Physical relationship between canine factor VIII coagulant activity and factor VIII-related antigen, *Proc. Soc. Exp. Biol. Med.* **153:**339.

Bergentz, S. E., and Nilsson, I. M., 1961, Effect of trauma on coagulation and fibrinolysis in dogs, *Acta Chir. Scand.* **122:**21.

Bertho, E., and Perez, A. P. Y., 1965, Etude comparée expérimentale de l'electrophorese des proteines et des électrolytes chez des chiens, *Can. Med. Assoc. J.* **92:**1210.

Birken, S., Wilner, G. D., and Canfield, R. E., 1975, Studies of the structure of canine fibrinogen, *Thromb. Res.* **7:**599.

Bloom, F., and Meyer, L. M., 1944, The morphology of the bone marrow cells in normal dogs, *Cornell Vet.* **34:**13.

Bonk, R. J., Swanson, M. B., and Malya, P. A. G., 1986, Automated synthetic substrate assays for coagulopathies of dogs, *Lab. Anim. Sci.* **36:**517.

Bruner, H. D., and Wakerling, G. E., 1937, The blood picture of the normal dog, *Proc. Soc. Exp. Biol. Med.* **36:**667.

Brunk, R. R., 1969, Standard values in the beagle dog: Haematology and clinical chemistry, *Food Cosmet. Toxicol.* **7:**141.

Buckwalter, J. A., Blythe, W. B., and Brinkhous, K. M., 1949, Effect of blood platelets on prothrombin utilization of dog and human plasma, *Am. J. Physiol.* **159:**316.

Byars, T. D., Ling, G. V., Ferris, N. A., and Keeton, K. S., 1976, Activated coagulation time (ACT) of whole blood in normal dogs, *Am. J. Vet. Res.* **37:**1359.

Cade, J. F., and Robinson, T. F., 1975, Coagulation and fibrinolysis in the dog, *Can. J. Comp. Med.* **39:**296.

Chignard, M., and Vargaftig, B. B., 1976, Dog platelets fail to aggregate when they form aggregating substances upon stimulation with arachidonic acid, *Eur. J. Pharmacol.* **38:**7.

Chignard, M., and Vargaftig, B. B., 1977, Synthesis of thromboxane A_2 by nonaggregating dog platelets challenged with arachidonic acid or with prostaglandin H_2, *Prostaglandins* **14:**222.

Ewald, B. H., and Sawyer, A., 1970, Automated haematological screening of the dog, *Lab. Anim. Med.* **20:**1103.

Feldman, B. F., Madewell, B. R., and O'Neill, S., 1981, Disseminated intravascular coagulation: Anti-thrombin, plasminogen, and coagulation abnormalities in 41 dogs, *J. Am. Vet. Med. Assoc.* **179:**151.

Hall, D. E., *Blood Coagulation and Its Disorders in the Dog,* Bailliere Tindall, London.

Hartmann, R. C., Conley, C. L., and Poole, E. L., 1952, Studies on the initiation of blood coagulation. III. The clotting properties of canine platelet free plasma, *J. Clin. Invest.,* **31:**685.

Heneghan, T., 1977, Haematological and biochemical variables in the greyhound, *Vet. Sci. Commun.* **1:**277.

Iatridis, S. G., Ferguson, J. H., Iatridis, P. G., and Mauldin, R., 1964, Thrombogenic properties of surface factor: Evidence for an anti-surface factor activity in canine plasma. *Thrombos. Haemorrh. (Stuttgaard)* **12:**1, 2, 35.

Ifran, M., 1961, Studies on the peripheral blood picture of the normal dog, *Ir. Vet. J.* **15:**65, 86, 110.

Jaques, W. E., Hampton, J. W., Bird, R. M., Bolten, K. A., and Randolph, B., 1960, Pulmonary hypertension and plasma in thromboplastin-antecedent deficiency in dogs, *Arch. Pathol.* **69:**248.

Johnson, G. J., Leis, L. A., Rao, G. H. R., and White, J. G., 1979, Arachidonate-induced platelet aggregation in the dog, *Thromb. Res.* **14:**147.

Johnson, G. J., Rao, G. H., Leis, L. A., and White, J. G., 1980, Effect of agents that alter cyclic AMP on arachidonate-induced platelet aggregation in the dog, *Blood* **55:**722.

Johnson, G. S., Benson, R. E., and Dodds, W. J., 1979, Ristocetin cofactor activity of purified canine factor VIII: Inhibition by plasma proteins, *Thromb. Res.* **15:**835.

Kaneko, J. J., and Smith, R., 1967, The estimation of plasma fibrinogen and its clinical significance in the dog, *Calif. Vet.* **21:**21.

Kraytman, M., 1973, Platelet size in thrombocytopenias and thrombocytosis of various origin, *Blood* **41:**587.

Lewis, J. H., 1963, Effects of epsilon aminocaproic acid (EACA) on survival of fibrinogen I^{131} and on fibrinolytic and coagulation factors in dogs, *Proc. Soc. Exp. Biol. Med.* **114:**777.

Lewis, J. H., and Ferguson, J. H., 1952, Fibrinolytic enzyme system of dog serum, *N. C. Med. J.* **13:**196.

Lewis, J. H., and Shirakawa, M., 1964, Effects of fibrinolytic agents and heparin on blood coagulation in dogs, *Am. J. Physiol.* **207:**1044.

Lumsden, J. H., Mullen, K., and McSherry, B. J., 1979, Canine hematology and biochemistry reference values, *Can. J. Comp. Med.* **43:**125.

Mayerson, H. S., 1930, The blood cytology of dogs, *Anat. Rec.* **47:**239.

Meyer, L. M., and Bloom, F., 1943, The bone marrow of normal dogs, *Am. J. Med. Sci.* **206:**637.

Michaelson, S. M., Scheer, K., and Gilt, S., 1966, The blood of the normal beagle, *J. Am. Vet. Med. Assoc.* **148:**532.

Morris, M. L., Stelton, N. J., Allison, J. B., and Green, D. F., 1940, Blood cytology of the normal dog, *J. Lab. Clin. Med.* **25:**353.

Porter, J. A., Jr., and Canaday, W. R., Jr., 1971, Hematologic values in mongrel and greyhound dogs being screened for research use, *J. Am. Vet. Med. Assoc.* **159:**1603.

Potkay, S., and Zinn, R. D., 1969, Effects of collection interval, body weight and season on the hemograms of canine blood donors, *Lab. Anim. Care* **19:**192.

Quick, A. J., and Hussey, C. V., 1951, On the comparative prothrombic activity of human and dog blood, *J. Clin. Invest.* **30:**903.

Rahn, B., and von Kaulla, K. N., 1964, Pharmacological induction of fibrinolytic activity in the dog, *Proc. Soc. Exp. Biol. Med.* **115:**359.

Rekers, P. E., and Coulter, M., 1948, A hematological and histological study of the bone marrow and peripheral blood of the adult dog, *Am. J. Med. Sci.* **216:**643.

Robinson, F. R., and Ziegler, R. F., 1968, Clinical laboratory values of beagle dogs, *Lab. Anim. Care* **18:**39.

Shively, J. N., Feldt, C., and Davis, D., 1969, Fine structure of formed elements in canine blood, *Am. J. Vet. Res.* **30:**893.

Smith, S. G., 1955, Evidence that the physiologic normal hemoglobin value for adult dog blood is 18 grams per 100 cc, *Am. J. Physiol.* **142:**476.

Takeuchi, T., and Takeda, Y., 1978, Physicochemical and biological properties of canine thrombin and prothrombin, *Thromb. Res.* **12:**635.

Usacheva, I. N., 1957, Peripheral blood and bone marrow indices in healthy dogs, *Bull. Exp. Biol. Med.* **44:**1133.

Van Loon, E. J., and Clark, B. B., 1943, Hematology of the peripheral blood and bone marrow of the dog, *J. Lab. Clin. Med.* **28:**1575.

Wagner, R. H., Brannan, W. M., Jr., and Brinkhous, K. M., 1955, Antiaccelerator (anticonvertin) activity of canine plasma and serum, *Soc. Exp. Biol. Med.* **89:**266.

Weiner, D. J., and Bradley, R. E., 1972, The hemogram and certain serum protein fractions in normal beagle dogs, *Vet. Med. Small Anim. Clin.* **67:**393.

Wilner, G. D., and Birken, S., 1975, Synthesis and radioimmunoassay of canine fibrinopeptide A, *Thromb. Res.* **7:**753.

Wintrobe, M. M., Schumaker, H. B., and Schmidt, W. J., 1933, Values for number, size and hemoglobin content of erythrocytes in normal dogs, rabbits and rats, *Folia Haematol.* **51:**502.

Zondag, A. C. P., Kolb, A. M., and Bax, N. M. A., 1985, Normal values of coagulation in canine blood, *Haemostasis* **15:**318.

Ferret

Atkinson, C. S., Press, G. A., Lyden, P., and Katz, B., 1989, Short communications: The ferret as an animal model in cerebrovascular research, *Stroke* **20:**1085.

May, R. M., 1986, News and views: Species conservation: The cautionary tale of the black-footed ferret, *Nature (London)* **320:**13.

May, R. M., 1989, News and views: Conservation biology: Black-footed ferret update, *Nature (London)* **339:**104.

Ryan, K. D., and Robinson, S. L., 1989, Prolactin or dopamine mediates the induction of puberty by long days in female ferrets, *Endocrinology* **125:**2605.

Ryan, K. D., Robinson, S. L., Tritt, S. H., and Zeleznik, A. J., 1988, Sexual maturation in the female ferret: Circumventing the gonadostat, *Endocrinology* **122:**1201.

Fox

Kennedy, A. H., 1935, A graphical study of the blood of normal foxes, *Can. J. Res.* **12:**796.

Spitzer, E. H., Coombes, A. I., and Wisnicky, W., 1941, Preliminary studies on the blood chemistry of the fox, *Am. J. Vet. Res.* **2:**193.

Mink

Bergman, R. K., Lodmell, D. L., and Hadlow, W. J., 1972, A technic for multiple bleedings or intravenous inoculations of mink at prescribed intervals, *Lab. Anim. Sci.* **22:**93.

Clark, R. A., Kimball, H. R., and Padgett, G. A., 1972, Granulocyte chemostaxis in the Chediak–Higashi syndrome of mink, *Blood* **39:**644.

Karstad, L., and Budd, J., 1967, Disseminated intravascular coagulation in mink, *Can. Vet. J.* **8:**239.

Kennedy, A. H., 1934, Cytology of the blood of normal mink and raccoon. I. Morphology of mink's blood, *Can. J. Res.* **12:**479.

Kennedy, A. H., 1934, Cytology of the blood of normal mink and raccoon. II. The numbers of the blood elements in the normal mink, *Can. J. Res.* **12:**484.

Kennedy, A. H., 1935, Cytology of the blood of normal mink and raccoon. III. Morphology and numbers of the blood elements in raccoon, *Can. J. Res.* **12:**495.

Kubin, R., and Mason, M. M., 1948, Normal blood and urine values for mink, *Cornell Vet.* **38:**79.

Law, R. G., and Kennedy, A. H., 1934, Nutritional anaemia in mink, *Can. Field Nat.* **48:**47.

Meyers, K. M., Holmsen, H., Seachord, C. L., Hopkins, G., and Gorham, J., 1979, Characterization of platelets from normal mink and mink with the Chediak–Higashi syndrome, *Am. J. Hematol.* **7:**137.

CHAPTER 20

SEALS

Harbor seal

Harp seal

1. INTRODUCTION

Pinnipedia are large, mostly marine carnivores that evolved from a primitive land Carnivora about 25 MYA. Their bodies are adapted for life in water, although they mate and give birth on land. They have greatly foreshortened extremities. The hands and feet have evolved into flippers. The hind flippers extend backward and act as a tail. The fore and powerful hind flippers propel their heavy, spindle-shaped bodies through the water or over rocky shores and beaches.

Pinnipedia are classified into two major families: Otarioidea, earless fur seals, sea lions, and walrus, and Phocoidea, earred seals such as the common harbor and harp varieties.

Seals are gregarious mammals, often living in large colonies. One bull may have a harem of 50–60 females. Each female usually produces one pup per year. The fur of most seals is valuable only when they are 2- or 3-day-old pups. To prevent eventual extinction, the gruesome annual harvest of baby seals must be reduced by at least one half.

Most Pinnipedia are confined to the oceans of the northern hemisphere. In summer, they breed and care for their young on shore; in the fall, they swim, often long distances, to warmer climates. A few seals live in the Southern hemisphere and some in fresh water, in particular the Caspian Sea and Lake Baikal. Their life expectancy is 12–30 years. They are hunted by man for blubber as well as pelts. Seals eat fish, squid, and other available small animals.

The harbor and harp seals are the most common species that inhabit the United States shores. Adults weigh 50–100 kg. The total world population of both species is probably not above 3 1/2 million. They are intelligent mammals, and many have been trained to perform in zoos and marinelands.

Blood studies on seals are limited to blood coagulation, cell counts, serum chemistries, and protein analysis. Geraci (1971) reported on the harp seal.

2. SOURCE

Blood samples from the three harbor seals (*Phoca vitulina*) weighing about 60 kg apiece were obtained at Mt. Desert Island Biological Laboratory from animals on which diving mechanisms were to be studied by Eugene Robin, M.D. Blood samples from two harp seals (*Pagophilus groenlandica*) weighing 80–90 kg were drawn at the Wometco Seaquarium by Jesse White, D.V.M. This blood was obtained by venipuncture into the plexus at the base of the flipper with a No. 19 needle.

3. TESTS CONDUCTED

3.1. Bleeding Times

On each of the three harbor seals, a nick was made on the edge of one flipper with sharp surgical scissors. Bleeding times were less than 2 min. Bleeding times were not done on harp seals.

3.2. General Coagulation Tests and Factors

Table 20.1 shows that the blood clotted in about 7 min in glass and 24 min in siliconized tubes. Clots retracted well and did not lyse. The SPTs were less than 20 sec, which is abnormal in humans. PTs with Simplastin, human brain, or RVV, APTTs, and Recal Ts were about the same in seals and humans. Thrombin times of the harbor seals were shorter than those of the harp seal or man. Seal TGTs were slightly longer than human, as shown in Table 20.2. All coagulation factors tested (I–XIII) assayed at levels similar to or higher than human (Table 20.3).

3.3. Fibrinolytic Enzyme System

Euglobulin lysis times were greater than 5 hr on all five animals. Streptokinase, urokinase, and staphylokinase activated harbor seal plasminogen. Anti–dog enzyme (anti-DE) was less than human.

3.4. Serum Proteins

Serum protein assays (Table 20.4) were different from human in that the β-globulin area, which was high, showed a split or double peak. Total protein, albumin, α_1, α_2, and γ-globulin all fell within human range.

3.5. Biochemical Tests

Biochemical tests were not performed on seal blood.

TABLE 20.1
General Coagulation Tests in Seals

	Harbor seal No.			Harp seal No.		All seals			Human	
Test	1 ♂	2 ♂	3 ♂	4 ♀	5 ♂	*N*	Mean	SD	Controls (6) (mean)	Normal range
Clot T (min)										
Glass	4	4	6	11	8	5	7	3.2	8	6–12
Silicone	10	15	25	39	28	5	24	11	26	20–59
Clot Retr	4/4	4/4	4/4	5/5	4/4	5	4–4	—	4/4	3–4+
Lys	0	0	0	0	0	5	0	0	0	0
SPT (sec)										
Human	17.0	18.0	17.2	15.2	17.8	5	17.0	1.1	42.8	20+
Seal	15.0	15.0	15.2	12.9	—	4	14.5	1.1	31.8	—
PT (sec)										
Simplastin	10.0	10.0	11.0	12.1	11.8	5	11.0	1.0	11.5	10–13
Hu Br	14.8	—	15.2	—	—	2	15.0	0.3	17.5	11–15
RVV	12.5	12.8	12.7	12.7	12.9	5	12.7	0.1	14.7	13–18
APTT (sec)	55.0	56.0	45.4	23.8	25.8	5	41.2	16	41.4	[a]
Recal T (sec)	145	160	180	128	160	5	155	19	132	90–180
Lys	0	0	0	0	0	5	0	0	0	0
Lys MCA	0	0	0	0	0	5	0	0	0	0
Thromb T (sec)										
Bov Thr	8.5	9.5	9.2	18.3	21.4	5	13.4	6.0	15.1	11–18

[a]Range was 45–55 sec for controls No. 1–3 and 24–34 sec for controls No. 4 and 5.

TABLE 20.2
Thromboplastin Generation Test in Seals

Generating mixture		Substrate clotting time (sec)[a]		
Human	Seal	Human	Harp	Harbor
H	—	**9.9**	10.2	—
—	Harp	14.4	**14.2**	—
H	—	**9.6**	—	10.6
—	Harbor	12.6	—	**12.6**

[a]Homologous systems are in **boldface**.

3.6. Miscellaneous Tests

Both harbor and harp seal plasmas clumped the standard staphylococcal preparation in dilutions of 1:2048.

3.7. Cellular Elements

HCTs and HGBs were high, and MCV and MCH were slightly above normal human range (Table 20.5). The WBC counts averaged $11.5 \times 10^3/mm^3$ and differential counts showed almost equal numbers of neutrophils and lymphocytes. Eosinophils were common and platelets plentiful.

3.8. Platelet Activities

Table 20.6 shows the platelet activities studied. Retention on glass beads was greater than 20%, as was the normal human. Seal platelets aggregated with ADP and collagen. They were not tested with other agents.

3.9. Platelet Ultrastructure

Electron micrographs were not prepared on seal platelets.

4. SUMMARY

These graceful diving animals certainly differ greatly from humans in their habitat, diet, and ability to hold their breath and swim to great depths (up to 200

TABLE 20.3
Coagulation Factors in Seals

Factor	Harbor seal No.			Harp seal No.		All seals			Normal human range
	1 ♂	2 ♂	3 ♂	4 ♀	5 ♂	N	Mean	SD	
I (mg/dl)	537	269	219	370	482	5	375	135	150–450
II (U/ml)	1.67	1.50	1.10	1.50	1.80	5	1.51	0.26	0.70–1.30
V (U/ml)	2.24	1.50	1.50	2.20	2.20	5	1.93	0.39	0.65–1.45
VII (U/ml)	3.00	2.0+	1.10	1.80	1.80	5	1.94	0.68	0.50–1.30
X (U/ml)	4.05	2.0+	2.0+	2.0+	2.0+	5	2.41	0.92	0.75–1.25
VII (U/ml)	4.00	2.0+	4.00	4.00	4.0+	5	3.60	0.89	0.75–1.40
IX (U/ml)	1.30	2.0+	1.50	2.0+	2.0+	5	1.76	0.34	0.65–1.70
XI (U/ml)	1.00	—	2.0+	2.0+	2.0+	4	1.75	0.50	0.75–1.40
XII (U/ml)	0.95	2.0+	2.0+	1.5	1.75	5	1.64	0.44	0.50–1.45
XIII (R)	16	8	8	16	16	5	13	4.4	4–16

TABLE 20.4
Serum Proteins by Paper Electrophoresis in Seals

Protein	Seals: Harbor No. 1	Seals: Harbor No. 2	Seals: Harp (No. 2)	Mean	Normal human range
TP (g/dl)	8.1	7.9	7.5	7.8	6.3–7.9
Albumin (g/dl)	3.2	3.2	3.3	3.2	3.2–4.4
α_1-Globulin (g/dl)	0.2	0.2	0.3	0.2	0.2–0.4
α_2-Globulin (g/dl)	0.5	0.7	0.4	0.5	0.4–1.0
β-Globulin[a] (g/dl)	2.6	2.9	2.4	2.6	0.5–1.0
γ-Globulin (g/dl)	1.6	0.9	1.1	1.2	0.6–1.8

[a]Seal serum showed a split or double β-globulin peak.

TABLE 20.5
Cellular Elements of the Peripheral Blood in Seals

Factor	Harbor seal No. 1 ♂	Harbor seal No. 2 ♂	Harbor seal No. 3 ♂	Harp seal No. 4 ♀	Harp seal No. 5 ♂	All seals N	All seals Mean	All seals SD	Normal human range
HCT (%)	44	57	41	55	53	5	50	7.1	37–52
HGB (g/dl)	17.8	23.0	17.2	18.8	20.0	5	19.4	2.3	12–18
RBC ($\times 10^6$/mm^3)	5.0	5.6	5.1	5.2	5.3	5	5.2	0.2	4.2–6.2
MCV (fl)	88	102	80	106	100	5	95	11	79–97
MCH (pg)	36	41	34	36	38	5	35	4.7	27–31
MCHC (g/dl)	40	40	42	34	38	5	39	3.0	32–36
WBC ($\times 10^3$/mm^3)	12.1	12.8	13.6	8.1	11.0	5	11.5	2.1	5–10
Neut (%)	66	43	27	45	58	5	48	15	55–75
Lymph (%)	32	53	56	47	40	5	46	9.8	20–40
Mono (%)	0	3	0	1	1	5	1	1.2	2–10
Eos (%)	2	1	15	6	1	5	5	6.0	0–3
Bas (%)	0	0	2	1	0	5	1	0.9	0–3
Plat C ($\times 10^3$/mm^3)	763	648	490	650	706	5	651	102	150–450

TABLE 20.6
Platelet Activities in Seals

Activity	Harbor seal No. 1 ♂	Harbor seal No. 2 ♂	Harbor seal No. 3 ♂	Harp seal No. 4 ♀	Harp seal No. 5 ♂	All seals N	All seals Mean	All seals SD	Normal human range
Plat Glass Ret Ind (%)	86	75	42	70	72	5	69	16	>20
Aggregation (%)									
ADP (10 μM)	100	90	—	100	100	4	98	5	70–100
Collagen (0.19 mg/ml)	100	100	—	100	100	4	100	0	80–100

m.) Limited studies of coagulation, protein, and cellular elements did not show any striking human vs. seal differences.

REFERENCE

Geraci, J. R., 1971, Functional haematology of the harp seal, *Pagophilus groenlandicus, Physiol. Zool.* **44:**162.

SUGGESTED READINGS

Bryden, M. M., and Lim, G. H. K., 1969, Blood parameters of the southern elephant seal *Mirounga leonina* in relation to diving, *Comp. Biochem. Physiol.* **28:**139.

Greenwood, A. G., Ridgway, S. H., and Harrison, R. J., 1971, Blood values in young gray seals, *J. Am. Vet. Med. Assoc.* **159:**571.

Lane, R. A. B., Morris, R. J. H., and Sheedy, J. W., 1972, A haematological study of the southern elephant seal, *Mirounga leonina* (Linn.), *Comp. Biochem. Physiol. A* **42:**841.

Lincoln, D. R., Edmunds, D. T., Gribble, T. J., and Schwartz, H. C., 1973, Studies on the hemoglobins of pinnipeds, *Blood* **41:**163.

Seal, U. S., Erickson, A. W., Siniff, D. B., and Hofman, R. J., 1970, Biochemical, population genetic, phylogenetic and cytological studies of Antarctic seal species, Symposium on Antarctic Ice and Water Masses, Tokyo, September 19, 1970, pp. 77–95.

Seal, U. S., Erickson, A. W., Siniff, D. B., and Cline, D. R., 1971, Blood chemistry and protein polymorphisms in three species of Antarctic seals (*Lobodon carcinophagus, Leptonychotes weddelli,* and *Mirounga leonina*), *Antarct. Res. Ser.* **18:**181.

Simpson, J. G., Gilmartin, W. G., and Ridgway, S. H., 1970, Blood volume and other hematologic values in young elephant seals (*Mirounga angustirostris*), *Am. J. Vet. Res.* **31:**1449.

CHAPTER 21

THE INDIAN ELEPHANT

1. INTRODUCTION

Since the extinction of the dinosaurs, elephants have been the largest land-dwelling animals. Proboscidea evolved as a separate order about 50 MYA, and its many species spread over most of the world. Today only two species survive: the Indian elephant (*Elephas maximus*) and the large-eared African elephant (*Loxodonta africana*). Adult elephants may weight up to 5000 kg (5 1/2 tons). Each has a proboscis, or trunk, used for breathing, for drinking by sucking up water and squirting it into the mouth, as a "hand" for tearing off food and conveying it to the mouth, and for carrying heavy loads. They have four enormous, pillarlike legs and feet with plantar surfaces that are thick, flexible, fatty pads that spread the weight evenly over a large area so that they leave only shallow footprints. Two upper incisors grow continuously and form the valuable ivory tusks. Elephants in captivity live 50–65 years. In the wild, their life expectancy is less due to disease and to slaughter by poachers collecting ivory. Elephants are herbivores, eating leaves, grasses, tree bark, and other vegetable matter in huge, often destructive quantities. Female elephants produce one offspring every 3 years. Gestation is about 20 months, followed by an almost equal period of nursing and nurturing during which the mother is not fertile.

Hematological observations on Proboscidea are largely confined to cell counts and morphology studies. Previous studies from this laboratory (Lewis, 1974) concerning blood coagulation are reprinted here with the kind permission of the editor of *Comparative Biochemistry and Physiology.*

2. SOURCE

Blood was obtained from nine elephants (*Elephas maximus*) owned by Ringling Brothers and Barnum & Bailey Circus. Blood from elephants No. 1–5 was obtained during the performing season in Pittsburgh, Pennsylvania, and from the others at winter quarters in Florida. We are indebted to J. Y. Henderson, D.V.M., and Mr. Axel Gautier for help in obtaining blood samples. All elephants in this study were adult, 25- to 30-year-old females except No. 5, which was a 3-year-old female. Blood was obtained by needle and syringe from a vein on the underside of the ear.

3. TESTS CONDUCTED

3.1. Bleeding Times

Bleeding times were not done on these valuable thick-skinned animals.

3.2. General Coagulation Tests and Factors

Table 21.1 lists the results found in general coagulation tests. The differences from human were not very great. Elephant blood clotted rapidly in glass and siliconized tubes. The clots retracted poorly and did not lyse. SPTs for both elephants and humans were less than 20 sec on elephant substrate. Prothrombin times with Simplastin, human brain, and RVV were similar to human. The APTTs were much shorter with elephant plasma than with human. The Recal Ts were about a minute, and the clots did not lyse alone or in 1% MCA, indicating the presence of factor XIII. Thrombin times were somewhat longer than human. The TGTs were slightly longer on human platelet-poor plasma (PPP) than on elephant PPP, as shown in Table 21.2.

Table 21.3 shows the results of coagulation factor assays. Factors I–XIII appeared in the high to high-normal human range. Remarkably, ATIII was hardly detectable.

3.3. Fibrinolytic Enzyme System

Lysis times of clots containing streptokinase or urokinase are shown in Table 21.4. Human plasma clots lysed in a very few minutes. Elephant clots were much slower to lyse, indicating either lack of reactivity to the kinases or lack of plasminogen.

3.4. Serum Proteins

Serum proteins were higher than human. By paper electrophoresis, the albumin appeared within the normal human range and the β-globulin was high and consistently split, as shown in Table 21.5.

3.5. Biochemical Tests

Table 21.6 lists the levels found in biochemical tests on five elephants. The total protein was in the high human range, while albumin was low. Cholesterol,

TABLE 21.1
General Coagulation Tests in the Indian Elephant

Test	Elephant No.						N	Mean	SD	Human	
	2 ♀	3 ♀	4 ♀	5 ♀	6 ♀	9 ♀				Controls (6) (mean)	Normal range
Clot T (min)											
Glass	5	8	6	6	6	9	6	6	1.5	10	6–12
Silicone	11	20	—	16	12	22	5	16	4.7	24	20–59
Clot Retr	2/2	2/2	2/–	2/2	2/2	2/2	5	2–2	—	4/4	3–4+
Clot Lys	0	0	0	0	0	0	6	0	0	0	0
SPT (sec)											
Human	62.4	66.8	65.4	73.2	88.4	58.6	6	69.1	11	32.8	20+
Elephant	15.4	12.8	13.0	13.4	15.2	14.1	6	14.0	1.1	16.8	—
PT (sec)											
Simplastin	10.2	10.3	10.2	10.1	10.8	11.4	6	10.5	0.5	11.8	10–13
Hu Br	17.2	16.8	18.4	16.8	—	—	4	17.3	0.8	18.0	11–15
RVV	16.8	14.4	15.8	16.0	15.7	15.8	6	15.8	0.8	15.7	13–18
APTT (sec)	21.4	19.8	20.4	22.2	22.6	25.8	6	22.0	2.2	28.5	24–34
Recal T (sec)	67	65	60	60	—	100	5	70	16	120	90–180
Lys	0	0	0	0	—	0	4	0	0	0	0
Lys MCA	0	0	0	0	—	0	4	0	0	0	0
Thromb T (sec)											
Bov Thr	44.4	36.0	38.4	41.2	—	26.4	5	37.3	6.8	15.2	11–18

TABLE 21.2
Thromboplastin Generation Test in the Indian Elephant

Generating mixture		Substrate Clot T (sec)[a]	
Human	Elephant	H	E
H	—	**9.8**	9.2
—	E2	12.0	**10.8**
—	E9	13.2	**10.8**

[a]Homologous systems are in **boldface**.

uric acid, chloride, and Na were lower than human. Ca, P, Alk Phos, and LDH were somewhat above human standards. Hydroxybutyric dehydrogenase (HBD) was high.

3.6. Miscellaneous Tests

Elephant plasma did not clump the standard staphylococcal preparation.

3.7. Cellular Elements

Table 21.7 shows that elephants had fewer but larger erythrocytes than do humans. These macrocytes contained plentiful hemoglobin. The total WBC count was slightly higher than that found in normal humans. The lymphocytes greatly outnumbered neutrophils, and many were binucleated. Eosinophils were numerous and showed large granules. Platelets were numerous. On stained smears, they appeared variable in size.

3.8. Platelet Activities

Table 21.8 shows that the platelet glass retention index was below the human range for elephant No. 3 only. Platelets aggregated moderately well with ADP, but three of the four showed biphasic curves, indicating some disaggregation. Additional ADP did not cause additional aggregation. Aggregation with bovine collagen was slightly less than that seen with human platelets.

TABLE 21.3
Coagulation Factors in the Indian Elephant

Factor	Elephant No.						N	Mean	SD	Normal human range
	2 ♀	3 ♀	4 ♀	5 ♀	8 ♀	9 ♀				
I (mg/dl)	428	375	456	400	428	332	6	403	44.5	150–450
II (U/ml)	1.67	1.93	1.78	1.98	1.60	1.80	6	1.79	0.2	0.70–1.30
V (U/ml)	2.07	2.20	2.07	2.35	1.70	1.80	6	2.03	0.2	0.65–1.45
VII (U/ml)	—	—	—	—	1.40	1.25	2	1.33	—	0.50–1.30
X (U/ml)	1.52	1.68	1.62	1.80	2.00	2.00	6	1.77	0.2	0.75–1.25
VIII (U/ml)	1.75	2.00	2.10	2.10	—	—	4	1.99	0.1	0.75–1.40
IX (U/ml)	1.40	1.50	1.75	1.90	—	—	4	1.64	0.2	0.65–1.70
XI (U/ml)	1.25	1.60	1.50	1.40	—	—	4	1.44	0.1	0.75–1.40
XII (U/ml)	1.75	2.00	1.70	1.75	—	—	4	1.80	0.1	0.50–1.45
ATIII (U/ml)	0.25	0.00	0.00	0.00	0.15	0.10	4	0.08	0.1	0.80–1.20
XIII (R)	64	64	16	32	32	16	6	37	22	4–16

TABLE 21.4
Effects of Streptokinase and Urokinase on Clot Lysis Times in the Indian Elephant[a]

Reagent	Clot lysis time (min)				
	2 ♀	3 ♀	4 ♀	5 ♀	H
Streptokinase					
250 U/ml	30+	27+	30+	30+	2
1000 U/ml	17½	17	17½	14½	2½
Urokinase					
100 U/ml	—	—	—	30	6
200 U/ml	30+	30+	30+	19½	4½
Buffer	30+	30+	30+	30+	30+

[a]Test system: 0.1 ml reagent + 0.1 ml elephant or human plasma + 0.1 ml 0.025 M $CaCl_2$.

TABLE 21.5
Serum Proteins by Paper Electrophoresis in the Indian Elephant

Protein	Elephant No.						*N*	Mean	SD	Normal human range
	2 ♀	3 ♀	4 ♀	5 ♀	6 ♀	9 ♀				
TP (g/dl)	9.0	8.6	8.7	7.8	7.6	7.6	6	8.2	0.6	6.3–7.9
Albumin (g/dl)	3.8	3.7	3.4	3.4	3.3	2.8	6	3.4	0.4	3.2–4.4
α_1-Globulin (g/dl)	0.8	0.7	0.6	0.7	0.6	0.6	6	0.7	0	0.2–0.4
α_2-Globulin (g/dl)	1.3	0.9	0.7	0.5	0.7	0.8	6	0.8	0.3	0.4–1.0
β-Globulin (g/dl)	1.1	1.5	1.8	1.4	1.2	2.4	6	1.6	0.5	0.5–1.0
γ-Globulin (g/dl)	2.0	1.8	2.2	1.8	1.8	1.0	6	1.8	0.4	0.6–1.8

[a]All elephant samples showed a split or double β-globulin peak.

TABLE 21.6
Biochemical Tests in the Indian Elephant

Substance	Elephant No.[a]								Normal human range
	2 ♀	3 ♀	4 ♀	5 ♀	6 ♀	*N*	Mean	SD	
TP (g/dl)	**8.9**	**8.8**	**8.9**	7.8	7.8	5	**8.4**	0.6	6.0–8.0
Albumin (g/dl)	*1.2*	*1.4*	*1.2*	*1.1*	*1.0*	5	*1.2*	0.1	3.5–5.0
Ca (mg/dl)	**10.8**	**11.2**	**10.6**	9.8	**10.7**	5	**10.6**	0.5	8.5–10.5
P (mg/dl)	**4.9**	**5.7**	**5.2**	**5.1**	**6.9**	5	**5.6**	0.8	2.5–4.5
Cholesterol (mg/dl)	*64*	*63*	*57*	*49*	*41*	5	*55*	9.8	150–200
Glucose (mg/dl)	82	95	94	105	**173**	5	110	36	65–110
Uric acid (mg/dl)	*0.2*	*0.2*	*0.2*	*0.2*	*0.3*	5	*0.22*	0	2.5–8.0
Creatinine (mg/dl)	1.4	**1.7**	**1.6**	**1.7**	1.3	5	**1.5**	0.2	0.7–1.4
T bilirubin (mg/dl)	0.1	0.1	0.1	0.1	0.1	5	0.1	0	0.1–1.4
BUN (mg/dl)	10	10	*8*	10	10	5	10	0.9	10–20
Chloride (meq/liter)	*81*	*83*	*82*	*83*	*81*	5	*82*	1.0	95–105
CO_2 (meq/liter)	25	*23*	25	30	27	5	26	2.7	24–32
K (meq/liter)	5	4.9	4.6	4.5	4.5	5	4.7	0.2	3.5–5.0
Na (meq/liter)	*126*	*126*	*124*	*126*	*130*	5	*126*	2.2	135–145
Alk Phos (IU/liter)	**101**	**110**	**102**	**91**	**222**	5	**125**	55	30–85
HBD (IU/liter)	**486**	**390**	**636**	**424**	**487**	5	**485**	94	14–185
CPK (IU/liter)	**113**	108	**140**	**143**	68	5	**114**	30	0–110
LDH (IU/liter)	**567**	**420**	**687**	**450**	**501**	5	**525**	106	60–200
SGOT (IU/liter)	20	11	13	16	26	5	17	6	0–41
SGPT (IU/liter)	10	*7*	*8*	*9*	*7*	5	*8*	1.3	10–50

[a]Values **higher** than human are in **boldface**, those *lower* than human in *italics*.

3.9. Platelet Ultrastructure

Figures 21.1 and 21.2 are TEMs of elephant platelets. The platelets are characterized by very large alpha granules and small dense bodies in vacuoles. Plentiful glycogen, an open canicular system, and microtubules are also visible.

4. SUMMARY

Elephant blood clotted rapidly, and the clots retracted only slightly. The APTT was much faster than human, and most coagulation factors were higher when assayed in systems designed for human plasma. ATIII was at a very low

TABLE 21.7
Cellular Elements of the Peripheral Blood in the Indian Elephant

Element	Elephant No.						N	Mean	SD	Normal human range
	3 ♀	5 ♀	6 ♀	7 ♀	8 ♀	9 ♀				
HCT (%)	49	29	35	39	35	33	6	36	6.9	37–52
HGB (g/dl)	16.4	9.7	11.6	13.4	11.6	10.6	6	12.2	2.4	12–18
RBC ($\times 10^6/mm^3$)	3.9	2.1	2.9	3.2	2.7	2.5	6	2.9	0.6	4.2–6.2
MCV (fl)	127	136	120	123	127	132	6	127	5.8	79–97
MCH (pg)	43	46	40	42	42	42	6	43	1.7	27–31
MCHC (g/dl)	34	28	33	34	33	32	6	32	2.4	32–36
WBC ($\times 10^3/mm^3$)	12.5	12.0	16.1	11.1	12.4	12.2	6	12.7	1.7	5–10
Neutr (%)	4	10	13	12	14	9	6	10	3.6	55–75
Lymph (%)	72	71	79	71	72	74	6	73	3.0	20–40
Mono (%)	11	5	0	1	0	2	6	3	4.3	2–10
Eos (%)	13	13	8	16	14	15	6	13	2.8	0–3
Plat C ($\times 10^3/mm^3$)	491	535	975	555	616	546	6	620	179	150–450

TABLE 21.8
Platelet Activities in the Indian Elephant

Activity	Elephant No. 2 ♀	3 ♀	5 ♀	6 ♀	N	Mean	SD	Normal human range
Plat Glass Ret Ind (%)	30	15	24	35	4	26	8.6	>20
Aggregation (%)								
ADP (10 μM)	100	55[a]	50[a]	45[a]	4	63	25	70–100
Collagen (0.19 mg/ml)	100	75	60	40	4	69	25	80–100

[a]Biphasic curve, showing aggregation followed by disaggregation.

level that would cause thrombotic tendency in humans. RBCs were larger but fewer than in humans. WBCs were slightly more plentiful and showed large numbers of eosinophils and lymphocytes, some of which were binucleated. Platelets were high in count, and aggregation was less than with human platelets. The platelets showed very large alpha granules.

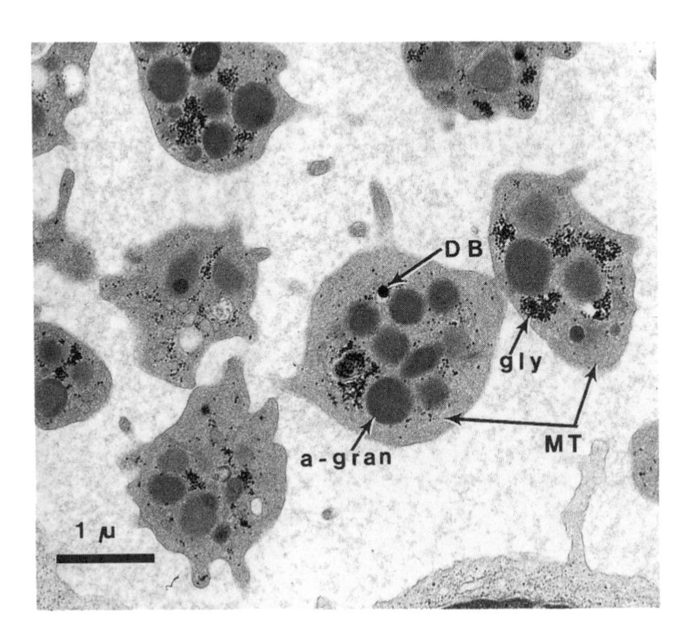

FIGURE 21.1. TEM of elephant platelets. Internal structures here and in Fig. 21.2: (a-gran) alpha granules; (DB) dense bodies; (gly) glycogen particles; (MT) microtubules; (OCS) open canalicular system.

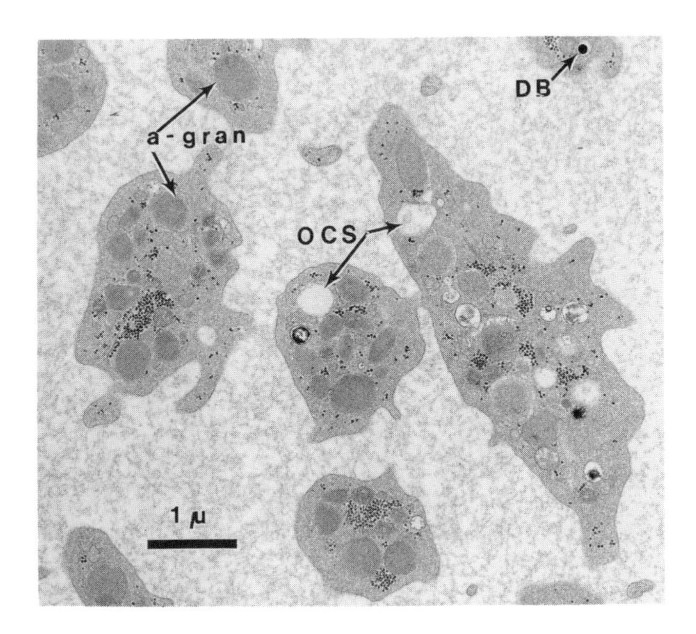

FIGURE 21.2. TEM of elephant platelets. Key in Fig. 21.1 caption.

REFERENCE

Lewis, J. H., 1974, Comparative hematology: Studies on elephants, *Elephas maximus*, *Comp. Biochem. Physiol. A* **49:**175.

SUGGESTED READINGS

Kleihauer, E., Buss, I. O., Luck, C. P., and Wright, P. G., 1965, Hemoglobins of adult and foetal African elephants, *Nature (London)* **207:**424.

Nirmalan, G., Nair, S. G., and Simon, K. J., 1967, Haematology of the Indian elephant (*Elephas maximus*), *Can. J. Physiol. Pharmacol.* **45:**985.

Schmitt, J., 1964, Hematological studies in elephants, *Vet. Med. Rev.* **2:**87.

Simon, K. J., 1961, Haematological studies in elephants, *Indian Vet. J.* **38:**241.

Young, E., 1967, Physiological values of the African elephant, *Loxodonta africana*, *Veterinarian* **4:**169.

CHAPTER 22

THE MANATEE

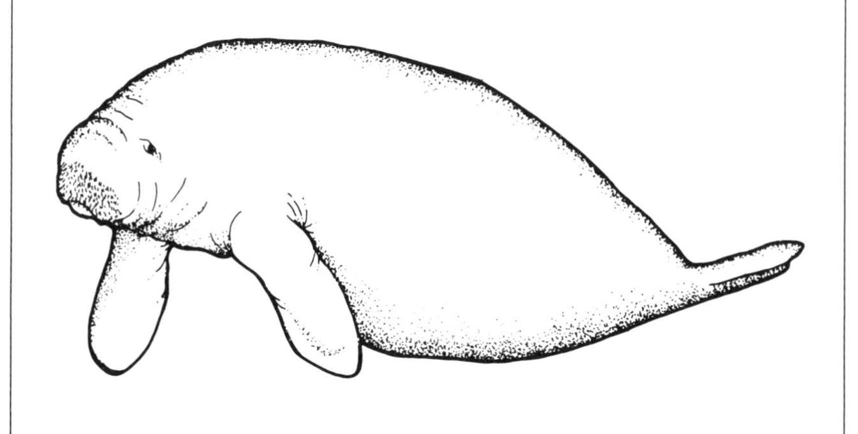

1. INTRODUCTION

The order Sirenia includes manatees and dugongs. The former live in the coastal waters of the southern United States, Central America, and northern South America. Dugongs inhabit the tropical shore waters of the Eastern hemisphere. Manatees are often called sea cows or mermaids and are large, ungainly mammals that weigh up to 600 kg and measure up to 4 m in length. They have broad heads, cleft upper lips, and a muzzle adorned with hairs or bristles. Their bodies are sparsely haired and roundish with front flippers and broad, flattened, rounded tails. They are said to be excellent, although sluggish, swimmers. They live in small groups in brackish or fresh water, lagoons, rivers, and estuaries. Sirenia are completely herbivorous, feeding on large quantities of water grasses or water hyacinths. Their gestation period is 11 months, and they usually produce one infant that nurses under water from teats near its mother's axillae.

Hematological studies on the manatee are rare. White *et al.* (1976) described the large erythrocytes found in these mammals. Medway and Black (1982) described blood counts and, with Bruss and Bengtson, some biochemical results (Medway *et al.*, 1982a). The only coagulation studies are those of Medway *et al.* (1982b).

2. SOURCE

Manatees No. 1 and 2 (*Trichechus manatus*) were studied at the Pittsburgh Zoo and Nos. 3–7 (*Trichechus manatus latirostris*) at Sea World in Orlando, Florida, through the courtesy of Daniel K. Odell, Ph.D., Michael Walsh, D.V.M., and Terry Campbell, D.V.M. The latter studies were approved by the U.S. Fish and Wildlife Service, Marine Mammal Section (PRT-684532). The manatees' official numbers were as follows: No. 3 = TM9005, No. 4 = TM8223, No. 5 = TM8321, No. 6 = TM8911, and No. 7 = TM8687. The author wishes to thank the staff of Sea World Research Laboratory for their generous help in collecting blood and performing blood counts and biochemical tests.

These heavy animals were each confined to a small tank and suspended in a large canvas sling. After the water level was lowered, each animal was raised with a special block and tackle and weighed, and other vital measurements were made. Manatees No. 3–7 weighed from 160 to 270 kg and were 2–3.5 m in length. Blood was collected from the vein at the base of the flipper with a No. 18

butterfly. Blood counts and biochemical tests were done at Sea World Research Laboratory.

3. TESTS CONDUCTED

3.1. Bleeding Times

Bleeding times were done on the first two animals only. Manatee skin is thick and tough, and the usual bleeding-time techniques do not produce any bleeding. Nicks about 1 cm deep and 1.5 cm long were made on a flipper edge with scissors. These bled for less than two min.

3.2. General Coagulation Tests and Factors

Table 22.1 shows that all manatee blood samples clotted well in glass or siliconized tubes; the clots retracted poorly and did not lyse. SPTs were done on the two manatees collected at the Pittsburgh Zoo and were short. Prothrombin times with Simplastin and RVV and APTTs were much shorter than those of human plasma. Recal Ts were short, and there was no lysis of the clot alone or in 1% MCA. Thrombin times with bovine thrombin and the two snake venoms were somewhat longer than human.

TGTs (Table 22.2) were performed on manatees No. 1 and 2 only. The substrate clotting times were longer than those usually found with human components (9.8–11.2 sec).

Coagulation factor assay results are shown in Table 22.3. Factors I, II, XI, Fletcher, and Fitzgerald were within the human normal range. Factors V, VII, X, VIII, IX, XII, XIII, and ATIII were high. Remarkably, protein C could not be detected in manatee blood. This is an activity test and should not be subject to species specificity. RCF appeared normal. Table 22.3 also shows that plasminogen (Plgn) was low and antiplasmin (Antipl) high by the kit assays.

3.3. Fibrinolytic Enzyme System

When manatees No. 1 and 2 were tested in the ^{131}I fibrin system (Table 22.4), their plasma showed the same fibrinolytic activity after adding streptokinase (SK) as did human plasma, suggesting that manatee plasminogen was similar to human. Activation with urokinase (UR) was moderately incomplete, and about one third of that with human plasma. With Staph K, activation was not greater than with buffer. Antitrypsin activity was lower than that of human plasma.

TABLE 22.1
General Coagulation Tests in the Manatee

Test	Manatee No.										Human Controls		
	1 ♂	2 ♂	3 ♀	4 ♀	5 ♀	6 ♂	7 ♀	N	Mean	SD	C (2) (Means)	C (5) (Means)	Normal range
Clot T (min)													
Glass	3	3	3	8	12	13	12	7	8	4.9	8	9	6–12
Silicone	6	7	5	13	25	20	14	7	13	7.6	24	21	20–59
Clot Retr	2/2	2/2	1/2	2/2	2/2	1/2	1/1	7	1–2	—	4/4	4/4	3–4+
Clot Lys	0	0	0	0	0	0	0	7	0	0	0	0	0
SPT (sec)													
Human	18.2	19.4	—	—	—	—	—	2	18.8	0.9	42.1	—	20+
Manatee	18.2	21.8	—	—	—	—	—	2	20.0	2.5	29.9	—	—
PT (sec)													
Simplastin	11.0	9.8	[a]	8.6	11.4	9.0	8.9	6	9.8	1.2	11.7	11.2	10–13
RVV	16.0	17.8	—	12.3	13.8	12.6	13.5	6	14.3	2.1	18.8	15.3	13–18
APTT (sec)	18.2	14.8	—	13.7	14.6	12.3	13.4	6	14.5	2.0	45.0	27.3	[b]
Recal T (sec)	75	75	—	63	78	58	65	6	69	8.1	125	180	90–180
Lys	0	0	—	0	0	0	0	6	0	0	0	0	0
Lys MCA	0	0	—	0	0	0	0	6	0	0	0	0	0
Thromb T (sec)													
Bov Thr	58.4	53.8	—	31.6	28.6	33.4	33.0	6	39.8	13	15.7	15.0	11–18
CHH	—	—	—	22.9	23.1	24.7	25.5	4	24.0	1.3	—	16.1	11–18
Atroxin	—	—	—	39.5	40.2	35.3	37.3	4	38.1	2.2	—	16.7	11–18

[a] Clot in citrate tube.
[b] Normal human range: controls No. 1 and 2, 45–55 sec; controls No. 3–7, 24–34 sec.

TABLE 22.2
Thromboplastin Generation Test in the Manatee

Generating mixture		Substrate Clot T (sec)[a]	
Human	Manatee	H	M
H	—	**9.8**	12.5
—	M1	17.0	**14.5**
—	M2	13.5	**13.5**

[a]Homologous systems are in **boldface**.

3.4. Serum Proteins

Table 22.5 shows the total protein (TP) and electrophoretic distribution of serum from manatees No. 1 and 2. TPs were slightly below the human range, but distribution followed the human pattern.

3.5. Biochemical Tests

Table 22.6 shows the results of biochemical tests done at the Sea World Research Laboratory in Florida. The P, creatinine, CO_2, and Na levels were above the human range. Cholesterol was 207 mg/dl, which is just above the acceptable top level (200 mg/dl) for humans. Alk Phos, LDH, γGT, SGPT, and CPK were also above the human range.

3.6. Miscellaneous Tests

Manatee plasmas (No. 1 and 2) clumped the standard staphylococcal preparation at dilutions of 1:128 and 1:2048, respectively.

3.7. Cellular Elements

Table 22.7 shows the blood cell parameters tested and the results obtained at the Sea World Laboratories. The manatees' HCTs and HGBs fell at the lower edge of the human range, and the RBC counts were below this range. Calcula-

TABLE 22.3
Coagulation Factors in the Manatee

Factor	Manatee						N	Mean	SD	Normal human range
	1 ♂	2 ♂	4 ♀	5 ♀	6 ♂	7 ♀				
I (mg/dl)	225	385	158	116	103	136	6	187	106	150–450
II (U/ml)	0.78	0.54	0.97	0.59	0.84	0.84	6	0.76	0.16	0.70–1.30
V (U/ml)	1.45	2.68	3.10	0.90	3.00	2.15	6	2.53	1.2	0.65–1.45
VII (U/ml)	1.80+	1.80+	4.96	3.48	4.06	3.91	4	4.10[a]	0.62	0.50–1.30
X (U/ml)	2.0+	2.0+	3.10	1.80	1.70	2.00	4	2.15[a]	0.65	0.75–1.25
VIII (U/ml)	2.0+	2.0+	8.75	6.25	12.00	8.00	4	8.75[a]	2.4	0.75–1.40
IX (U/ml)	2.0+	2.0+	6.00	4.50	3.95	4.00	4	4.61[a]	0.96	0.65–1.70
XI (U/ml)	2.0+	2.0+	1.50	1.12	2.20	1.48	4	1.57[a]	0.45	0.75–1.40
XII (U/ml)	2.0+	2.0+	9.00	9.00	10.00	7.25	4	8.81[a]	1.1	0.50–1.45
Flet (U/ml)	—	—	1.35	1.45	1.45	1.35	4	1.40	0.06	0.50–1.50
Fitz (U/ml)	—	—	1.75	0.99	1.60	1.70	4	1.51	0.35	0.50–1.50
ATIII (U/ml)	—	1.25	—	1.37	—	—	2	1.31	0.7	0.80–1.20
Prot C (U/ml)	—	—	[b]	[b]	[b]	[b]	4	[b]	0	0.60–1.40
Plgn (U/ml)	—	—	0.20	0.17	0.17	0.17	4	0.18	0.01	0.80–1.20
Antipl (U/ml)	—	—	1.82	1.84	1.89	1.86	4	1.85	0.03	0.80–1.20
RCF (U/ml)	—	—	0.87	0.74	0.69	0.71	4	0.75	0.08	0.50–1.50
XIII (R)	64	64	—	—	—	—	2	64	0	4–16

[a]Mean of manatees No. 4–7. [b]Less than 0.01 U/ml.

TABLE 22.4
Fibrinolytic Activity with Streptokinase, Urokinase, Staphylokinase, and Antitrypsin in the Manatee

	^{131}I Fibrin lysed		
	Plasma		
	Manatee No.		
Reagent	1 ♂	2 ♂	Human
Buffer	3%	5%	5%
SK (1000 U/ml)	100%	81%	84%
UK (400 U/ml)	12%	11%	34%
Staph K (2%)	5%	4%	90%
Antitrypsin	38%	35%	100%

tions show that the RBCs were large and contained high levels of HGB. The WBC counts were within the human range. The numbers of lymphocytes were greater than those of neutrophils in three of the five animals. Monocytes were rare, and eosinophils and basophils were not seen. Platelet counts were not remarkable. Platelets were slightly smaller than their human counterparts.

3.8. Platelet Aggregation

The blood samples were maintained for 6–8 hr at room temperature from the time of venipuncture to study in Pittsburgh. Table 22.8 shows that manatee

TABLE 22.5
Serum Proteins by Paper Electrophoresis in the Manatee

	Manatee No.		
Protein	1 ♂	2 ♂	Normal human range
TP (g/dl)	5.9	5.0	6.3–7.9
Albumin (g/dl)	3.9	3.8	3.2–4.4
α_1-Globulin (g/dl)	0.4	0.3	0.2–0.4
α_2-Globulin (g/dl)	0.4	0.3	0.4–1.0
β-Globulin (g/dl)	0.3	0.2	0.5–1.0
γ-Globulin (g/dl)	0.9	0.4	0.6–1.8

TABLE 22.6
Biochemical Tests in the Manatee

	Manatee No.[a]								
Substance	3 ♀	4 ♀	5 ♀	6 ♂	7 ♀	*N*	Mean[a]	SD	Normal human range
TP (g/dl)	7.0	7.1	7.2	7.5	7.2	5	7.2	1.9	6.0–8.0
Albumin (g/dl)	4.8	4.8	4.9	5.1	5.1	5	4.9	1.5	3.5–5.0
Ca (mg/dl)	9.2	9.8	9.3	9.3	9.1	5	9.3	2.7	8.5–10.5
P (mg/dl)	**5.1**	**4.6**	4.4	**5.9**	**6.3**	5	**5.3**	0.8	2.5–4.5
Cholesterol (mg/dl)	**224**	176	167	**290**	177	5	**207**	52	150–200
Triglyceride (mg/dl)	65	99	69	116	107	5	91	23	45–200
Glucose (mg/dl)	*49*	*58*	69	69	65	5	*62*	8.5	65–110
Creatinine (mg/dl)	**3.0**	**2.8**	**2.2**	**2.4**	**2.3**	5	**2.5**	0.3	0.7–1.4
T bilirubin (mg/dl)	*<0.1*	*<0.1*	*<0.1*	*<0.1*	*<0.1*	5	*<0.1*	0	0.1–1.4
Fe (mg/dl)	114	*67*	133	127	121	5	112	26	80–160
TIBC (mg/dl)	382	306	335	392	329	5	349	37	250–420
BUN (mg/dl)	13.8	*9.6*	13.2	10.5	16.7	5	12.8	2.8	10–20
Chloride (meq/liter)	98	98	100	95	95	5	98	2.1	95–105
CO_2 (meq/liter)	**44**	**45**	**41**	**43**	**44**	5	**43**	1.5	24–32
K (meq/liter)	4.5	4.7	4.9	4.7	4.3	5	4.6	2.2	3.5–5.0
Na (meq/liter)	**149**	**150**	**147**	**149**	**146**	5	**148**	1.6	135–145
Alk Phos (IU/liter)	**123**	**100**	**130**	**109**	**161**	5	**124**	23	30–85
CPK (IU/liter)	**431**	**701**	**206**	**622**	**231**	5	**438**	223	0–110
LDH (IU/liter)	**327**	**268**	170	**224**	**265**	5	**251**	58	60–200
GGT (IU/liter)	**62**	**66**	**68**	**68**	**57**	5	**64**	4.7	<44
SGOT (IU/liter)	9	7	3	10	5	5	6.8	3.1	0–110
SGPT (IU/liter)	*5*	*3*	*3*	*9*	*6*	5	*5.2*	2.5	10–50

[a]Values **higher** than human are in **boldface**, those *lower* than human in *italics*.

platelets aggregated with ADP at 20 or 10 μM and with collagen and arachidonic acid (Arach A). One of the four animal platelet-rich plasmas (PRPs) did not show aggregation with ristocetin (Risto). Thrombin, CHH, and pig plasma had almost no aggregating effect on manatee platelets.

3.9. Platelet Ultrastructure

The TEMs of manatee platelets were unsatisfactory, probably because of the delay of 6–8 hr before fixation. Unfortunately, it was not possible to complete fixation before return to Pittsburgh.

TABLE 22.7
Cellular Elements of the Peripheral Blood in the Manatee

Element	Manatee No. 3 ♀	4 ♀	5 ♀	6 ♂	7 ♀	*N*	Mean	SD	Normal human range
HCT (%)	38	36	36	40	41	5	38	2.5	37–52
HGB (g/dl)	12.3	11.3	11.9	13.2	13.3	5	12.4	0.8	12–18
RBC ($\times 10^6/mm^3$)	3.0	2.7	2.9	3.2	3.7	5	3.0	0.2	4.2–6.2
MCV (fl)	127	130	131	126	133	5	129	2.9	79–97
MCH (pg)	41	42	42	42	44	5	42	1.1	27–31
MCHC (g/dl)	33	32	32	33	33	5	33	0.6	32–36
WBC ($\times 10^3/mm^3$)	10.1	6.9	8.0	11.1	12.7	5	9.8	2.3	5–10
Neutr (%)	31	69	54	44	37	5	47	15	55–75
Lymph (%)	69	28	44	55	60	5	51	16	20–40
Mono (%)	0	3	2	1	3	5	2	1.3	2–10
Eos (%)	0	0	0	0	0	5	0	0	0–3
Bas (%)	0	0	0	0	0	5	0	0	0–3
Plat C ($\times 10^3/mm^3$)	397	381	244	364	348	5	347	60	150–450
MPV (μM^3)	6.5	6.2	6.4	6.0	6.0	5	6.2	0.2	5.6–10.4

TABLE 22.8
Platelet Aggregation in the Manatee

Reagent	Manatee No. 4 ♀	5 ♀	6 ♂	7 ♀	*N*	Mean	SD	Human: Controls (4) (Mean)	Human: Normal range
Aggregation (%)									
ADP (20 µM)	67	52	70	88	4	69	15	84	70–100
(10 µM)	76	63	58	79	4	69	10	74	70–100
(5 µM)	0	—	0	—	2	0	0	38	50–100
(2.5 µM)	3	—	0	—	2	1.5	2.1	18	40–100
Collagen (0.19 mg/ml)	100	93	85	85	4	91	7.2	100	80–100
Risto (0.9 mg/ml)	60	0	51	68	4	45	31	100	80–100
Risto ½ (0.45 mg/ml)	0	0	0	0	4	0	0	5	<18 000
Arach A (0.5 mg/ml)	65[a]	42[a]	38[a]	36[a]	4	45[a]	13	85	60–100
Bov Thr (1.0 U/ml)	—	Clot	—	—	0	—	—	—	60–100
(0.6 U/ml)	0	—	—	0	2	0	0	Clot	60–100
(0.5 U/ml)	0	—	—	0	2	0	0	—	60–100
(0.4 U/ml)	0	0	0	0	4	0	0	80	60–100
CHH (0.05 mg/ml)		—	0	—	2	0.5	0.7	51	60–100
Pig Pl (1 : 10)	2	13	13	—	3	9	6.4	90	60–100

[a]Biphasic curve, showing aggregation followed by disaggregation.

4. SUMMARY

Blood from seven manatees was studied. The first two, studied at the Pittsburgh Zoo, were the West Indian manatee (*Trichechus manatus*), and the rest, studied at Sea World in Florida, were the Florida manatee *Trichechus manatus latirostris*. No differences were detected between the two subspecies. Clotting times were fast and clot retraction poor. Prothrombin times with Simplastin or RVV, APTTs, and Recal Ts were rapid. Thrombin times were somewhat longer than human. TGTs were longer than human for reasons not determined. Protein C was not detectable.

Manatee plasminogen activated fully with SK, but only slightly with UK or Staph K. The serum proteins were not remarkable. The biochemical tests showed only slight differences from the human ranges. Erythrocytes were large and contained high levels of HGB. WBCs showed predominance of lymphocytes and rare or absent monocytes, eosinophils, and basophils. Platelets were numerous and slightly smaller than human. They aggregated with ADP, collagen, and Arach A. Three of the four manatee PRPs tested reacted to Risto. Thrombin, CHH, and pig plasma had almost no aggregating activity with manatee platelets.

REFERENCES

Medway, W., and Black, D. J., 1982, Hematology of the West Indian manatee (*Trichechus manatus*), *Vet. Clin. Pathol.* **11:**11.

Medway, W., Bruss, M. L., Bengston, J. L., and Black, D. J., 1982a, Blood chemistry of the West Indian manatee (*Trichechus manatus*), *J. Wildl. Dis.* **18:**229.

Medway, W., Dodds, W. J., Moynihan, A. C., and Bonde, R. K., 1982b, Blood coagulation of the West Indian manatee, *Cornell Vet.* **72:**120.

White, J. R., Harkness, D. R., Isaacks, R. E., and Duffield, D. D., 1976, Some studies on blood of the Florida manatee, *Trichechus manatus latirostris, Comp. Biochem. Physiol. A* **55:**413.

SUGGESTED READINGS

Dierauf, L. A., ed., 1990, *CRC Handbook of Marine Mammal Medicine: Health, Disease, and Rehabilitation,* Boca Raton, CRC Press.

Farmer, M., Weber, R. E., Bonaventura, J., Best, R. C., and Domning, D., 1979, Functional properties of hemoglobin and whole blood in an aquatic mammal, the Amazonian manatee (*Trichechus inunguis*), *Comp. Biochem. Physiol. A* **62:**231.

Fawcett, D. W., 1942, A comparative study of blood-vascular bundles in the Florida manatee (*Trichechus latirostris*) and in certain cetaceans and edentates, *J. Morphol.* **71:**104.

Marsh, H., Spain, A. V., and Heinsohn, G. E., 1978, Minireview: Physiology of the dugong, *Comp. Biochem. Physiol. A* **61:**159.

O'Shea, T. J., and Rathbun, G. B., 1985, Tolerance of West Indian manatees to capture and handling, *Biol. Conserv.* **33:**335.

Quiring, D. P., and Harlan, C. F., 1953, On the anatomy of the manatee, *J. Mammal.* **34:**192.

CHAPTER 23

THE THOROUGHBRED RIDING HORSE

1. INTRODUCTION

The order Perissodactyla, odd-toed, hoofed mammals, includes horses, asses, zebras, rhinoceroses, and tapirs. A small primitive horse, *Eohippus,* differentiated from the early ungulates in the Eocene Age (40 MYA). By the beginning of the Ice Age, large, single-toed horses could be found on every continent except Australia. Horses actually walk on tiptoe; only one toe—the remaining middle one—supports the weight. This toe has evolved a hoof, which consists of a curved covering of horn composed mostly of keratin that protects the distal digit. Horses have long tails made up of heavy coarse hairs that can swish away irritating insects. Their front teeth are sharp for cropping grass; farther back in the jaw, they have large, flat molars for grinding their foodstuff, grass and hay. A mare produces and nurtures a foal about once every 2 years. Riding horses are almost always thoroughbreds, the so-called "hot-blooded" strain, which results from crossbreeding with an Arabian strain. Many of our wild horses, including the mustang, have Arabian blood inherited from the horses brought in by the invading Spaniards.

The literature shows a relative abundance of inconsistent observations concerning the clotting of horse blood. Bell *et al.* (1955) describe a "haemophilioid" state marked by a long clotting time and low levels of factor VIII. Sjølin (1956, 1957) first described a lack of factor IX, but later indicated that there was a lack of factor XII. Abilgaard and Link (1965) found low levels of factors XI and XII and human-range levels of factors V, VIII, and IX. They did not detect changes after exercise. Descriptions of cellular elements are well summarized in veterinary textbooks.

2. SOURCE

Blood was obtained from eight thoroughbred horses (*Equus caballus*) at a local riding academy. Horses No. 1 and 2 were stallions, Nos. 4 and 7 mares, and the others geldings (G). They weighed between 350 and 500 kg. Blood was easily withdrawn by a multisyringe technique from a large jugular vein. The horses were not anesthetized, but were gently soothed while being held by their trainer.

3. TESTS CONDUCTED

3.1. Bleeding Times

Bleeding times were not done on these thoroughbred horses.

3.2. General Coagulation Tests and Factors

Table 23.1 shows that the clotting times in new glass tubes were longer than human in five of the six horses tested. Surprisingly, the clotting times in siliconized tubes were not longer than human. The clots retracted and did not lyse. SPTs were faster than human in four of the six. Prothrombin times with Simplastin, horse brain, or RVV were about the same on horse as on human plasma. The APTTs and Recal Ts were consistently long. Horse plasma clotted very slowly with bovine thrombin and moderately slowly with Atroxin.

The long thrombin times appeared to be due to a heat-stable, non-heparin-like inhibitor. An equivolume mixture of horse and human plasma clotted in 31 sec. This inhibitory effect was not removed by heating the horse plasma to 56°C for 5 min or by adsorbing oxalated horse plasma with barium sulfate, which removes heparin.

Table 23.2 shows the TGT results. Recalcified mixtures of horse platelets, barium-treated plasma, and serum formed much less coagulant activity than did a similar mixture of human materials. Differences between human and horse substrate clotting times were not significant.

The coagulation factor assays shown in Table 23.3 indicated that horse factors V, VII, and X were higher than human. Factors I, II, VIII, IX, XII, ATIII, and RCF fell within the human range. Factor XIII appeared to be about half the human level. Factors XI and Fletcher were lower in assay than those of humans.

3.3. Fibrinolytic Enzyme System

Table 23.4 shows that horse plasminogen did not activate over a wide range of streptokinase (SK) levels. With high concentrations of urokinase (UK) as activator, fibrinolysis was observed but was slower than human.

3.4. Serum Proteins

Table 23.5 shows that total protein and electrophoretic distribution on paper electrophoresis gave values that fell within the human range.

TABLE 23.1
General Coagulation Tests in the Thoroughbred Riding Horse

Test	Horse No.									Human	
	3 G[a]	4 ♀	5 G[a]	6 G[a]	7 ♀	8 G[a]	*N*	Mean	SD	Controls (3) (mean)	Normal range
Clot T (min)											
Glass	20	21	15	7	15	19	6	16	5.0	—	6–12
Silicone	40	36	30	25	25	24	6	30	6.7	—	20–59
Clot Retr	3/4	3/3	2/3	3/4	4/4	4/4	6	2–4	—	—	3–4+
Clot Lys	0	0	0	0	0	0	6	0	0	—	0
SPT (sec)											
Human	18.2	33.4	10.8	16.4	36.1	9.4	6	20.7	11	38.6	20+
Horse	19.4	31.4	13.0	17.8	39.4	13.2	6	22.4	11	40.7	—
PT (sec)											
Simplastin	8.9	9.0	9.1	9.1	9.1	9.0	6	9.0	0.1	10.7	10–13
Horse Br	16.8	17.4	17.8	15.4	15.2	16.8	6	16.6	1.1	16.8	—
RVV	15.8	17.4	16.4	16.2	14.8	15.2	6	15.9	0.9	16.4	13–18
APTT (sec)	68.8	57.2	48.2	83.9	89.5	72.5	6	69.9	16	30.2	24–34
Recal T (sec)	300	330	315	300	330	345	6	320	18	145	90–180
Lys	0	0	0	0	0	0	6	0	0	0	0
Lys MCA	0	0	0	0	0	0	6	0	0	0	0
Thromb T (sec)											
Bov Thr	60+	60+	60+	60+	60+	60+	6	60+	0	15.4	11–18
Atroxin	30.4	28.4	26.2	31.4	30.4	28.2	6	29.2	1.9	15.4	11–18

[a]G, gelding (castrated male).

TABLE 23.2
Thromboplastin Generation Test in the Thoroughbred Riding Horse

Generating mixture		Substrate Clot T (sec)[a]	
Human	Horse	H	Ho
H	—	**11.1**	10.1
—	Ho1	23.3	**21.6**
—	Ho3	23.6	**19.8**
—	Ho4	22.8	**25.7**

[a]Homologous systems are in **boldface**.

3.5. Biochemical Tests

Table 23.6 shows the results of biochemical tests on the eight thoroughbred horses. Compared to human, cholesterol and uric acid were low and Ca, Alk Phos, HBD, CPK, LDH, and SGOT were high.

3.6. Miscellaneous Tests

Horse plasma in dilutions up to 1:512 clumped a standard preparation of killed staphylococci (Newman D_2C coagulase–negative). Ethanol and protamine gel tests were negative on horse plasma.

3.7. Cellular Elements

Cellular elements of horse blood are shown in Table 23.7. Horse red cells were small (MCV = 49 fl) and numerous and sedimented very rapidly. Leukocytes and platelet counts fell within the human range. The eosinophils had very large granules.

3.8. Platelet Activities

Over 50% of horse platelets were retained when blood was passed through a standard glass bead column (Table 23.8). Horse platelets aggregated with ADP, collagen, thrombin, and CHH. Aggregation was minimal or absent with ristocetin (Risto).

TABLE 23.3
Coagulation Factors in the Thoroughbred Riding Horse

Factor	Horse No.									
	3 G	4 ♀	5 G	6 G	7 ♀	8 G	N	Mean	SD	Normal human range
I (mg/dl)	325	285	285	230	255	255	6	273	33	150–450
II (U/ml)	1.41	1.34	1.25	1.47	1.23	1.34	6	1.34	0.09	0.70–1.30
V (U/ml)	2.65	2.85	2.50	2.55	2.73	2.73	6	2.67	0.13	0.65–1.45
VII (U/ml)	6.50	8.00	9.00	6.00	5.80	6.10	6	6.90	1.3	0.50–1.30
X (U/ml)	2.40	2.00	2.10	2.10	2.10	1.70	6	2.10	0.2	0.75–1.25
VIII (U/ml)	1.25	0.92	1.20	0.90	0.85	0.78	6	0.98	0.19	0.75–1.40
IX (U/ml)	0.63	0.68	0.58	0.52	0.61	0.61	6	0.61	0.06	0.65–1.70
XI (U/ml)	0.30	0.33	0.35	0.33	0.31	0.22	6	0.31	0.05	0.75–1.40
XII (U/ml)	1.00	0.66	0.84	0.84	0.78	0.58	6	0.78	0.15	0.50–1.45
Flet (U/ml)	0.20	0.21	0.19	0.21	0.06	0.15	6	0.17	0.06	0.50–1.50
ATIII (U/ml)	1.17	0.75	0.90	1.20	1.10	1.00	6	1.02	0.17	0.80–1.20
RCF (U/ml)	1.02	1.08	0.87	0.53	0.92	0.72	6	0.86	0.2	0.50–1.50
XIII (R)	4	4	4	4	8	4	6	5	1.6	4–16

TABLE 23.4
Effects of Streptokinase and Urokinase on Clot Lysis Times in the Thoroughbred Riding Horse[a]

Reagent	Clot lysis time (min) Horse No. 3 G	4 ♀	5 G	Human
Buffer	360+	360+	360+	360+
Streptokinase				
1000 U/ml	360+	90	360+	2
125 U/ml	360+	360+	360+	2
31.25 U/ml	360+	360+	360+	3¼
Urokinase				
1000 U/ml	4½	6	4½	2
500 U/ml	90	90	90	3½
250 U/ml	360+	360+	360+	5½
125 U/ml	360+	360+	360+	360+

[a]Test system: 0.1 ml reagent + 0.1 ml horse or human plasma + 0.1 ml Simplastin 1 : 100 in 0.025 $CaCl_2$.

3.9. Platelet Ultrastructure

Figures 23.1 and 23.2 are TEMs of horse platelets. The usual structures recognized in other mammals can be seen. Internally, microtubules, an open canalicular system, alpha granules, and small glycogen particles were evident. Dense bodies were scarce.

TABLE 23.5
Serum Proteins by Paper Electrophoresis in the Thoroughbred Riding Horse

Protein	Horse No. 1 ♂	2 ♂	3 G	4 ♀	5 G	6 G	7 ♀	8 G	*N*	Mean	SD	Normal human range
TP (g/dl)	7.4	6.7	6.5	6.7	6.2	4.4	6.5	7.0	8	6.4	0.9	6.3–7.9
Albumin (g/dl)	3.4	3.3	3.1	2.8	2.6	1.8	2.5	2.9	8	2.8	0.5	3.2–4.4
α_1-Globulin (g/dl)	0.6	0.5	0.9	1.3	1.1	0.7	1.0	0.9	8	0.9	0.3	0.2–0.4
α_2-Globulin (g/dl)	0.7	0.2	0.8	1.0	1.1	0.8	1.1	1.3	8	0.9	0.3	0.4–1.0
β-Globulin (g/dl)	0.9	0.7	0.6	0.5	0.4	0.4	0.6	0.5	8	0.6	0.2	0.5–1.0
γ-Globulin (g/dl)	1.8	1.9	1.1	1.1	0.9	0.6	1.2	1.2	8	1.2	0.4	0.6–1.8

TABLE 23.6
Biochemical Tests in the Thoroughbred Riding Horse

Substance	Horse No.[a]								N	Mean[a]	SD	Normal human range
	1 ♂	2 ♂	3 G	4 ♀	5 G	6 G	7 ♀	8 G				
TP (g/dl)	6.9	7.3	8.4	7.9	6.9	6.7	7.0	7.3	8	7.3	0.6	6.0–8.0
Albumin (g/dl)	*3.4*	3.7	3.8	3.9	3.9	4.0	3.9	3.6	8	3.8	0.2	3.5–5.0
Ca (mg/dl)	**12.2**	**12.9**	**12.2**	**12.4**	**12.8**	**12.2**	**12.3**	**12.3**	8	**12.4**	0.3	8.5–10.5
P (mg/dl)	3.9	3.1	3.0	2.9	3.9	3.6	3.8	3.2	8	3.4	0.4	2.5–4.5
Cholesterol (mg/dl)	*86*	*125*	*105*	*115*	*110*	*110*	*99*	*110*	8	*108*	11	150–200
Glucose (mg/dl)	67	86	**112**	89	**135**	100	110	110	8	101	21	65–110
Uric acid (mg/dl)	*0.6*	*0.6*	*0.5*	*1.0*	*0.2*	*0.6*	*0.6*	*0.9*	8	*0.6*	0.2	2.5–8.0
Creatinine (mg/dl)	**1.5**	**1.5**	**1.9**	**1.6**	**1.6**	**1.5**	1.4	**2.0**	8	**1.6**	0.2	0.7–1.4
T bilirubin (mg/dl)	0.8	1.1	1.2	1.4	1.2	1.3	**1.6**	1.0	8	1.2	0.2	0.1–1.4
BUN (mg/dl)	**26**	18	20	**25**	20	19	20	**24**	8	**22**	3.0	10–20
Chloride (meq/liter)	98	102	101	**107**	104	104	103	**110**	8	104	3.6	95–105
CO_2 (meq/liter)	27	27	25	24	26	26	28	27	8	26	1.3	24–32
K (meq/liter)	**5.6**	4.5	3.1	3.2	3.9	3.6	4.2	4.4	8	4.1	0.8	3.5–5.0
Na (meq/liter)	138	141	138	140	141	141	140	140	8	140	0.9	135–145
Alk Phos (IU/liter)	**320**	**196**	**187**	**228**	**270**	**220**	**140**	**175**	8	**187**	57	30–85
HBD (IU/liter)	**696**	**488**	**942**	**468**	**890**	**903**	**978**	**467**	8	**729**	221	14–185
CPK (IU/liter)	**128**	**160**	**169**	**187**	**291**	**275**	**129**	**148**	8	**186**	63	0–110
LDH (IU/liter)	**378**	**429**	**828**	**397**	**696**	**765**	**821**	**459**	8	**599**	202	60–220
SGOT (IU/liter)	**352**	**287**	**300+**	**300+**	**300+**	**300+**	**300+**	**300+**	8	**305**	20	0–41
SGPT (IU/liter)	15	12	17	12	15	17	17	14	8	15	2.1	10–50

[a]Values **higher** than human are in **boldface**, those *lower* than human in *italics*.

TABLE 23.7
Cellular Elements of the Peripheral Blood in the Thoroughbred Riding Horse

Element	Horse No.											
	1 ♂	2 ♂	3 G	4 ♀	5 G	6 G	7 ♀	8 G	*N*	Mean	SD	Normal human range
HCT (%)	41	40	37	32	37	39	37	41	8	38	3.0	37–52
HGB (g/dl)	15.5	14.0	15.5	13.4	15.6	16.3	14.4	16.7	8	15.2	1.1	12–18
RBC ($\times 10^6/mm^3$)	8.7	8.3	7.6	6.6	7.8	7.9	7.1	8.1	8	7.8	0.7	4.2–6.2
MCV (fl)	47	48	49	48	48	50	52	51	8	49	1.7	79–97
MCH (pg)	18	17	20	20	20	21	20	21	8	20	1.4	27–31
MCHC (g/dl)	38	44	42	52	42	42	52	51	8	46	5.5	32–36
WBC ($\times 10^3/mm^3$)	8.0	9.5	7.4	5.7	7.7	6.5	6.0	7.3	8	7.3	1.2	5–10
Neutr (%)	53	54	66	79	70	60	72	56	8	64	9.5	55–75
Lymph (%)	40	42	22	19	28	39	22	38	8	31	9.5	20–40
Mono (%)	6	2	0	0	1	1	0	0	8	1	2.1	2–10
Eos (%)	1	1	9	2	1	0	6	5	8	3	3.2	0–3
Bas (%)	0	1	3	0	0	0	0	1	8	1	1.1	0–3
Plat C ($\times 10^3/mm^3$)	500	146	279	197	252	184	161	158	8	235	117	150–450

TABLE 23.8
Platelet Activities in the Thoroughbred Riding Horse

Activity	Horse No.						N	Mean	SD	Human	
	3 G	4 ♀	5 G	6 G	7 ♀	8 G				Control	Normal range
Plat Glass Ret Index (%)	63	60	61	66	61	70	6	64	4	—	>20
Aggregation (%)											
ADP (10 μM)	100	100	100	100	100	95	6	99	2	95	70–100
(5 μM)	100	100	100	85	100	75	6	93	11	90	50–100
Collagen (0.19 mg/ml)	100	100	100	100	100	100	6	100	0	90	80–100
Risto (0.9 mg/ml)	0	0	20[a]	20[a]	35	0	6	13	15	100	80–100
Bov Thr (0.4 U/ml)	100	100	100	100	100	100	6	100	0	100	60–100
CHH (0.05 mg/ml)	100	100	100	100	100	100	6	100	0	75	60–100

[a]Biphasic curve, showing aggregation followed by disaggregation.

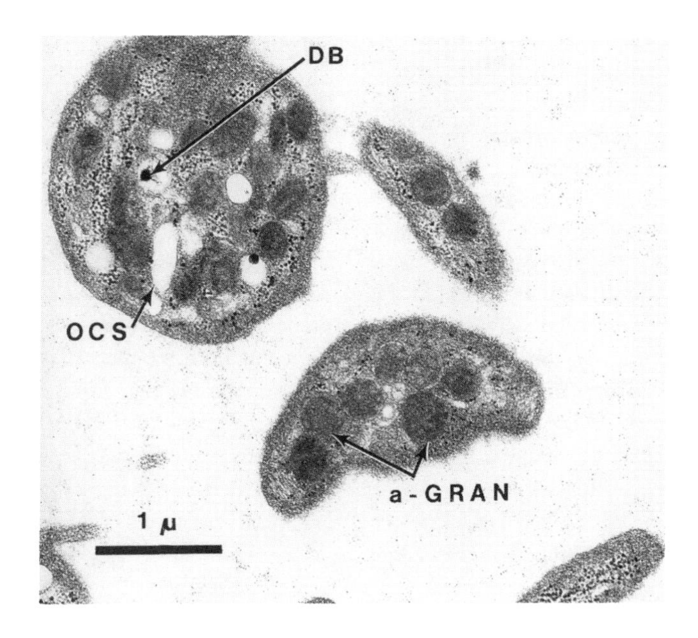

FIGURE 23.1. TEM of horse platelets. Internal structures here and in Fig. 23.2: (a-gran) alpha granules; (DB) dense body; (MT) microtubules; (OCS) open canalicular system.

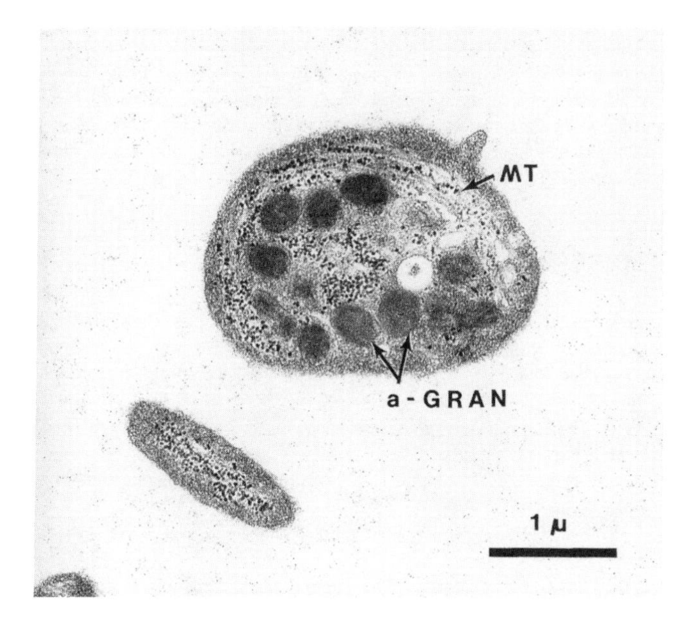

FIGURE 23.2. TEM of horse platelets. Key in Fig. 23.2 caption.

4. SUMMARY

Horse blood showed slower clotting in glass, and longer APTTs and Recal Ts, than any other mammals of this study with the exception of those from order Cetacea. The combination of slow clotting in glass and normal clotting (by human standards) in silicone suggests an abnormality in the contact pathway and may be due to a combination of low levels of Fletcher factor and factor XI. The long thrombin time appears to be due to a heat-stable, non-heparin-like inhibitor. Equine plasminogen was activated with UK, but not with SK. Erythrocyte and leukocyte counts fell in line with values given in veterinary hematology textbooks. Granules in the eosinophils were very much larger than those seen in humans, an observation previously made by Sonoda (1972). Equine platelets aggregated with ADP, collagen, thrombin, and snake venom (CHH). Only half the horses' platelets reacted with Risto, and then only minimally. Platelet ultrastructure was similar to human. Lack of platelet aggregation with Risto is suggestive of a von Willebrand's type of disorder, but this suggestion is not supported by the "normal" levels of factor VIII and RCF.

REFERENCES

Abildgaard, C. F., and Link, R. P., 1965, Blood coagulation and hemostasis in thoroughbred horses, *Proc. Soc. Exp. Biol. Med.* **119:**212.

Bell, W. N., Tomlin, S. C., and Archer, R. K., 1955, The coagulation mechanisms of the blood of the horse with particular reference to its "haemophilioid" status, *J. Comp. Pathol.* **65:**255.

Sjølin, K. E., 1956, Lack of Christmas factor in horse plasma, *Nature (London)* **178:**153.

Sjølin, K. E., 1957, Coagulation defect in horse plasma, *Proc. Soc. Exp. Biol. Med.* **94:**818.

Sonoda, M., 1972, Electron microscopy of eosinophils in the peripheral blood of clinically healthy horses, *Acta Haematol. Jpn.* **35:**39.

SUGGESTED READINGS

Allen, B. V., and Archer, R. K., 1973, Studies with normal erythrocytes of the English thoroughbred horse, *Equine Vet. J.* **5:**135.

Allen, B. V., and Archer, R. K., 1976, Some hematological values in English thoroughbred horses, *Vet. Rec.* **98:**195.

Archer, R. K., 1959, The normal haemograms and coagulograms of the English thoroughbred horse, *J. Comp. Pathol.* **69:**390.

Archer, R. K., 1965, The effect of excitation and exertion on the circulating blood of horses, *Vet. Rec.* **77:**689.

Archer, R. K., and Flute, P. T., 1959, Heparin and thromboplastin generation in the horse, *Nature (London)* **183:**235.

Archer, R. K., and Miller, W. C., 1959, The interpretation of haematological examinations in thoroughbred horses, *Vet. Rec.* **71:**273.

Barkhan, P., Tomlin, C., and Archer, R. K., 1957, Comparative coagulation studies on horse and human blood, *J. Comp. Pathol.* **67:**358.
Brenon, H. C., 1956, Erythrocyte and haemoglobin studies in thoroughbred racing horses, *J. Am. Vet. Med. Assoc.* **128:**343.
Brenon, H. C., 1958, Further erythrocyte and haemoglobin studies in thoroughbred racing horses, *J. Am. Vet. Med. Assoc.* **133:**102.
Campbell, M. D., Bellamy, J. E., and Searey, G. P., 1981, Determination of plasma fibrinogen concentration in the horse, *Am. J. Vet. Res.* **42:**100.
Catling, S. J., 1975, Studies on haematology of exertion in horses, *Br. J. Haematol.* **30:**123.
Dodds, W. J., 1975, Normal coagulation in the horse, in: *Proceedings of the First International Symposium on Equine Hematology* (H. Kitchen and J. D. Krehbiel, eds.), pp. 202, American Association of Equestrian Practices,
Fantl, P., and Marr, A. G., 1958, The coagulation of horse blood, *J. Physiol.* **142:**197.
Fegler, G., 1948, Haemoglobin concentration, haematocrit value, and sedimentation rate of horse blood, *Q. J. Exp. Physiol. Cogn. Med. Sci.* **34:**129.
Finocchio, E. J., Cottman, J. R., and Osbaldiston, G. W., 1970, Platelet counts in horses, *Cornell Vet.* **60:**518.
Gardikas, C., Kallinikou, M., and Kallinikos, G., 1965, Observations on horse blood coagulation, *Scand. J. Haematol.* **2:**31.
Gentry, P. A., Woodbury, F. R., and Black, W. D., 1978, Comparative study of blood coagulation tests in the horse and pony, *Am. J. Vet. Res.* **39:**333.
Hansen, M. F., and Todd, A. C., 1951, Preliminary report on the blood picture of the Arabian horse, *J. Am. Vet. Med. Assoc.* **118:**886.
Hansen, M. F., Todd, A. C., Kelley, G. W., and Hull, F. E., 1950, Studies on the haematology of the thoroughbred horse. I. Mares in foal, *Am. J. Vet. Res.* **11:**296.
Hansen, M. F., Todd, A. C., Cawein, M., and McGee, W. R., 1950, Studies on the haematology of the thoroughbred horse. III. Stallions, *Am. J. Vet. Res.* **11:**397.
Hansen, M. F., Todd, A. C., and McGee, W. R., 1950, Blood picture of lactating and non-lactating thoroughbred mares, *Vet. Med.* **45:**228.
Hansen, M. F., Todd, A. C., Kelley, G. W., and Cawein, M., 1951, Studies on the haematology of the thoroughbred horse, *Am. J. Vet. Res.* **12:**31.
Jeffcott, L. B., 1977, Clinical haematology of the horse, in: *Comparative Clinical Haematology* (R. K. Archer and L. B. Jeffcott, eds.), p. 161, Blackwell, Oxford.
Knill, L. M., McConaughy, M. S., Camarena, B. S., and Day, M., 1969, Hemogram of the Arabian horse, *Am. J. Vet. Res.* **30:**295.
Littlejohn, A., 1968, Packed cell volume, haemoglobin and plasma electrolyte studies in horses. I. Mean values in clinically normal horses, *Br. Vet. J.* **124:**530.
MacLeod, J., Ponder, E., Aitken, G. J., and Brown, R. B., 1947, The blood picture of the thoroughbred horse, *Cornell Vet.* **37:**305.
Meyers, K. M., Lindner, C., and Grant, B., 1979, Characterization of the equine platelet aggregation response, *Am. J. Vet. Res.* **40:**260.
Moore, V. A., Haring, C. M., and Cady, B. J., 1904, The clinical examination of the blood of the horse and its value to the veterinarian, *Proc. Am. Vet. Med. Assoc.* **41:**284.
Ollenforf, P., 1959, Defects in and variability of the thromboplastin system in horse plasma, *Thromb. Diath. Haemorrh.* **4:**45.
Rawlings, C. A., Byars, T. D., Van Noy, M. K., and Bisgard, G. E., 1975, Activated coagulation test in normal and heparinized ponies and horses, *Am. J. Vet. Res.* **36:**711.
Scott, E. A., Sandler, G. A., and Byars, T. D., 1979, Warfarin: Effects on anticoagulant, hematologic and blood enzyme values in normal ponies, *Am. J. Vet. Res.* **40:**142.

Stewart, G. A., and Steel, J. D., 1974, Haematology of the fit racehorse, *J. S. Afr. Vet. Assoc.* **45:**287.

Stewart, G. A., Riddle, C. A., and Salmon, P. W., 1977, Haematology of the racehorse: A recent study of thoroughbreds in Victoria, *Aust. Vet. J.* **53:**353.

Stewart, J., and Holman, H. H., 1940, The "blood picture" of the horse, *Vet. Rec.* **52:**157.

Stolpe, J., and Wiesner, E., 1970, Fibrinogen status of galloping and trotting horses subject to epistaxis, *Arch. Exp. Vet. Med.* **24:**903.

Trum, B. F., 1952, Normal variances in horse blood due to breed, age, lactation, pregnancy and altitude, *Am. J. Vet. Res.* **13:**514.

Tufvesson, G., 1952, Lymphangitis in horses. IV. A study of blood coagulability, *Nord. Veterinaermed.* **4:**1.

Yousef, M. K., Burk, C., and Dill, D. B., 1971, Biochemical properties of the blood of three equines, *Comp. Biochem. Physiol. B* **39:**279.

CHAPTER 24

PIGS

Standard pig

Mini-pig

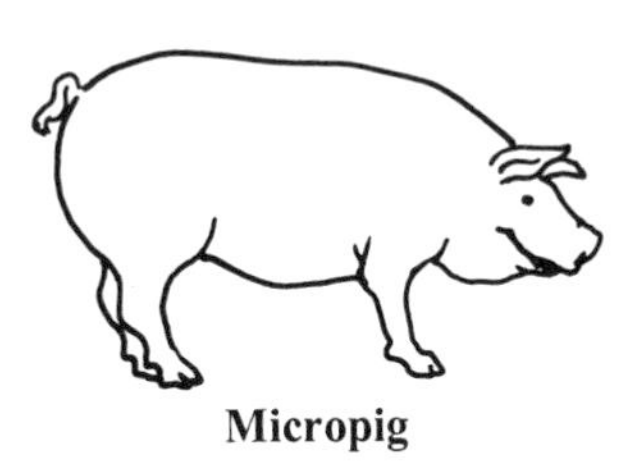

Micropig

1. INTRODUCTION

Turning to Artiodactyla, the order of even-toed, hoofed mammals, our first example studied is the pig. Most pigs today are the domesticated descendants of the ancestral pig (*Sus scrofa*). A few feral animals of the same suborder still exist: the European wild boar, the warthog, and the closely related hippopotamus. For some years, pigs have been used as models for experimental surgery and other research studies because they are the right size, cheap, and adaptable to laboratory housing. Unfortunately, they often outgrow the experiments. Standard pigs continue to grow as long as food is available and may weigh as much as 160–180 kg by the time they reach full adulthood. For this reason, smaller pigs have been sought. Selective crossbreeding of small pigs from both the Hanford and Yucatan strains has resulted in animals that are reproducibly stunted. The Charles River Breeding Laboratories has developed two strains—Mini-pigs (30–50 kg) and Micropigs (20–40 kg)—from the original Yucatan swine. These two strains are not completely inbred, but each strain has a limited gene pool because they are interbred from a limited number of stock animals.

A good deal of material has been published about coagulation in pigs because an excessive bleeding tendency in one cohort was described years ago (Hogan *et al.*, 1941; Bogart and Muhrer, 1942) and has been studied in detail (see the Suggested Readings). This has been identified as a porcine version of human von Willebrand's disease (vWD). Radin *et al.* (1986) described hematological and serum biochemical values in Yucatan miniature swine.

2. SOURCE

Blood was obtained from standard pigs at the Central Animal Facility, University of Pittsburgh, through the courtesy of Paul H. Bramson, D.V.M. Standard pigs were young Yorkshires (Nos. 1–4) and Chestershires (Nos. 5–9) and weighed about 50 kg. Pigs No. 1–6 were female and No. 7–9 were male. Blood from Mini-pigs and Micropigs was obtained at the Charles River Breeding Farm, Windham, Maine, through the courtesy of Mr. Jace Trickey and W. Judd, D.V.M. Mini-pigs No. 10–12 (female) and No. 13 and 14 (male) weighed 30–40 kg; Micropigs No. 15–17 (female) and No. 18–20 (male) weighed 20–30 kg. All pigs were housed indoors and fed vegetables and chow. Blood samples were taken from females without anesthesia and from the more ferocious males after ketamine injections. Blood was obtained from the vena cava or femoral vein by external puncture. All pigs were breeders or used for different experimental purposes, and our samples were obtained before other bleedings or procedures.

Blood for blood counts and chemistries was transferred to the Presbyterian Hospital Clinical Laboratories (for standard pigs) or to the Northern Diagnostic Laboratories, Portland, Maine, an affiliate of Maine Medical Center (for Minipigs and Micropigs). Blood for coagulation, platelet, and electron-microscopic studies was transported as quickly as possible to the Central Blood Bank Laboratories in Pittsburgh, Pennsylvania.

3. TESTS CONDUCTED

3.1. Bleeding Times

Bleeding times using the Simplate method on the forelegs of six standard pigs were shorter than 3 min with a mean of 105 sec.

3.2. General Coagulation Tests and Factors

Table 24.1 shows the results of general coagulation tests in the three sizes of pigs. The blood clotted in 6–9 min in new glass and slightly longer in siliconized tubes. Clots retracted and did not lyse. SPTs performed on human or pig barium-treated plasma were within the normal human range. The prothrombin times with Simplastin or RVV and the APTTs were about the same in the three types of pigs and in humans. Pig brain clotted pig plasma more rapidly than it did human plasma. Pig Recal Ts were somewhat shorter than those of human plasma. The clots were solid and did not lyse alone or in 1% MCA. Thrombin times with bovine thrombin, CHH, and Atroxin were all longer than human.

The TGTs (Table 24.2) used human or pig generating mixtures to clot human or pig platelet-poor plasma (PPP) substrates. Both animal and human mixtures clotted the human substrate somewhat more rapidly than that of the pig. Pig results did not suggest any plasma coagulation factor or platelet factor 3 deficiency.

Coagulation factor assay results (Table 24.3) were much the same for the three sizes of pigs. Fibrinogen and factors II, XI, Fletcher, ATIII, protein C (A), antiplasmin (Antipl), and XIII fell in the normal human range. The others, except for protein S and plasminogen (Plgn), were high or very high. Protein S was measured by an immunoassay, and the antibody apparently did not react to pig protein. Plgn was measured by streptokinase (SK) activation. Pig plasma was not sensitive to SK; therefore, no activity would be expected.

TABLE 24.1
General Coagulation Tests in Pigs

	Standard pigs			Mini-pigs			Micropigs			All pigs (mean)	Human		
											Controls[a]		Normal range
											S (4)	M (3)	
	N	Mean	SD	N	Mean	SD	N	Mean	SD		(Means)		
Clot T (min)													
Glass	6	6	3.8	4	9	1.9	4	7	1.7	7	6	7	6–12
Silicone	6	21	9.8	4	21	3.0	4	12	2.9	18	22	30	20–59
Clot Retr	6	3–4	0	4	4–4	0	4	4–4	0	4–4	4/4	4/4	3–4+
Clot Lys	6	0	0	4	0	0	4	0	0	0	0	0	0
SPT (sec)													
Human	6	27.9	5.9	—	—	—	—	—	—	27.9	44.8	45.5	20+
Pig	4	32.6	6.4	—	—	—	—	—	—	32.6	22.4	—	—
PT (sec)													
Simplastin	6	12.6	1.1	4	14.2	3.0	4	14.1	3.6	13.5	12.4	12.7	10–13
Pig Br	6	14.6	2.5	—	—	—	—	—	—	14.9	24.8	—	—
RVV	6	16.8	2.5	4	15.8	2.6	4	13.7	1.8	15.9	19.8	15.4	13–18
APTT (sec)	6	23.0	5.4	3	22.0	0.2	4	24.5	2.2	23.2	28.8	25.6	24–34
Recal T (sec)	6	90	22	4	86	7.5	4	107	15	94	130	135	90–180
Lys	6	0	0	4	0	0	4	0	0	0	0	0	0
Lys MCA	6	0	0	4	0	0	4	0	0	0	0	0	0
Thromb T (sec)													
Bov Thr	6	61+	—	4	42.2	6.1	4	40.2	3.3	50.0	14.9	16.0	11–18
CHH	2	32	3.1	4	34.2	7.7	4	32.8	5.0	33.3	16.2	13.8	11–18
Atroxin	4	52	6.3	4	39.8	2.4	4	36.4	2.6	48.0	15.1	14.0	11–18

[a]Controls for standard (S) and Mini-pigs and Micropigs (M).

TABLE 24.2
Thromboplastin Generation Test in Pigs

Generating mixture		Substrate Clot T (sec)[a]	
Human	Pig	Human	Pig
H1	—	**8.6**	12.7
—	P1	8.9	**11.2**
—	P2	8.5	**9.5**

[a]Homologous systems are in **boldface**.

3.3. Fibrinolytic Enzyme System

Pig euglobulin lysis time was greater than 4 hr. Trace activation of Plgn occurred with SK and urokinase and none with staphylokinase. Antipl was about the same as that from human.

3.4. Serum Proteins

Serum proteins, done on the standard pigs, did not differ greatly from human (Table 24.4).

3.5. Biochemical Tests

Results of biochemical tests (Table 24.5) showed little variation among the three sizes of pigs. Most values were similar to those found in humans. Pig plasma was high in P, uric acid, and the liver enzyme LDH. The cholesterol level was below the human range.

3.6. Miscellaneous Tests

Pig plasma in dilutions up to 1:2048 aggregated the standard staphylococcal preparation.

Pig plasma showed a remarkable ability to aggregate normal human platelets. This effect was shared with porpoise, cow, goat, and sheep, plasmas (see Chapters 18 and 26). It has been recognized for many years (Forbes and Prentice,

TABLE 24.3
Coagulation Factors in Pigs

Factor	Standard pigs			Mini-pigs			Micropigs			All pigs	Normal human
	N	Mean	SD	N	Mean	SD	N	Mean	SD	(mean)	range
I (mg/dl)	6	348	78	4	251	104	4	310	67	309	150–450
II (U/ml)	6	0.68	0.3	4	0.43	0.1	4	0.80	0.6	0.64	0.70–1.30
V (U/ml)	6	2.80	0.6	4	5.15	0.8	4	7.20	3.3	4.73	0.65–1.45
VII (U/ml)	6	1.92	0.4	4	1.67	0.8	4	2.75	1.7	2.09	0.50–1.30
X (U/ml)	6	2.54	0.9	4	1.06	0.3	4	1.75	0.9	1.89	0.75–1.25
VIII (U/ml)	6	6.42	0.8	4	5.88	1.0	4	5.95	1.0	6.13	0.75–1.40
IX (U/ml)	6	3.20	0.3	4	4.10	1.4	4	5.73	1.5	4.18	0.65–1.70
XI (U/ml)	6	1.67	0.3	4	1.16	0.5	4	1.35	0.1	1.43	0.75–1.40
XII (U/ml)	6	5.65	1.4	4	3.98	0.7	4	4.95	1.2	4.95	0.50–1.45
Flet (U/ml)	6	2.00	0.5	4	1.40	0.4	2	1.25	0.1	1.67	0.50–1.50
Fitz (U/ml)	6	3.05	0.3	4	3.00	0.6	2	4.65	0.4	3.30	0.50–1.50
ATIII (U/ml)	6	1.09	0.2	3	1.11	0.05	2	1.18	0.01	1.11	0.80–1.20
Prot C (A) (U/ml)	2	0.70	0.1	3	0.56	0.2	2	0.94	0.12	0.71	0.60–1.40
Prot S (I) (U/ml)	2	0.10	0	3	0.07	0.01	2	0.07	0.01	0.08	0.60–1.40
Plgn (U/ml)	2	0.50	0.1	2	0.08	0.02	2	0.10	0.01	0.08	0.80–1.20
Antipl (U/ml)	2	1.20	0.1	2	1.32	0.01	2	1.22	0.04	1.25	0.80–1.20
XIII (R)	6	15	9.4	4	12	4.6	4	16	0	15	4–16

TABLE 24.4
Serum Proteins by Paper Electrophoresis in Pigs

Protein	Standard pig No. 1 ♀	2 ♀	3 ♀	4 ♀	N	Mean	SD	Normal human range
TP (g/dl)	5.4	7.4	6.5	5.6	4	6.2	0.9	6.3–7.9
Albumin (g/dl)	2.3	3.4	2.7	2.4	4	2.7	0.5	3.2–4.4
α_1-Globulin (g/dl)	1.3	1.5	1.8	1.3	4	1.5	0.2	0.2–0.4
α_2-Globulin (g/dl)	0.4	0.6	0.6	0.5	4	0.5	0.2	0.4–1.0
β-Globulin (g/dl)	0.7	1.0	0.6	0.7	4	0.8	1.0	0.5–1.0
γ-Globulin (g/dl)	0.7	0.9	0.8	0.7	4	0.8	1.0	0.6–1.8

1973; de Gaetano, 1974). This activity is associated with the high-molecular-weight fraction of pig factor VIII complex (the vW factor).

3.7. Cellular Elements

Table 24.6 illustrates the blood counts found in the pigs. The RBC size (MCV) and hemoglobin content (MCH) were lower than human and the WBC count slightly higher. Otherwise, most of the values fell within the human range. Pig platelets appeared to be about the same in number and size as human platelets.

3.8. Platelet Activities

Table 24.7 shows that more than 20% of pig platelets were retained after passage of blood through a glass bead column. This percentage was within the human range. The aggregation of pig platelets in platelet-rich plasma (PRP) was very different from that seen with human PRP. With ADP, the curves were biphasic; that is, the platelets aggregated and then spontaneously disaggregated. Pig platelets aggregated with collagen almost as completely and irreversibly as did human. No aggregation was seen with ristocetin (Risto), arachidonic acid (Arach A), thrombin, CHH, pig and sheep plasmas, all of which aggregate human platelets.

To test the effects of plasmas on aggregation, human and pig platelets were concentrated by gentle centrifugation and then suspended in homologous or heterologous PPP and their aggregation with Risto tested. Table 24.8 shows that human platelets suspended in pig or human PPP aggregated completely with

TABLE 24.5
Biochemical Tests in Pigs

Substance	Standard pigs			Mini-pigs			Micropigs			All pigs (mean)[a]	Normal human range
	N	Mean[a]	SD	*N*	Mean[a]	SD	*N*	Mean[a]	SD		
TP (g/dl)	4	6.2	0.7	5	7.8	0.5	5	7.7	0.7	7.3	6.0–8.0
Albumin (g/dl)	—	—	—	5	4.7	0.2	5	4.7	0.5	4.7	3.5–5.0
Ca (mg/dl)	4	10.2	1.9	5	10.0	0.4	5	10.1	0.6	10.1	8.5–10.5
P (mg/dl)	4	**7.1**	0.8	5	**6.2**	0.5	5	**6.7**	0.8	**6.6**	2.5–4.5
Cholesterol (mg/dl)	4	*124*	28	5	*66*	12	5	*84*	49	*89*	150–200
Triglyceride (mg/dl)	—	—	—	5	*43*	23	5	78	60	61	45–200
Glucose (mg/dl)	4	**138**	46	5	69	11	5	*64*	15	87	65–110
Uric acid (mg/dl)	4	**>10**	—	5	**>10**	—	5	**>10**	—	**>10**	2.5–8.0
Creatinine (mg/dl)	4	1.2	0.4	5	**1.7**	0.2	5	**1.5**	0.1	**1.5**	0.7–1.4
T bilirubin (mg/dl)	4	0.15	<1	5	0.1	<1	5	0.2	<1	0.15	0.1–1.4
BUN (mg/dl)	4	15.3	6.2	5	18.7	3.4	5	**21.1**	4.9	18.6	10–20
Chloride (meq/liter)	4	98	6.5	5	102	1.7	5	100	2.2	100	95–105
CO_2 (meq/liter)	4	25.3	5.6	5	26.3	3.9	5	26.1	3.4	26	24–32
K (meq/liter)	4	4.5	0.3	5	4.6	0.7	5	4.7	0.8	4.6	3.5–5.0
Na (meq/liter)	4	139	7.6	5	139	1.3	5	142	4.4	140	135–145
Alk Phos (IU/liter)	—	—	—	5	33	8.7	5	42	8.6	38	30–85
LDH (IU/liter)	4	**423**	49	5	**509**	249	5	**563**	97	**505**	60–200
SGOT (IU/liter)	4	**54**	3.8	5	40	4.6	5	**43**	8.0	**45**	0–41
SGPT (IU/liter)	—	—	—	5	37	4.9	5	**58**	11	48	10–50

[a]Values **higher** than human are in **boldface**, those *lower* than human in *italics*.

TABLE 24.6
Cellular Elements of the Peripheral Blood in Pigs

Elements	Standard pigs			Mini-pigs			Micropigs			All pigs (mean)	Normal human range
	N	Mean	SD	*N*	Mean	SD	*N*	Mean	SD		
HCT (%)	4	38	2.1	5	36	2.5	5	36	1.7	37	37–52
HGB (g/dl)	4	14.4	0.6	5	12.6	0.7	5	12.4	0.6	13.0	12–18
RBC ($\times 10^6/mm^3$)	4	6.5	0.6	5	5.3	0.2	5	5.3	0.4	5.6	4.2–6.2
MCV (fl)	4	58	7.0	5	68	3.9	5	69	3.5	66	79–97
MCH (pg)	4	22	2.2	5	24	0.9	5	23	1.5	23	27–31
MCHC (g/dl)	4	38	0.8	5	35	0.7	5	34	0.7	36	32–36
WBC ($\times 10^3/mm^3$)	4	21.5	2.1	5	8.5	1.3	5	14.8	3.1	14.5	5–10
Neutr (%)	4	45	4.2	5	52	1.5	5	69	4.3	56	55–75
Lymph (%)	4	47	5.6	5	40	12	5	25	3.5	37	20–40
Mono (%)	4	2	<1	5	4	3	5	4	3	3	2–10
Eos (%)	4	3	1	5	3	3	5	1	1	2	0–3
Bas (%)	4	1	<1	5	1	1	5	1	1	1	0–3
Plat C ($\times 10^3/mm^3$)	9	580	115	4	337	108	6	379	46	465	150–450
MPV (μM^3)	2	6.8	0.2	4	7.4	0.8	4	7.8	0.8	7.2	5.6–10.4

TABLE 24.7
Platelet Activities in Pigs

Activity	Standard pigs			Mini-pigs			Micropigs			All pigs (mean)	Human	
	N	Mean	SD	*N*	Mean	SD	*N*	Mean	SD		Control[a]	Normal human
Plat Glass Ret Ind (%)	5	32	5	—	—	—	—	—	—	32	39	>20
Aggregation (%)												
ADP (20 μM)	1	56[b]	—	3	51[b]	7	4	46[b]	18	49[b]	100	70–100
(10 μM)	3	50[b]	2	3	44[b]	7	5	44[b]	16	46[b]	100	70–100
(5 μM)	1	42[b]	—	3	31[b]	2	4	29[b]	15	31[b]	100	50–100
(2.5 μM)	1	28[b]	—	3	16[b]	1	4	15[b]	7	23[b]	73	40–100
Collagen (0.19 mg/ml)	3	74	4	3	63	6	5	75	24	71	100	80–100
Risto (0.9 mg/ml)	3	5	0	3	4	1	5	4	3	<5	100	80–100
Risto ½ (0.45 mg/ml)	3	3	0	3	4	2	5	5	3	<5	2	<18
Arach A (0.5 mg/ml)	3	0	0	3	5	2	5	6	2	<5	100	60–100
Bov Thr (0.4 U/ml)	3	0	0	3	6	2	5	6	3	<5	100	60–100
CHH (0.05 mg/ml)	3	0	0	3	4	1	5	8	5	<5	100	60–100
Pig Pl[c] (1 : 10)	1	0	—	3	6	4	5	10	2	<5	100	60–100
Sh Pl[c] (1 : 10)	1	5	—	3	8	1	5	4	1	<5	100	60–100
Hu Pl (1 : 10)	1	2	—	1	2	0	3	4	1	<5	0	<10

[a]Controls for Mini-pigs and Micropigs.
[b]Biphasic curve, showing aggregation followed by disaggregation.
[c]Pooled frozen, stored at −70°C.

TABLE 24.8
Aggregation of Washed Platelet Resuspended in Platelet-Poor Plasma

Suspension (ml)				
Washed platelets		PPP		Aggregation by
Human	Pig	Human	Pig	Risto (0.9 mg/ml)
0.1	—	0.3	—	100%
0.1	—	—	0.3	100%
—	0.1	0.3	—	0%
—	0.1	—	0.3	0%

Risto. Obversely, pig platelets suspended in human or pig PPP did not aggregate with Risto.

3.9. Platelet Ultrastructure

Figures 24.1 and 24.2 are TEMs of pig platelets in the resting state. The striking and unique finding in pig platelets was the freckled alpha granules. Such granules were not seen in any other mammal we have studied. These freckles appeared to be cross sections of internal bands or fingers. These pig platelets showed most of the common structural elements seen in other platelets: circumferential bands of microtubules, an open canalicular system, mitochondria with cristae, and glycogen particles. Dense bodies were rare and usually found in vacuoles.

4. SUMMARY

Various hematological parameters were studied in three sizes of pigs: standard, Mini-pig, and Micropig. Bleeding times were less than 3 min. Blood clotted in 6–9 min in glass and slightly longer in siliconized tubes. Clots retracted and did not lyse. Prothrombin times using Simplastin or RVV and APTTs were about the same in the three sizes of pigs and in the human controls. Prothrombin times with pig brain were shorter in pigs than in humans. Thrombin times were much longer on porcine plasma than on human plasma. Coagulation factors I, II, XI, ATIII, Prot C, Antipl, and XIII assayed at near-human levels and factors V, VII, VIII, IX, X, XII, Flet, and Fitz at higher than human levels. Prot S and Plg assayed very low. Pig serum proteins and biochemistries were not

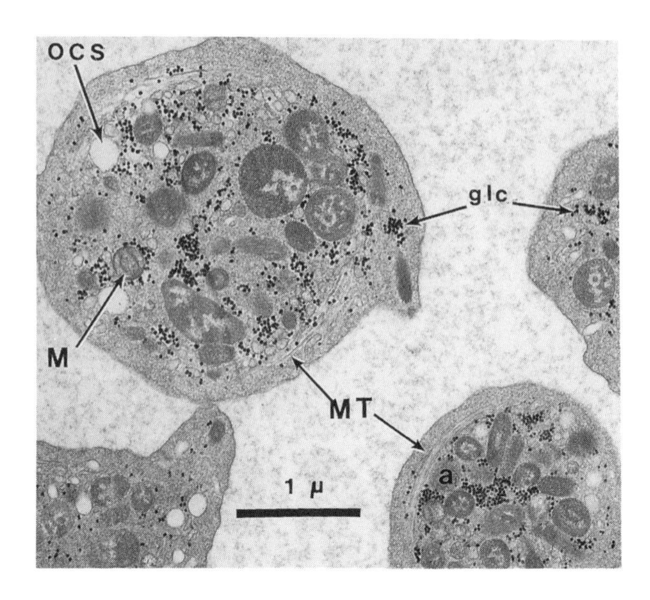

FIGURE 24.1. TEM of pig platelets. Note here and in Fig. 24.2 that almost all the alpha granules contain "freckles," round internal structures unique to pig platelets. Internal structures here and in Fig. 24.2: (F-Gran) "freckled" alpha granules; (glc) glycogen; (M) mitochondria; (MT) microtubules; (OCS) open canalicular system.

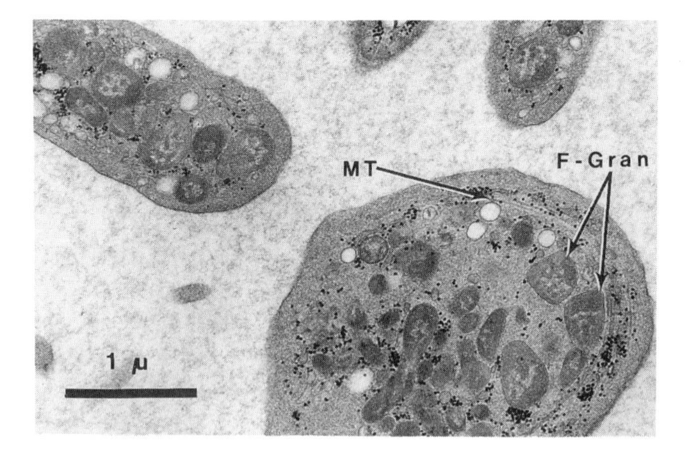

FIGURE 24.2. TEM of pig platelets. Key in Fig. 24.1 caption.

greatly different from human or among the three sizes of pigs. Pig erythrocytes were slightly smaller with less hemoglobin than their human counterparts. Leukocyte counts were slightly higher than humans. Platelet count and size were not remarkable. Pig platelets aggregated with ADP and collagen but *not* with Risto, Arach A, or thrombin. TEMs showed the usual internal structure and a remarkable, distinctive finding: The alpha granules showed "freckles," round internal structures, suggesting that internal bands or fingers had been cross-sectioned. Except for the marked differences in platelet reactivity, pigs should be excellent research models.

REFERENCES

Bogart, R., and Muhrer, M. E., 1942, The inheritance of a hemophelia-like condition in swine, *J. Heredity* **33:**59.

de Gaetano, G., Donati, M. B., and Vermylen, J., 1974, Evidence that human platelet aggregating activity in porcine plasma is a property of von Willebrand factor, *Thromb. Diath. Haemorrh.* **32:**549.

Forbes, C. D., and Prentice, C. R. M., 1973, Aggregation of human platelets by purified porcine and bovine antihaemophilic factor, *Nature (London)* **241:**149.

Hogan, Muhree, M. L., and Bogart, R., 1941, A hemophilia-like disease in swine, *Proc. Soc. Exp. Biol. Med.* **48:**217.

Radin, M. J., Weiser, M. G., and Fettman, M. J., 1986, Hematologic and serum biochemical values for Yucatan miniature swine, *Lab. Anim. Sci.* **36**(4)**:**425.

SUGGESTED READINGS

Belcher, T. E., and Gunstone, M. J., 1969, Fibrinolysis, coagulation, and haematological findings in normal large white/Wessix cross pigs, *Br. Vet. J.* **125:**74.

Bowie, E. J., Owen, C. A., Jr., Zollman, P. E., Thompson, J. H., Jr., and Fass, D. N., 1973, Tests of hemostasis in swine: Normal values and values in pigs affected with von Willebrand's disease, *Am. J. Vet Res.* **34:**1405.

Brinkhous, K. M., Read, M. S., Rodman, N. F., and Mason, R. G., 1969, The need of fibrinogen for thrombin-induced aggregation of porcine platelets, *J. Lab. Clin. Med.* **73:**1000.

Chan, J. Y. S., Owen, C. A., Jr., Bowie, E. J. W., Didisheim, P., Thompson, J. H., Jr., Muhrer, M. L., and Zollman, P. E., 1968, Von Willebrand's disease "stimulating factor" in porcine plasma, *Am. J. Physiol.* **214:**1219.

Cornell, C. N., and Muhrer, M. E., 1964, Coagulation factors in normal and hemophiliac-type swine, *Am. J. Physiol.* **206:**926.

Cornell, C. N., Copper, R. G., Kahn, R. A., and Grab, S., 1969, Platelet adhesiveness in normal and bleeder swine as measured in a celite system, *Am. J. Physiol.* **216:**1170.

Coulter, D. B., Ewan, R. C., Swenson, M. J., Atherne, F. X., and Wyllie, D., 1970, Plasma and erythrocytic concentration of electrolytes in blood of fetal and maternal swine, *Am. J. Vet. Res.* **31:**1179.

Crookshank, H. R., Smalley, H. E., and Steel, E., 1975, Hematological parameters of American-Essex swine, *J. Anim. Sci.* **40:**190.

Cross, M. J., 1962, The nature of the thromboplastic activity appearing during the clotting of pig's plasma, *Thromb. Diath. Haemmorh.* **8:**472.

Cross, M. J., 1962, The separation of thromboplastin formed from pig's plasma in the thromboplastin generation test, *Thromb. Diath. Haemmorrh.* **8:**485.

Fass, D. N., Bowie, E. J. W., Owen, C. A., and Mann, K. G., 1975, Stability of porcine factor VIII, *Thromb. Res.* **6:**109.

Fass, D. N., Brockway, W. J., Owen, C. A., and Bowie, E. J. W., 1976, Factor VIII (Willebrand) antigen and ristocetin–Willebrand factor in pigs with von Willebrand's disease, *Thromb. Res.* **8:**319.

Fraser, A. C., 1938, A study of the blood of pigs, *Vet. J.* **94:**3.

Gerloff, J., and von Kaulla, K. N., 1972, Prevention of dog-serum induced aggregation of pig platelets, *Proc. Soc. Exp. Biol. Med.* **141:**298.

Gregor, G., 1979, Haematological and biochemical studies on Mini-mini-lewe miniature swine. I. Blood picture and serum proteins, *Z. Versuchstierkd.* **21:**92.

Kinlough-Rathbone, R. L., Chahil, A., and Mustard, J. F., 1974, Divalent cations and the release reaction of pig platelets, *Am. J. Physiol.* **226:**235.

Kopitar, M., Stegnar, M., Accetto, B., and Levez, D., 1974, Isolation and characterization of plasminogen activator from pig leucocytes, *Thromb. Diath. Haemorrh.* **31:**72.

Kristjansson, F. K., 1960, Genetic control of blood serum proteins in swine, *Can. J. Genet. Cytol.* **2:**295.

Kristjansson, F. K., 1960, Inheritance of a serum protein in swine, *Science* **131:**1681.

Kristjansson, F. K., 1961, Genetic control of three haptoglobins in pigs, *Genetics* **46:**907.

Mandel, L., and Travnicek, J., 1982, Haematology of conventional and germfree miniature Minnesota piglets. I. Blood picture, *Z. Versuchstierkd.* **24:**299.

Mertz, E. T., 1942, The anomaly of a normal Duke's and a very prolonged saline bleeding time in swine suffering from an inherited bleeding disease, *Am. J. Physiol.* **136:**360.

Muller, E. R., Ullrey, D. E., Ackermann, I., Schmidt, D. A., Luecke, R. W., and Hoefer, J. A., 1961, Swine haematology from birth to maturity. II. Erythrocyte population, size, and haemoglobin concentration, *J. Anim. Sci.* **20:**890.

Nafstad, H. J., and Nafstad, I., 1968, An electron microscopic study of normal blood and bone marrow in pigs, *Pathol. Vet.* **5:**451.

Olson, D. J., Brockway, W. J., Fass, D. N., Magnuson, M. A., and Bowie, E. J. W., 1975, Evaluation of ristocetin–Willebrand factor assay and ristocetin-induced platelet aggregation, *Am. J. Clin. Pathol.* **63:**210.

Olson, D. J., Brockway, W. J., Fass, D. N., Bouwie, E. J. W., and Mann, K. G., 1977, Purification of porcine and human ristocetin–Willebrand factor, *J. Lab. Clin. Med.* **89:**1278.

Wu, Q. Y., Drouet, L., Carrier, J. L., Rothschild, C., Berard, M., Rouault, C., Caen, J., and Meyer, D., 1987, Differential distribution of von Willebrand factor in endothelial cells: Comparison between normal pigs and pigs with von Willebrand disease, *Arteriosclerosis* **7:**47.

CHAPTER 25

THE CAMEL FAMILY

1. INTRODUCTION

The second group of animals of the order Artiodactyla studied is the family Camelidae, which originated in North America in the early Eocene Age (40–50 MYA). About 2 MYA, groups migrated over land bridges to Eurasia and South America. Just 5000 years ago, at the end of the glacial period, Camelidae became extinct in North America. The group in Eurasia developed into suborder Camelus and those in South America into suborder Lama. There are several species. *Camelus dromedarius* is an entirely domesticated species of large, one-humped animals bred for riding or carrying loads and traveling long distances in desert areas without food or water. They are taller than horses and have long, slender legs with padded, flexible hooves that keep them from sinking in the sand. They walk like a boat on waves, swaying from side to side. Their bushy eyebrows, long eyelashes, and nostrils that can be closed keep out the desert sands. The hump stores fat that when metabolized produces water and CO_2; this metabolic water can replace water from external sources to maintain life. For centuries, the one-humped dromedary was used to transport materials across arid areas, particularly in Africa. The two-humped Bactrian camel has long hair and lives in the colder regions of Asia. Camels are said to be "unpleasant, smelly, dull, and stubborn." Even so, the females provide milk, the animals provide skins for tents and covers, and, in emergencies, their meat can sustain human life.

The South American Camelidae are smaller, up to 75 kg, and have developed into several species: the guanaco (*Lama guanicoe*), the domestic llama (*Lama guanicoe gama*), the alpaca (*Lama guanicoe pacos*), and the vicuna (*Lama vicuna*). These animals usually live in the high mountains, where they develop luxurious wool coats that are sheared annually. In addition, they are used for transport and recently have become popular as pets.

Camelidae produce one offspring after a gestation period of approximately 1 year. This offspring is nursed and nurtured for up to a year. Camels live for 20–24 years. Even if a camel produces one offspring every 2 years for life, careful husbandry is required to maintain the population.

Animals of the family Camelidae are the only mammals who have oval erythrocytes. The unusual ellipsoid appearance of the stained or fresh erythrocyte has been known for over a hundred years (Gulliver, 1875; Ponder *et al.*, 1928). Scanning electron microscopy (Jain and Keeton, 1974) showed camel and llama cells to be wafer-thin and elliptical.

Results of coagulation and hematological studies performed in this laboratory (Lewis, 1976) on two camels, one guanaco, and one llama are reprinted here with the kind permission of the editor of *Comparative Biochemistry and Physiology*.

2. SOURCE

The two camels, both female and each weighing approximately 600 kg, and the llama (75 kg) were young animals owned by the Pittsburgh Zoo. We are indebted to J. Swart, D.V.M., for aid in obtaining blood samples by jugular venipuncture. Blood from the guanaco (60 kg) was obtained through the courtesy of J. Y. Henderson, D.V.M., Ringling Brothers and Barnum & Bailey Circus, Sarasota, Florida.

3. TESTS CONDUCTED

3.1. Bleeding Times

Bleeding times were not done on these hairy animals.

3.2. General Coagulation Tests and Factors

Table 25.1 lists the results found in general tests of coagulation. The blood clotted in times comparable to human blood. The clots retracted well and did not

TABLE 25.1
General Coagulation Tests in Camelidae

	Camel No.				Human	
Test	1 ♀	2 ♀	Guanaco	Llama	Control[a]	Normal range
Clot T (min)						
Glass	10	9	10	15	8	6–12
Silicone	25	33	30	55	24	20–59
Clot Retr	4/4	4/4	1/1	4/4	4/4	3–4+
Clot Lys	0	0	0	0	0	0
SPT (sec)						
Human	—	120+	—	120+	41.8	20+
Homolog	120+	120+	—	120+	25.4	—
PT (sec)						
Simplastin	10	9.2	10.2	9.1	12	10–13
RVV	—	14.4	19.2	14.0	19.8	13–18
APTT (sec)	47	50	36.4	26.4	50.4	45–55
Recal T (sec)	105	109	158	101	127	90–180
Lys	0	0	0	0	0	0
Lys MCA	0	0	0	0	0	0
Thromb T (sec)						
Bov Thr	13.1	26.1	24.8	25.9	15.1	11–18

[a]Control for llama.

TABLE 25.2
Thromboplastin Generation Test in Camelidae

Generating mixture			Substrate Clot T (sec)[a]		
Human	Camel (mean)	Guanaco	H	C	G
H	—	—	**92**	10.2	10.4
—	C (2)	—	10.8	**10.2**	10.4
—	—	G	9.8	10.4	**9.8**

[a]Homologous systems are in **boldface**.

lyse. The prothrombin times and APTTs were slightly shorter than human and the thrombin times slightly longer.

Results with the TGT test on camel and guanaco were very similar to human (Table 25.2).

Table 25.3 shows the coagulation factor assay results. Most factors (I, II, VII, XI, XII, ATIII, XIII) fell within or close to the human range. Factor V and VIII were higher than human in one camel, the quanaco, and the llama.

3.3. Fibrinolytic Enzyme System

Whole blood and recalcified clots of camel, guanaco, and llama blood did not lyse. Euglobulin fractions were not prepared. Camel and guanaco plas-

TABLE 25.3
Coagulation Factors in Camelidae

	Camel No.					
Factor	1 ♀	2 ♀	Mean	Guanaco	Llama	Normal human range
I (mg/dl)	280	237	259	318	237	150–450
II (U/ml)	1.30	1.19	1.25	1.10	1.03	0.70–1.30
V (U/ml)	0.93	2.14	1.54	2.0+	1.70	0.65–1.45
VII (U/ml)	1.20	1.04	1.12	1.30	0.87	0.50–1.30
X (U/ml)	—	0.50	0.50	1.37	0.89	0.75–1.25
VIII (U/ml)	2.0+	2.0+	2.0+	2.0+	7.10	0.75–1.40
IX (U/ml)	0.50	0.65	0.58	1.50	2.50	0.65–1.70
XI (U/ml)	0.90	0.60	0.75	1.25	1.35	0.75–1.40
XII (U/ml)	2.0+	0.75	1.38	1.25	1.75	0.50–1.45
ATIII (U/ml)	—	1.05	1.05	0.60	1.28	0.80–1.20
XIII (R)	—	16	16	8	16	4–16

minogens were not activated with streptokinase. Urokinase and staphylokinase were not tested.

3.4. Serum Proteins

In the three animals for which results are shown in Table 25.4, the total proteins and γ-globulins were slightly below the human range.

3.5. Biochemical Tests

As shown in Table 25.5, Ca, P, and Na levels in camel No. 1 and the llama were above the human range. The enzymes Alk Phos and SGOT were very high in both animals, as were CPK and LDH in camel No. 1.

3.6. Miscellaneous Tests

Plasmas from camel, guanaco, and llama were tested for their ability to clump staphylococci, and each was positive up to a dilution of 1:256.

3.7. Cellular Elements

The ellipitical erythrocytes of Camelidae contain large amounts of hemoglobin in their small volumes. On stained smears or scanning electron microscopy, they appeared flat, but by phase-contrast microscopy, they appeared like small boats. Figures 25.1 and 25.2 are SEMs of mixtures of human and llama or camel RBCs. The human RBCs looked like large doughnuts; the llama or camel

TABLE 25.4
Serum Proteins by Paper Electrophoresis in Camelidae

Protein	Camel No. 2 ♀	Guanaco	Llama	Normal human range
TP (g/dl)	4.8	5.4	6.0	6.3–7.9
Albumin (g/dl)	3.4	3.1	3.5	3.2–4.4
α_1-Globulin (g/dl)	0.2	0.5	0.1	0.2–0.4
α_2-Globulin (g/dl)	0.4	0.7	0.6	0.4–1.0
β-Globulin (g/dl)	0.6	0.7	1.4	0.5–1.0
γ-Globulin (g/dl)	0.2	0.4	0.5	0.6–1.8

TABLE 25.5
Biochemical Tests in Camelidae

Substance	Camel No. 1 ♀[a]	Llama[a]	Normal human range
Ca (mg/dl)	**11.3**	**10.9**	8.5–10.5
P (mg/dl)	**8.2**	**8.6**	2.5–4.5
Cholesterol (mg/dl)	*45*	*45*	150–200
Glucose (mg/dl)	**205**	106	65–110
Uric acid (mg/dl)	*0.25*	*<1.0*	2.5–8.0
Creatinine (mg/dl)	1.0	**1.7**	0.7–1.4
T bilirubin (mg/dl)	0.2	0.1	0.1–1.4
BUN (mg/dl)	10	**23.0**	10–20
Chloride (meq/liter)	105	**112**	95–105
CO_2 (meq/liter)	30	27	24–32
K (meq/liter)	4.2	**5.6**	3.5–5.0
Na (meq/liter)	**152**	**154**	135–145
Alk Phos (IU/liter)	**>350**	**110**	30–85
CPK (IU/liter)	**230**	—	0–110
LDH (IU/liter)	**570**	—	60–200
SGOT (IU/liter)	**147**	**213**	0–41

[a]Values **higher** than human are in **boldface**, those *lower* than human in *italics*.

erythrocytes were smaller, flat ovals. Table 25.6 shows the RBC parameters. The oral RBCs were small (MCV ±50) with low hemoglobin (MCH ±25). Total WBC counts were higher than those found in humans, but differential counts were remarkable only for the high numbers of eosinophils found in camel No. 2 and llama. Platelets were small and numerous.

3.8. Platelet Activities

Table 25.7 shows that platelets from camel No. 2 and the llama aggregated with ADP, collagen, and ristocetin (Risto). Thrombin did not aggregate camel platelets.

3.9. Platelet Ultrastructure

Figure 25.3 shows platelets from the llama. The appearance and internal structure were similar to those of human platelets.

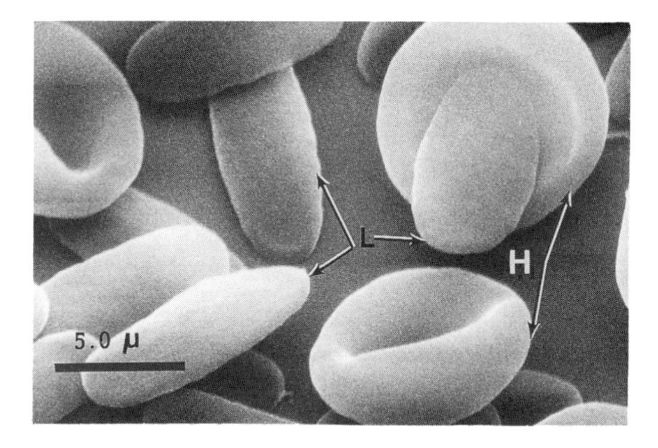

FIGURE 25.1. SEM of a mixture of llama and human erythrocytes.

4. SUMMARY

Blood samples from two camels, one guanaco, and one llama were studied. All samples clotted promptly; the clots retracted and did not lyse. Prothrombin times and APTTs were shorter than human. Plasma thromboplastin was generated effectively. Coagulation factor assays fell in the high to high-normal ranges as compared to human. Serum proteins were similar to human. Biochemical tests showed high levels of Ca, P, Na, and the enzymes Alk Phos, CPK, LDH, and SGOT.

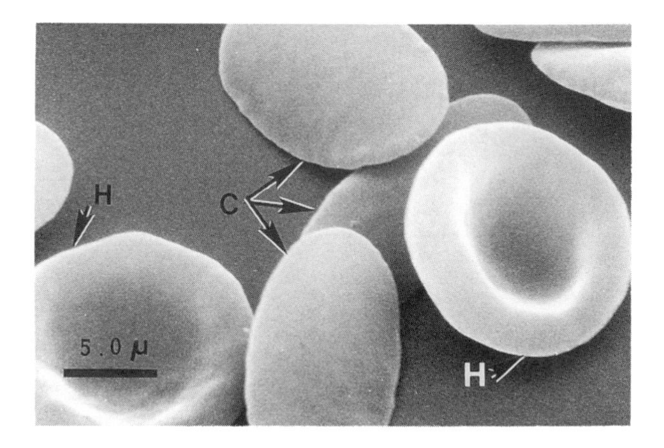

FIGURE 25.2. SEM of a mixture of camel and human erythrocytes.

TABLE 25.6
Cellular Elements of the Peripheral Blood in Camelidae

Element	Camel[a]	Guanaco	Llama	Normal human range
HCT (%)	30	32	32	37–52
HGB (g/dl)	13.6	16.4	16.3	12–18
RBC (× 10^6/mm^3)	5.9	7.6	5.9	4.2–6.2
MCV (fl)	51	40	54	79–97
MCH (pg)	23	22	28	27–31
MCHC (g/dl)	45	51	51	32–36
WBC (× 10^3/mm^3)	16.8	17.5	13.9	5–10
Neutr (%)	53	93	55	55–75
Lymph (%)	38	5	35	20–40
Mono (%)	1	0	0	2–10
Eos (%)	6	2	10	0–3
Bas (%)	1	0	0	0–3
Plat C (× 10^3/mm^3)	260	520	680	150–450

[a]Mean of camels No. 1 and 2.

Erythrocytes were flat ellipsoids containing large amounts of hemoglobin (MCHC > 50 g/dl). White cells did not differ measurably from human.

Platelets from Camelidae aggregated with ADP, collagen, or Risto. Aggregation with thrombin (bovine), and snake venom was essentially absent. By electron microscopy, the animal platelets showed the usual ultrastructural components: microtubules, alpha granules, dense bodies, open canalicular system, and glycogen particles.

TABLE 25.7
Platelet Activities in Camelidae

Activity	Camels (mean of 2)	Llama	Human	
			Control	Normal range
Plat Glass Ret Ind (%)	26	68	—	>20
Aggregation (%)				
ADP (10 μM)	80	100	100	70–100
Collagen (0.19 mg/ml)	70	50	90	80–100
Risto (0.9 mg/ml)	40	100	100	80–100
Bov Thr (0.5 U/ml)	0	—	—	60–100
CHH (0.05 mg/ml)	0	50	80	60–100
Bov Fib (0.1 %)	—	0	100	60–100

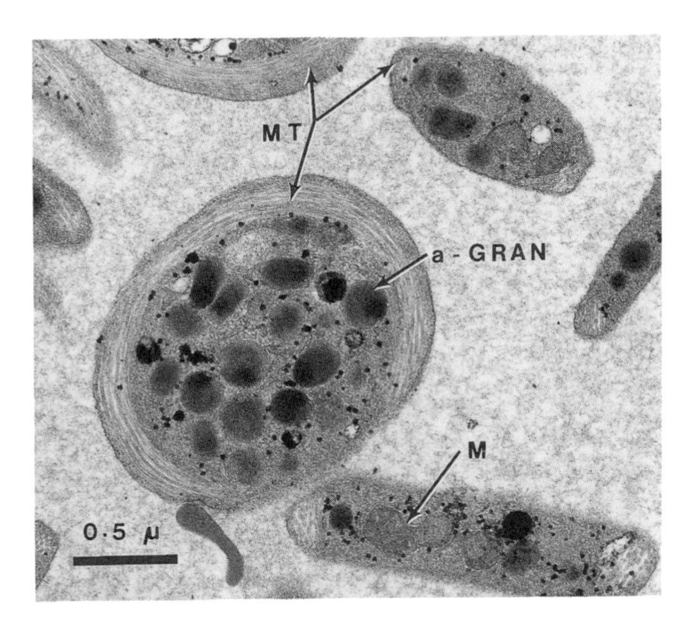

FIGURE 25.3. TEM of llama platelets. Internal structures: (MT) circumferential microtubules; (a-GRAN) alpha granules; (M) mitochondrion.

REFERENCES

Gulliver, G., 1875, Observations on the size and shape of the red corpuscles, *Proc. Zool. Soc. London* **1875:**474.

Jain, N. C., and Keeton, K. S., 1974, Morphology of camel and llama erythrocytes as viewed with the scanning electron microscope, *Br. Vet. J.* **130:**288.

Lewis, J. H., 1976, Comparative hematology—studies on Camelidae, *Comp. Biochem. Physiol. A* **55:**367.

Ponder, E., Yeager, J. F., and Charipper, H. A., 1928, Haematology of the Camelidae, *Zool. Sci. (N. Y. Zool. Soc.),* **11:**1.

SUGGESTED READINGS

Banerjee, S., Bhattacharjee, R. C., and Singh, T. I., 1962, Haematological studies in the normal adult Indian camel (*Camelus dromedarius*), *Am. J. Physiol.* **203:**1185.

Bokori, J., 1974, Contribution to the haemograms of the buffalo and of the camel, *Acta Vet. Hung.* **24:**73.

Cornelius, C. E., and Kaneko, J. J., 1962, Erythrocyte life span in the guanaco, *Science* **137:**673.

Elamin, F. M., and Saha, N., 1980, Blood protein polymorphism in the one-humped camel (*Camelus dromedarius*) in the Sudan, *Anim. Blood Groups Biochem. Genet.* **11:**39.

Kohli, R. N., 1963, Cellular micrometry of camel's blood, *Indian Vet. J.* **40:**134.
Kraft, H.; 1957, Untersuchungen über das Blutbild der Camelen, *Tierarztl. Wochenschr.* **70:**371.
Kushner, H. F., 1938, Composition of the blood of camels in relation to working ability of these mammals, *C. R. (Dokl.) Acad. Sci. USSR N.S.* **18:**681.
Little, A., McKenzie, A. J., Morris, R. J. H., Roberts, J., and Evans, J. V., 1970, Blood electrolytes in the Australian camel, *Aust. J. Exp. Biol. Med. Sci.* **48:**17.
Perk, K., 1963, The camel's erythrocyte, *Nature (London)* **200:**272.
Ponder, E., Yeager, J. F., and Charipper, H. A., 1929, Studies in comparative hematology I. Camelidae, *Q. J. Exp. Physiol.* **19:**115.
Soliman, M. K., and Shaker, M., 1967, Cytological and biochemical studies on the blood of adult she camels, *Indian Vet. J.* **44:**989.
Soni, B. K., and Aggarwala, A. C., 1958, Studies on the physiology of the camel (*Camelus dromedarius*). I. Cellular blood constituents, *Indian Vet. J.* **35:**209.
Yagil, R., Sod-Moriah, U. A., and Meyerstein, N., 1974, Dehydration and camel blood. I. Red blood cell survival in the one-humped camel *Camelus dromedarius, Am. J. Physiol.* **226:**298.
Yagil, R., Sod-Moriah, U. A., and Meyerstein, N., 1974, Dehydration and camel blood. II. Shape, size, and concentration of red blood cells, *Am. J. Physiol.* **226:**301.

CHAPTER 26

THE HOLLOW-HORNED DOMESTIC RUMINANTS

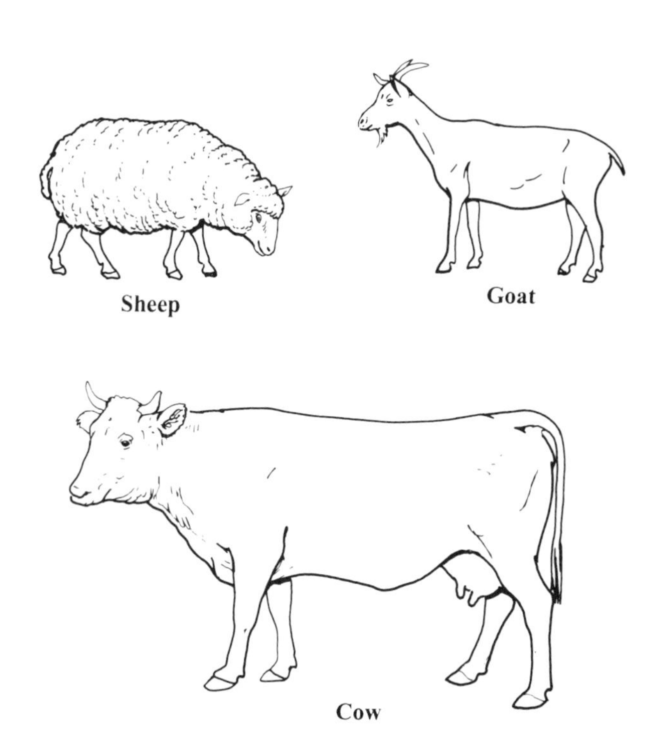

1. INTRODUCTION

In tropical areas, early man developed from hunter–gatherer to stockman and farmer about 9000 years ago. In temperate zones, this major ecological change did not take place until after the last Ice Age—about 5000 years ago.* Man cultivated his own grains and vegetables and provided food and shelter for animals that, in turn, provided valuable goods for man. Ruminants—cows, goats, and sheep—were among the earliest animals to be domesticated and became important in providing milk, meat, and hides for clothing and tents. The ruminants are large quadrupeds with four-chambered stomachs. Grass or hay is cut off with the front incisors, partially ground with the molars, and swallowed. Later it is regurgitated, reground to a fine mix (chewed cud), and reswallowed into the second section of the stomach, from which it is digested. The females are usually pregnant or producing milk for an offspring or for the farmer.

Veterinary testbooks describe the cellular elements, proteins, and blood chemistries of various strains of cows, goats, and sheep in health and disease.

There are many coagulation studies in cattle. An acquired bleeding disorder that occurred in cattle after ingestion of spoiled sweet clover was described by Schofield (1922). Roderick (1931) blamed a fall in prothrombin as the cause of the bleeding. Campbell and Link (1941) isolated and identified the causative agent in spoiled clover as bishydroxy-coumarin. The fascinating story of the synthesis of a number of coumarin-related compounds and their abilities to inhibit the production of factors II, VII, X, and IX by blocking vitamin K activity in the liver has been described by Link (1959). Some of these derivatives are extremely valuable for the treatment and prevention of human thrombotic diseases and as rat and mouse poisons.

Two hereditary hemorrhagic disorders have been recognized in ruminants: afibrinogenemia in goats and factor XI deficiency or dysfunction in cows [see Chapter 27 (Section 3)]. Material from a study on goats (Lewis, 1976) is reproduced here with the kind permission of the editor of the *American Journal of Veterinary Research*.

2. SOURCE

Blood was obtained from seven lactating cows (Jerseys) (*Bos taurus*), weighing 250–400 kg, at the Department of Veterinary Science, Penn State

*The end of the Ice Age has been ascribed to times varying from 5,000 to 18,000 years ago.

University, State College, Pennsylvania. Three non-pregnant female and three male goats (*Capra musimon*), weighing 45–60 kg, were purchased from a local animal dealer and housed in the animal laboratories of the University of Pittsburgh Medical School. Blood samples were obtained from the jugular veins of cows and goats. Ten pregnant sheep (*Ovis aegagras*) weighing 60–80 kg, were from three different strains: Polipay, Dorset, and Shropshire. The sheep were to be the subjects of obstetrical research, and preliminary blood samples from the vena cava or jugular veins were obtained for us by Anthony Batteli, from the animal research facility of Magee Womens Hospital, Pittsburgh, Pennsylvania.

3. TESTS CONDUCTED

3.1. Bleeding Times

Bleeding times were not done on cattle or sheep. On goats, using a new No. 10 Bard-Parker blade, small cuts were made on the less hairy inguinal areas lateral to the femoral artery. Two cuts (right and left) were made on each of four goats. The bleeding times were 3 1/2, 2, 2, 2, 6, 4 1/2, 3, and 3 1/2 min (mean: 3.34 min).

3.2. General Coagulation Tests and Factors

Table 26.1 shows the results obtained in six cows, six goats, and ten sheep. Coagulation times were short in glass or siliconized tubes, and the clots retracted promptly and well and did not lyse. SPTs were long (normal for humans). Prothrombin times on cow plasma with Simplastin, cow brain, and RVV and the APTTs were all long. On goat and sheep, the prothrombin times with Simplastin and RVV and the APTTs were shorter and essentially within the human range. Goat brain was not prepared. Sheep brain homogenate in the prothrombin time test gave about the same times for sheep and human plasmas. Recal Ts were longer than human in the cow, but shorter in goat and sheep. Thrombin times in cow and goat were longer than human. Sheep plasma clotted in the thrombin, CHH, and Atroxin time tests slightly faster than human. The TGT (Table 26.2) was done on goat and sheep materials. The results with goat were longer than human, while with sheep ingredients they were just above the human range.

Coagulation factor results are shown in Table 26.3. Cow II and VII, goat II, and sheep II, VII, and X assayed below 0.5 U/ml. Cow V and VIII, goat V, VIII, IX, and XII, and sheep V, VIII, and IX assayed above 2.0 U/ml.

TABLE 26.1
General Coagulation Tests in the Hollow-Horned Domestic Ruminants

	Cow			Goat			Sheep			Human	
Test	*N*	Mean	SD	*N*	Mean	SD	*N*	Mean	SD	Controls (3) (mean)	Normal range
Clot T (min)											
Glass	6	7	1.3	6	6	0.5	9	7	2.1	8	6–12
Silicone	6	11	1.2	6	19	4.4	9	12	2.6	30	20–59
Clot Retr	6	3–4	—	6	4–4	—	9	4–4	—	4–4	3–4+
Clot Lys	6	0	0	6	0	0	9	0	0	0	0
SPT (sec)											
Human	6	29.0	2.7	6	50.6	0.8	9	23.9	0.4	44.6	20+
Homolog	6	65.0	6.6	6	33.4	0.4	5	27.3	3.7	60+	—
PT (sec)											
Simplastin	6	22.6	0.9	6	11.7	0.5	10	13.6	1.2	11.1	10–13
Homolog Br	6	32.3	2.3	—	—	—	5	20.2	0.8	[a]	—
RVV	6	24.7	3.4	3	18.5	1.1	9	21.7	2.8	16.1	13–18
APTT (sec)	6	59.6	8.9	6	32.4	3.3	10	26.3	4.8	33.9	24–34
Recal T (sec)	6	235	30	6	94	3.1	6	98	6.1	128	90–180
Lys	6	0	0	6	0	0	6	0	0	0	0
Lys MCA	6	0	0	6	0	0	6	0	0	0	0
Thromb T	6										
Bov Thr	6	21.8	11	6	27.0	5.0	10	15.3	1.5	15.5	11–18
CHH	—	—	—	—	—	—	5	13.0	0.7	14.7	11–18
Atroxin	—	—	—	—	—	—	5	13.6	0.2	15.1	11–18

[a]On human plasma: cow brain, 26.8 sec; sheep brain 20.9 sec.

TABLE 26.2
Thromboplastin Generation Test in the Hollow-Horned Domestic Ruminants

Generating mixture			Substrate Clot T (sec)[a]		
Human	Goat	Sheep	H	G	S
	Means				
H (3)	—	—	**9.6**	10.1	11.4
—	G (6)	—	16.1	**18.2**	—
—	—	S (5)	11.5	—	**11.4**

[a]Homologous systems are in **boldface**.

3.3. Fibrinolytic Enzyme System

The fibrinolytic system in the ruminants was activated poorly or not at all by streptokinase (SK) (100 and 500 U/ml) or urokinase (UK) (50 U/ml). Staphylokinase (Staph K) (2%) activated both goat and sheep plasminogen to about the same extent as human.

3.4. Serum Proteins

Table 26.4 shows total protein and paper electrophoretic distribution in the ruminants. Cow electrophoresis usually did not discriminate α_1 and α_2, but

TABLE 26.3
Coagulation Factors in the Hollow-Horned Domestic Ruminants

	Cow			Goat			Sheep			Normal human range
Factor	*N*	Mean	SD	*N*	Mean	SD	*N*	Mean	SD	
I (mg/dl)	6	180	20	6	336	66	10	272	79	150–450
II (U/ml)	6	0.42	0.1	6	0.62	0.2	10	0.62	0.1	0.70–1.30
V (U/ml)	6	2.0+	—	6	6.45	1.4	10	2.0+	—	0.65–1.45
VII (U/ml)	6	0.19	0.1	6	1.63	0.2	10	0.36	0.1	0.50–1.30
X (U/ml)	6	1.09	0.3	6	1.75	0.2	10	0.44	0.1	0.75–1.25
VIII (U/ml)	6	2.0+	—	6	12.8	2.4	10	2.0+	—	0.75–1.40
IX (U/ml)	6	0.76	0.1	6	3.33	1.8	10	2.0+	—	0.65–1.70
XI (U/ml)	6	1.01	0.1	6	1.48	0.3	8	0.98	0.3	0.75–1.40
XII (U/ml)	6	0.97	0.2	6	2.47	0.3	9	0.98	—	0.50–1.45
ATIII (U/ml)	6	0.93	0.5	6	0.91	0.2	4	1.05	0.1	0.80–1.20
XIII (R)	6	9.3	5.5	6	24	8.8	6	19	11	4–16

TABLE 26.4
Serum Proteins by Paper Electrophoresis in the Hollow-Horned Domestic Ruminants

	Cow			Goat			Sheep			Normal human range
Protein	*N*	Mean	SD	*N*	Mean	SD	*N*	Mean	SD	
TP (g/dl)	6	7.2	0.6	6	7.4	1.4	7	7.5	1.1	6.3–7.9
Albumin (g/dl)	6	2.9	0.2	4	4.0	0.6	7	4.1	0.8	3.2–4.4
α_1-Globulin (g/dl)	6	1.1	0.1	4	0.2	0.2	7	0.2	0.2	0.2–0.4
α_2-Globulin (g/dl)	[a]	[a]	[a]	4	0.7	0.3	7	0.7	0.3	0.4–1.0
β-Globulin (g/dl)	6	1.1	0.2	4	0.5	0.1	7	0.5	0.1	0.5–1.0
γ-Globulin (g/dl)	6	2.1	0.4	4	2.0	0.5	7	2.0	0.4	0.6–1.8

[a]There is a single α_1–α_2-globulin peak in cows.

showed a single or run-together α peak. The γ-globulins were high in all three ruminants.

3.5. Biochemical Tests

Table 26.5 shows the results obtained in seven cows, three goats, and one sheep. The albumin, cholesterol, glucose, and uric acid were below the human range. P and SGOT were high.

3.6. Miscellaneous Tests

Staphylococcal clumping did not occur when cow, goat, or sheep plasma was mixed with the standard staphylococcal preparation. The reason for this result is unknown.

3.7. Cellular Elements

Table 26.6 shows the blood counts found in these groups of animals. All have low HCTs, high red counts, and moderate HGBs. These data mean that the RBC indices calculate to give the ruminants low MCVs and MCHs. In fact, goats have the smallest erythrocytes of all common vertebrates. Figure 26.1 is an SEM of a mixture of goat and human erythrocytes. The goat cells are very much smaller than human RBCs [see Chapter 28 (Section 1)].

TABLE 26.5
Biochemical Tests in the Hollow-Horned Domestic Ruminants

Substance	Cow			Goat			Sheep		Normal human range
	N	Mean[a]	SD	*N*	Mean[a]	SD	*N*	Mean[a]	
TP (g/dl)	7	7.7	0.6	3	6.7	0.5	1	7.3	6.0–8.0
Albumin (g/dl)	7	*2.4*	0.7	3	*2.7*	0.2	1	*3.0*	3.5–5.0
Ca (mg/dl)	7	9.5	0.6	3	9.6	0.4	1	10.3	8.5–10.5
P (mg/dl)	7	**5.4**	2.0	3	**7.1**	0.8	1	**5.2**	2.5–4.5
Cholesterol (mg/dl)	7	*128*	50	3	*76*	5.5	1	*92*	150–200
Glucose (mg/dl)	7	*59*	13	3	*55*	9.5	1	*40*	65–110
Uric acid (mg/dl)	7	*1.3*	0.4	—	—	—	1	*Low*	2.5–8.0
Creatinine (mg/dl)	7	1.1	0.2	3	*0.6*	0.1	1	0.8	0.7–1.4
T bilirubin (mg/dl)	7	0.2	0.1	3	0.2	0.1	1	1.0	0.1–1.4
BUN (mg/dl)	7	15	5.1	3	16	0.1	1	*8.7*	10–20
Chloride (meq/liter)	7	98	2.7	3	**108**	0.8	1	103	95–105
CO_2 (meq/liter)	7	30	4.1	3	25	21	1	*21*	24–32
K (meq/liter)	7	5.0	0.6	3	4.6	0.7	1	4.9	3.5–5.0
Na (meq/liter)	7	143	3.5	3	**146**	0	1	142	135–145
Alk Phos (IU/liter)	7	65	44	3	**173**	37	1	37	30–85
HBD (IU/liter)	6	**1803**	140	—	—	—	—	—	14–185
CPK (IU/liter)	6	36	13	—	—	—	—	—	0–110
LDH (IU/liter)	6	**1032**	324	—	—	—	—	—	60–200
SGOT (IU/liter)	7	**94**	24	3	**107**	23	1	**52**	0–41
SGPT (IU/liter)	6	26	6.1	—	—	—	—	—	10–50

[a]Values **higher** than human are in **boldface**, those *lower* than human in *italics*.

TABLE 26.6
Cellular Elements of the Peripheral Blood in the Hollow-Horned Domestic Ruminants

Elements	Cow			Goat			Sheep			Normal human range
	N	Mean	SD	*N*	Mean	SD	*N*	Mean	SD	
HCT (%)	6	34	2.9	6	34	4.9	9	37	4.4	37–52
HGB (g/dl)	6	12.3	1.2	3	12.7	1.5	9	10.3	1.1	12–18
RBC ($\times 10^6$/mm^3)	6	6.5	0.6	5	14.5	2.9	9	9.9	1.1	4.2–6.2
MCV (fl)	6	53	1.6	5	23	2.2	9	37	6.4	79–97
MCH (pg)	6	19	0.5	3	8	0.4	9	11	1.6	27–31
MCHC (g/dl)	6	36	0.5	3	34	1.5	9	28	5.6	32–36
WBC ($\times 10^3$/mm^3)	6	7.9	1.4	6	13.3	2.7	9	7.9	2.6	5–10
Neutr (%)	6	34	4.5	5	44	6.7	9	31	5.2	55–75
Lymph (%)	6	53	4.1	5	51	11	9	58	5.1	20–40
Mono (%)	6	5	2.1	5	2	0.4	9	1	0.9	2–10
Eos (%)	6	7	1.4	5	2	0.9	9	9	3.7	0–3
Bas (%)	6	1	0.8	5	1	1	9	1	0.9	0–3
Plat C ($\times 10^3$/mm^3)	6	350	150	7	551	93	9	479	122	150–450

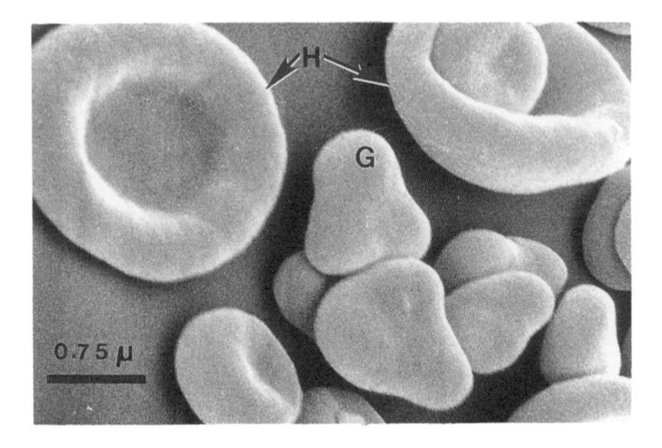

FIGURE 26.1. SEM of mixed goat (G) and human (H) erythrocytes.

Ruminant white cell counts were not remarkable, and the three species all had slightly greater numbers of lymphocytes than of neutrophils. Platelet counts were in the high human range, and the platelets were small.

3.8. Platelet Activities

As Table 26.7 shows, the platelet glass retention for all the ruminants was within the normal human range. Aggregation with ADP was also within the human range. With collagen, aggregation was less than human in cows and sheep. With ristocetin (Risto), it was very low in cows and goats and low-normal in sheep. Arachidonic acid was tested only on sheep and was low-normal. Thrombin and CHH aggregated goat and sheep platelets. Pig plasma was tested only on sheep and had little activity.

3.9. Platelet Ultrastructure

Figures 26.2, 26.3, and 26.4 show TEMs of ruminant platelets. The three species were very similar. All showed marginal bundles of microtubules, an open canalicular system, alpha granules, dense bodies (usually in vacuoles), and fine glycogen particles. An occasional mitochondrion with cristae was seen.

TABLE 26.7
Platelet Activities in the Hollow-Horned Domestic Ruminants

Activity	Cow			Goat			Sheep			Human	
	N	Mean	SD	*N*	Mean	SD	*N*	Mean	SD	Control[a]	Normal range
Plat Glass Ret Ind (%)	6	56	8.8	4	64	10	3	46	6.8	41	>20
Aggreation (%)											
ADP (20 μM)	—	—	—	—	—	—	4	75	17	85	70–100
(10 μM)	6	87	11	4	98	5.0	4	75	23	87	70–100
(5 μM)	—	—	—	—	—	—	3	77	29	80	50–100
(2.5 μM)	—	—	—	—	—	—	3	63	28	28	40–100
Collagen (0.19 mg/ml)	6	55	23	4	80	17	4	10	3.7	76	80–100
Risto (0.9 mg/ml)	4	0	0	4	12	0	4	43	25	83	80–100
Risto ½ (0.45 mg/ml)	—	—	—	—	—	—	4	10	2.5	6	<18
Arach A (0.5 mg/ml)	—	—	—	—	—	—	4	49	95	80	60–100
Bov Thr (0.3 U/ml)	—	—	—	4	100	0	4	67	22	Clot	60–100
(0.2 U/ml)	—	—	—	—	—	—	4	9	9.1	Clot	60–100
CHH (0.05 mg/ml)	—	—	—	1	100	0	4	75	13	80	60–100
Pig Pl (1 : 10)	—	—	—	—	—	—	4	11	8.9	83	60–100

[a]Control for sheep.

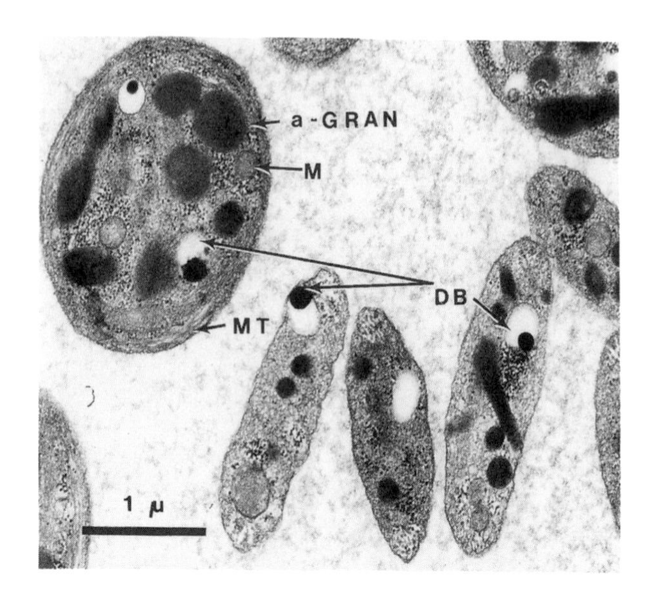

FIGURE 26.2. TEM of cow platelets. Internal structures here and in Figs. 26.3 and 26.4: (a-GRAN) alpha granules; (DB) dense bodies; (gly) glycogen particles; (M) mitochondria; (MT) microtubules.

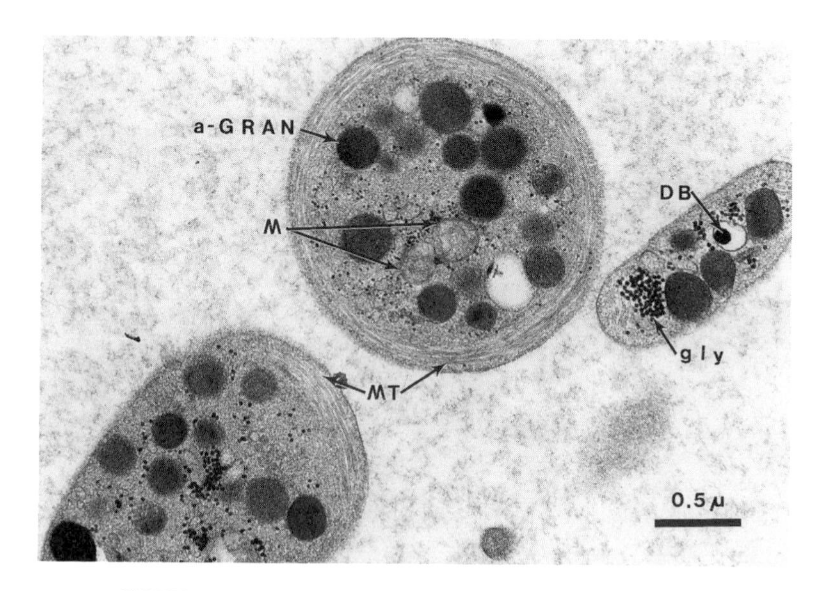

FIGURE 26.3. TEM of goat platelets. Key in Fig. 26.2 caption.

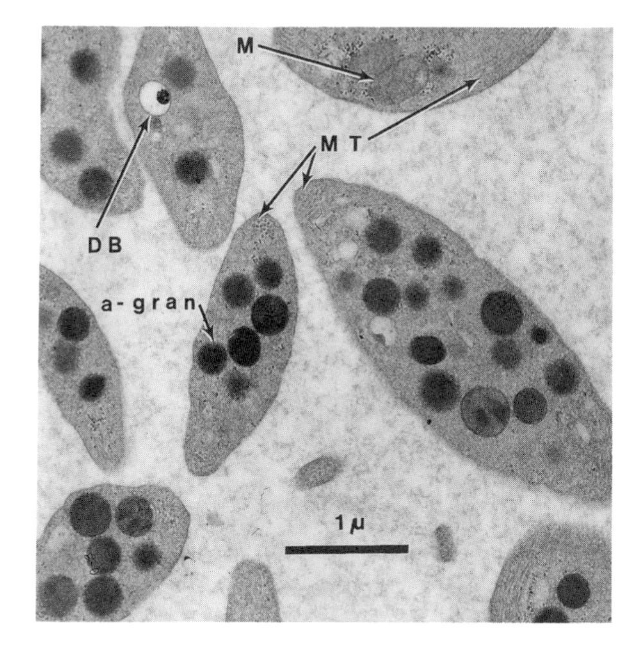

FIGURE 26.4. TEM of sheep platelets. Key in Fig. 26.2 caption.

4. SUMMARY

Coagulation of ruminant blood was prompt; clots retracted and did not lyse. PTs, APTTs, and thrombin times were short. Cow II and VII, goat II, and sheep II, VII, and X were low, and other factors were moderate to high. The fibrinolytic systems did not activate with SK or UK, but did with Staph K. Ruminant plasma did not clump staphylococcal preparations.

Erythrocytes were small and numerous. Platelets appeared similar to human on electron microscopy, but aggregations were dissimilar, with little response to Risto and pig plasma where tested.

REFERENCES

Campbell, H. A., and Link, K. P., 1941, Studies on the hemorrhagic sweet clover disease. IV. The isolation and crystallization of the hemorrhagic agent, *J. Biol. Chem.* **138:**21.

Lewis, J. H., 1976, Comparative hematology: Studies on goats, *Am. J. Vet. Res.* **37:**601.

Link, K. P., 1959, The discovery of dicumarol and its sequels, *Circulation* **19:**97.

Roderick, L. M., 1931, A problem in the coagulation of the blood: "Sweet clover" disease of cattle, *Am. J. Physiol.* **96:**413.

Schofield, F. W., 1922, A brief account of a disease in cattle simulating hemorrhagic septicemia due to feeding sweet clover, *Can. Vet. Rec.* **3:**74.

SUGGESTED READINGS

Ruminants: General

Fraser, A. C., 1929–1930, A study of the blood of cattle and sheep in health and disease, *Univ. Cambridge Inst. Anim. Pathol. Rep. Dir.* **1:**114.

Cattle

Alencar, R. A., de Fo Penha, A. M., and Cintra, L. C., 1971, Blood picture of acclimatized Dutch red-and-white cattle, *Biologico* **37:**272.

Casillas, G., Simonetti, C., Vasquez, C., and Pavlovsky, A., 1976, Physical, chemical and immunological studies on bovine factor VIII, *Haemostasis* **5:**1.

Clawson, A. B., 1914, Some results of blood counting on cattle, *Am. Vet. Rev.* **45:**527.

Culbertson, R., Jr., Abilgaard, C. F., Harrison, J., Jettcoat, K., and Osburn, B. L., 1979, Ontogeny of bovine hemostasis, *Am. J. Vet. Res.* **40:**1402.

Dimock, W. W., and Thompson, M. C., 1906, Clinical examination of the blood of normal cattle, *Am. Vet. Rev.* **30:**553.

Ferguson, L. C., Irwin, M. R., and Beach, B. A., 1945, On variation in the blood cells of healthy cattle, *J. Infect. Dis.* **76:**24.

Heyns, H., 1971, The effect of breed on the composition of the blood, *J. Agric. Sci.* **76:**563.

Holman, H. H., 1955, The blood picture of the cow, *Br. Vet. J.* **111:**440.

Jones, W. H., Hughes, C. D., Swenson, M. J., and Underbjerg, G. K. L., 1956, Plasma prothrombin time and haematocrit values of blood of dairy cattle, *Proc. Soc. Exp. Biol. Med.* **91:**14.

Kimeto, B. A., 1976, Ultrastructure of blood platelets in cattle with East Coast fever, *Am. J. Vet. Res.* **37:**443.

Marzec, U., Johnston, G. G., and Bernstein, E. F., 1975, Platelet function in calves: A study of adhesion, aggregation, release, clotting factor activity and life span, *Trans. Am. Soc. Artif. Internal Organs* **21:**581.

Penny, R. H. C., Scofield, A. M., and Cembrowicz, H., 1966, Haematological values for the clinically normal bull, *Br. Vet. J.* **122:**239.

Ryan, G. M., 1971, Blood values in cows: Erythrocytes, *Res. Vet. Med.* **12:**572.

Ryan, G. M., 1971, Blood values in cows: Leucocytes, *Res. Vet. Med.* **12:**576.

Straub, O. C., 1981, *Advances in Veterinary Medicine,* Vol. 36, *Bovine Hematology,* p. 64, Paul Parey, New York.

Sutton, R. H., and Hobman, B., 1975, The value of plasma fibrinogen estimations in cattle, *N. Z. Vet. J.* **23:**21.

White, W. J., Greenshields, J. E. R., and Chubaty, W., 1954, The effect of feeding sweet clover silage on the prothrombin time of blood of cattle, *Can. J. Agric. Sci.* **34:**601.

Wiesner, E., Rex, J. O., and Wiesner, B., 1968, Blood coagulation in cattle fed sweet clover. II. Feeding experiments, *Am. J. Vet. Med.* **23:**788.

Winqvist, G., 1954, Morphology of the blood and the hemopoietic organs in cattle under normal and some experimental conditions, *Acta Anat. Suppl.* **21**:1.

Zajdel, M., Wagrzynowicz, Z., and Jeljaszewicz, J., 1975, Action of staphylothrombin on bovine fibrinogen, *Thromb. Res.* **6**:501.

Goats

Bhalla, N. P., Bhalla, R. C., and Sharma, G. L., 1966, Haematological values in healthy hill goats, *Indian J. Vet. Sci.* **36**:33.

Bhargava, S. C., 1980, Haematological studies in goats, *Indian Vet. J.* **57**:485.

Fletcher, W. S., Rogers, A., and Donaldson, S. S., 1964, The use of the goat as an experimental animal, *Lab. Anim. Care.* **14**:65.

Gonzaga, A. C., and Novilla, M. N., 1962, The mean corpuscular hemoglobin concentration in native goats, *Philipp. J. Vet. Med.* **1**:105.

Holman, H. H., and Dew, S. M., 1964, The blood picture of the goat. II. Changes in erythrocytic shape, size and number associated with age, *Res. Vet. Sci.* **5**:274.

Holman, H. H., and Dew, S. M., 1965, The blood picture of the goat. IV. Changes in coagulation times, platelet counts and leucocyte numbers associated with age, *Res. Vet. Sci.* **6**:510.

Houchin, O. B., Graham, W. R., Jr., Peterson, V. E., and Turner, C. W., 1939, The chemical composition of the blood of the dairy goat, *J. Dairy Sci.* **22**:241.

Mews, A., and Mowlem, A., 1982, Normal haematological and biochemical values in the goat, *Dairy Goat J.* **60**:147.

Vaidya, M. B., Vaghari, P. M., and Patel, B. M., 1970, Haematological constituents of the blood of goats, *Indian Vet. J.* **47**:642.

Wilkins, J. H., and Hodges, R. E. D. H., 1962, Observations on normal goat blood, *R. Army Vet. Corps. J.* **33**:7.

Sheep

Becker, D. E., and Smith, S. E., 1946, A chemical and morphological study of normal sheep blood, *Cornell Vet.* **36**:25.

Fantl, P., and Ward, H. A., 1960, Clotting activities of maternal and foetal sheep blood, *J. Physiol.* **150**:607.

Gajewski, J., and Povar, M. L., 1971, Blood coagulation values of sheep, *Am. J. Vet. Res.* **32**:405.

Holman, H. H., 1944, Studies on the haematology of sheep. I. The blood picture of healthy sheep, *J. Comp. Pathol.* **54**:26.

Holman, H. H., 1945, Studies on the haematology of sheep. IV. Erythrocytic and thrombocytic pictures, and variations in physical attributes, *J. Comp. Pathol. Ther.* **55**:146.

Hudson, A. E. A., and Osborn, J. C., 1954, A note on certain blood values of adult sheep, *Vet. Med.* **49**:423.

Jones, D. C. L., and Krebs, J. S., 1972, Hematologic characteristics of sheep, *Am. J. Vet. Res.* **33**:1537.

Linton, J. H., Goplen, B. P., Bell, J. M., and Jaques, L. B., 1963, Dicoumarol studies. II. The prothrombin time response of sheep to various levels of contamination in low coumarin sweet clover varieties, *Can. J. Anim. Sci.* **43**:353.

Loly, W., Israels, L. G., Bishop, A. J., and Israels, E. D., 1971, A comparative study of adult and fetal sheep fibrinogen, sulf-fibrinogen and fibrinogen degradation products, *Thromb. Diath. Haemmorrh.* **26:**526.

Morag, M., and Robertson Smith, D., 1971, The day to day variations in certain blood values and in blood coagulation times of healthy ewes, *Refu. Vet.* **28:**74.

Osbaldiston, G. W., and Hoffman, M. W., 1971, Blood coagulation values in normal sheep and in two mutant strains with hyperbilirubinemia, *Can. J. Comp. Med.* **35:**150.

Reda, H., and Hathout, A. F., 1957, The haematological examination of the blood of normal sheep, *Br. Vet. J.* **113:**251.

Rees Evans, E. T., 1968, Sickling phenomenon in sheep, *Nature (London)* **217:**74.

Smith, M. L., Lee, R., Sheppard, S. J., and Farris, B. L., 1978, Reference ovine serum chemistry values, *Am. J. Vet. Res.* **39:**321.

Stubbs, E. L., and Boyer, C. I., 1954, Studies of sheep blood, *Vet. Ext. Q. Univ. Pa.* **136:**20.

Tillman, P., Carson, S. N., and Talken, L., 1981, Platelet function and coagulation parameters in sheep during experimental vascular surgery, *Lab. Anim. Sci.* **31:**263.

Ullrey, D. E., Miller, E. R., Long, C. H., and Vincent, B. H., 1965, Sheep hematology from birth to maturity. I. Erythrocyte population, size and hemaglobin concentration, *J. Anim. Sci.* **24:**135.

Winter, H., 1964, The myelogram of normal sheep, *J. Comp. Pathol. Ther.* **74:**457.

PART III

COMPARATIVE STUDIES

CHAPTER 27

COMPARATIVE HEMOSTASIS

1. RESULTS OF ANIMAL TESTS

1.1. Bleeding Times

The bleeding time is the time measured with a stopwatch between the making of a small incision in the skin and the moment the bleeding stops. Duke in 1910 was the first to describe, in humans, a bleeding time involving a cut on the fingertip or earlobe. Ivy in 1941 standardized the procedure to the making of a shallow cut with a new knife blade on the ventral surface of the forearm with a sphygmomanometer cuff on the upper arm inflated to 30–40 mm Hg. Mielka *et al.* (1969) developed a template to standardize the length and depth of the cut. Currently, most bleeding times done in humans involve a perpendicular cut on the cleaned volar surface of the forearm in an area without visible veins. A commercially available template, Simplate (**OT**), is most frequently used.

For animals, the template may be used if hair or feathers are sparse. Hair may be removed by applying the depilatory Nair. This is best done 1 day before the test so that any inflammation can subside. If no skin areas are available, the tongue may be used by making small cuts on the edge with short, sharp scissors.

Table 27.1 lists the 35 species on which bleeding times were performed, the number of animals of each species tested, the mean bleeding time, the range, the instrument used, the use of a sphygmomanometer cuff, the use of Nair, and the site of the test. Animals not listed were not tested. Only the baboon bleeding time range was longer than the human range of 3–9½ min.

1.2. Clotting Time, Clot Retraction, Clot Lysis, and Clot Recalcification Time

The whole blood clotting time (Clot T) (see Chapter 2) measured in glass or silicone-coated tubes is an old test that originated in the early years of this century. It serves as an index of the capability of shed blood to form a fibrinous clot. That clotting does not necessarily play an essential role in hemostasis is evidenced by the long clotting times but lack of excessive bleeding associated with certain coagulation deficiencies or defects (D/Ds) in humans (D/Ds of factors XII, Fletcher, Fitzgerald) and other mammals (D/D of factor XI in bats and of factor XII in Cetacea) as well as the long clotting times in many premammalian vertebrates. Clot T tests are of special value because extraneous materials are not purposefully introduced; thus, problems of species specificity are not involved. The surface on which the blood is clotted is important. Glass serves to activate the early contact system in mammalian plasma. The siliconized surface is more like the nautral endothelial surface than is glass. Various plastic surfaces are avoided for clotting tests because they may inhibit clot retraction (Clot Retr).

The only extraneous material involved in the recalcification time (Recal T) is added calcium, and optimum concentration was used whenever possible. The Recal T is dependent on platelet/thrombocyte concentration, which in turn is dependent on the speed and duration of centrifugation used in preparing the plasma. This is standardized at 2000 *g* for 10 min. The Recal clots are observed for clot lysis (Clot Lys) and for Clot Lys in 1% monochloroacetic acid (MCA). Lack of dissolution in MCA indicates the presence of factor XIII (fibrin-stabilizing factor) in the clot.

The major source of error in these tests was the introduction of tissue fluid into the blood during the phlebotomy. Almost any tissue fluid contains tissue factor or thromboplastin and will accelerate blood clotting.

Obviously, blood obtained by drawing it from large animals with large blood vessels using large needles and discarding the first blood to be shed was most likely to be free of tissue factor. For small animals, small needles were often necessary, and the amount of blood obtained was very limited; thus, it was necessary to use many animals to accomplish the studies described herein.

In normal human subjects, whole blood Clot T was 6–12 min in glass and 20–59 min in siliconized tubes. The clots retracted 3+ to 4+ and did not lyse. Recal T was 90–180 sec. Recal T clots did not dissolve in 1% MCA or alone.

In patients, short Clot Ts or Recal Ts sometimes reflected intravascular activation of coagulation. More frequently they related to technical problems with venipuncture. Long Clot Ts or Recal Ts were due to deficiencies or defects (D/D) of any coagulation factor except VII or XIII or were due to the presence of an inhibitor. Inhibitors included anti-VIII, anti-V, anti-X, heparin, or a high titer-blocking inhibitor (lupus anticoagulant). Decreased clot retraction was found in association with platelet abnormalities such as thrombocytopenia or thromocytopathia (including Glanzmann's thrombasthenia). Clot retraction was also dependent upon RBC/plasma volume ratios, being low in polycythemia and high in

TABLE 27.1
Bleeding Times

Species	N	Mean (min)	Range (min)	Method: Instrument[a]	Cuff	Nair	Site
Opossum	9	1	3/4–1 1/4	K	–	–	Abdominal area
Hedgehog	2	3 1/2	2 1/2–4	K	–	–	Abdominal area
Baboon	22	5 1/4	2 1/2–10 1/4	S	+	+	Forearm
Man	100	3 3/4	3–8	S	+	0	Forearm
Monkey	4	2 1/2	2–3	S	–	–	Forearm
Armadillo	4	3	2–4	K	–	–	Abdominal area
Rabbit	6	3 3/4	3 1/4–4 1/2	S	–	–	Ear
	6	2 3/4	2–4	Sc	–	–	Ear margin
Guinea pig	20	4	1–6	S	+	+	Forearm
Rat							
S-D	8	3 1/4	2–5	S	–	–	Abdominal area
Wistar	6	3	1–5	S	+	+	Thigh
Porpoise	1	2	2	K	–	–	Flipper
Cat	6	3 1/2	3–4	S	–	+	Forearm
Dog	6	3 1/4	2–6	S	–	–	Inguinal area
Seal	3	2	2	Sc	–	–	Flipper
Manatee	2	2	2	Sc	–	–	Flipper
Pig	6	1 3/4	0–3	S	–	–	Foreleg
Goat	4	3 1/2	2–6	K	–	–	Foreleg
Chicken	6	3/4	0–1 1/2	S	–	[b]	Breast
Turkey	3	2 1/2	2–3	K	–	–	Thigh
Loggerhead turtle	2	3 1/2	3–4	K	–	–	Toe
Snake	4	1 1/4	0–2	K	–	–	Abdominal area
Alligator	2	1	0–2	K	–	–	Abdominal area
Amphiuma	2	1/2	1/2	Sc	–	–	Abdominal area
Necturus	3	0	0	Sc	–	–	Abdominal area
Frog	4	1/2	0–1	Sc	–	–	Abdominal area
Bullfrog	6	1 1/2	0–2	Sc	–	–	Thigh
Giant toad	4	1 1/2	1/2–2	K	–	–	Thigh
Goosefish	4	1 1/4	1/2–2	K	–	–	Abdominal area
Mullet	1	1	—	S	–	–	Abdominal area
Porgy	1	1 1/2	—	S	–	–	Abdominal area
Sculpin	5	1 1/2	0–3 3/4	K	–	–	Abdominal area
Sea robin	1	1 1/2	—	S	–	–	Abdominal area
Spadefish	1	1	—	S	–	–	Abdominal area
Black-tipped shark	1	4	—	K	–	–	Abdominal area
Dogfish	6	1 1/4	0–2 3/4	K	–	–	Abdominal area
Skate	2	<1	—	K	–	–	Abdominal area
Lamprey	6	1	0–2	K	–	–	Abdominal area

[a]Instrument: (K) knife; (S) Simplate; (Sc) surgical scissors. [b]Plucked.

anemia. Recal Clot Lys in 1% MCA was found in lack of factor XIII or presence of anti-factor XIII. Positive whole blood Clot Lys or Recal Clot Lys was found only after severe stress, trauma, sudden death, transplant surgery, or in severe liver disease.

Table 27.2 shows the Clot Ts, Clot Retr, Clot Lys, and Recal Ts, Lys, and MCA Lys for most of the animals on which the tests were done. The Clot Ts in glass or siliconized tubes are the means for the animals tested. Clot Retr was graded visually (0–4+), and the highest was reported. After 24 hr, each tube was examined, and if the blood had clotted and the clot was completely gone, lysis was recorded. The number of animal samples that lysed vs. the number observed is entered (e.g., $^2/_5$ = 2 of 5 chimpanzee samples). Lys MCA was recorded at 1 hr after the clot was placed in MCA.

Mammalia: Glass Clot Ts were long with porpoise and borderline with horse, llama, and bat blood. Silicone Clot Ts were long only for porpoise blood. Clot Retr was decreased in hedgehog, elephant, manatee, and guanaco. This decrease did not appear to be due to a low platelet count. The only mammalian Clot Lys seen was in two chimpanzees (Clot T had received ketamine for anesthesia) and one armadillo Recal T.

Aves: Glass Clot Ts of blood from domestic fowl were variable and usually longer than human. Except for the turkey, silicone Clot Ts were within the human range. Clot Retr was poor or absent. Recal Ts were long. The clots did not lyse alone or in MCA.

Reptilia: Clot Ts of blood from the large marine turtles were very long. Often no clot was seen at 24 hr in the siliconized tubes. Clot Retr was poor, and Clot Lys was not seen. Clotting of alligator blood was slower than human, but the clots did retract well and did not lyse. The addition of calcium did not accelerate reptilian clotting.

Amphibia: Clot Ts of amphibians were variable, and the clots frequently lysed. At least one sample each from frog, bullfrog, and toad had a long Clot T, suggesting that all anuran Clot Ts might have been long if the blood had been obtained by a perfect technique.

Osteichthyes: About half the fish had blood that clotted promptly, while the other Clot Ts were much longer (see also Table 6.1). This difference seemed to be a species characteristic rather than dependent on technique. The Arctic grayling and trout blood samples were obtained from relatively large fish living in the same lake at 2°C. Clot Ts of five grayling were all very rapid (1 min), while those of the four trout were much slower (7–27 min). Blood from goosefish, tautogs, toadfish, and sturgeons all clotted slowly.

Chondrichthyes: Blood from elasmobranchs clotted very slowly or not at all unless tissue fluid or calcium was added. A granular appearance due to thrombocyte aggregation was often noted at about 15 min. Despite this phenomenon, no gel formation was noted in the great majority of samples. The few that did clot

TABLE 27.2[a]
Clotting Times, Clot Retraction, Clot Lysis, and Recalcification Times

		Clot T (min)		Clot Retr[b]			Recalcification		
Animal	*N*	G	S	G	S	Clot Lys[b]	Time (sec)[c]	Lys	Lys MCA
Echidna	3	3	9	2	3	0	150	0	0
Opossum	10	6	26	3	4	0	93	0	0
Quokka	1	6	14	4	4	0	56	0	0
Wallaby	1	5	14	3	3	0	66	0	0
Hedgehog	2	3	15	1	1	0	60	0	0
Bat	5	13	25	3	4	0	348	0	0
Baboon	7	6	21	4	4	0	280	0	0
Chimpanzee	5	3	3	2	4	2/5	73	0	0
Man									
Low	100+	6	20	3	3	0	90	0	0
High	100+	12	59	4	4	0	180	0	0
Monkey	10	5	11	3	4	0	111	0	0
Armadillo	6	6	33	2	4	0	200	1/6	0
Rabbit	6	4	20	3	4	0	56	0	0
Guinea pig	11	3	7	3	4	0	50	0	0
Hamster	7	5	8	4	4	0	97	0	0
Rat									
S-D	6	3	9	3	4	0	85	0	0
Wistar	6	4	8	3	4	0	180	0	0
Porpoise	12	37	133	3	4	0	1800	0	0
Cat	8	6	30	3	4	0	81	0	0
Dog	6	4	16	3	4	0	99	0	0
Ferret	6	2	3	1	3	0	66	0	0
Fox	2	6	25	3	4	0	104	0	0
Mink	3	4	8	5	5	0	67	0	0
Raccoon	3	4	14	2	3	0	87	0	0
Seal	5	7	24	4	4	0	155	0	0
Elephant	6	6	16	2	2	0	70	0	0
Manatee	7	8	13	1	2	0	69	0	0
Horse	6	16	30	3	4	0	651	0	0
Pig	14	7	18	4	4	0	94	0	0
Camel	2	10	29	4	4	0	107	0	0
Guanaco	1	10	30	1	1	0	158	0	0
Llama	1	15	55	4	4	0	101	0	0
Cow	6	7	11	3	4	0	235	0	0
Goat	6	6	19	4	4	0	94	0	0
Sheep	9	7	12	4	4	0	98	0	0
Chicken	9	8	8	0	2	0	355	0	0
Duck	4	16	19	0	0	0	453	0	0
Goose	6	12	22	2–3	2–3	0	1510	0	0

(*continued*)

TABLE 27.2[a] *(Continued)*

Animal	N	Clot T (min) G	Clot T (min) S	Clot Retr[b] G	Clot Retr[b] S	Clot Lys[b]	Recalcification Time (sec)[c]	Recalcification Lys	Recalcification Lys MCA
Pigeon	2	25	32	0	3	0	600+	0	0
Turkey	8	61	69	1	2	0	600+	0	0
Turtles									
Green	1	420	1440	1	0	0	600+	0	0
Loggerhead	2	120	1440	5	5	0	600+	0	0
PSE	11	33	141	0	2	0	543	0	0
Miscellaneous	4	9	13	4	4	0	400	0	0
Snake	5	61	120+	3	4	0	300+	0	0
Alligator	6	20	135	4	4	0	363	0	0
Amphiuma	3	4	9	5	5	3/3	117	3/3	?[d]
Cryptobranchus	4	6	18	2	3	1/4	103	1/3	0
Necturus	3	5	7	4	4	0	123	1/3	0
Salamander	1	3	3	4	4	0	72	0	0
Frog	5	32	94	2	4	5/5	185	2/5	0
Bullfrog	14	19	96	5	5	14/14	146	8/8	?[d]
Toad	3	20	26	2	3	0	53	0	0
Giant toad	5	14	120+	2	3	4/5	129	4/5	0
Arctic grayling	5	1	1	4	4	0	—	—	—
Arctic trout	4	7	27	4	4	1/3	—	—	—
Cobia	1	12	120	4	4	0	[c]	0	0
Goosefish	4	44	120+	2	2	0	[c]	0	0
Sculpin	4	6	22	4	4	0	75	0	0
Sturgeon	7	52	456	2	2	3/7	[c]	0	0
Tautog	1	30	120	4	4	0	43	0	0
Toadfish	1	30	240	4	4	0	[c]	0	0
Dogfish	7	60–120	600+	3	[b]	[b]	[c]	7/7	0
Guitarfish	1	60	120	1	1	1/1	[c]	1/1	?[d]
Shark									
Black-tipped	1	1440+	1440+	[b]	[b]	[b]	[c]	1/1	0
Blue	2	240+	240+	[b]	[b]	[b]	[c]	1/1	0
Mako	1	240+	240+	[b]	[b]	—	—	—	—
Skate	3	240+	240+	[b]	[b]	[b]	[c]	0	0
Lamprey	3	120	280	1	2	0	450	0	0

[a]More fish are listed in Table 6.1. [b]Not clotted at 24 hr.
[c]More than 120 sec. [d]Clot lysed before MCA test.

were lysed by 24 hr. Calcium, sea water, and tissue extracts had marked clot-promoting effects, and at 24 hr, the clots were either retracted or lysed.

One 2.4-kg dogfish was given 12 meq $CaCl_2$ intravenously. The clotting time shortened from over 60 min to 8½ min. After the animal was returned to the tank, oozing from the second venipuncture site was observed. This was the only time "excessive bleeding" was observed in any of the animals.

Cyclostomata: Lamprey blood clotted slowly. The clots retracted poorly and did not lyse.

1.3. Prothrombin Time with Simplastin, Homologous Brain, and Russell's Viper Venom, and Activated Partial Thromboplastin Time

Prothrombin times (PTs) with Simplastin and Russell's viper venom (RVV), and activated partial thromboplastin times (APPTs), were done on 33 species of mammals and 35 species of premammals. Prothrombin times with homologous brain (Homolog Br) or other tissues were done whenever possible.

Prothrombin times are shown in Table 27.3. The almost consistent results on human plasma with Simplastin (10.7–13.8 sec) show that there was little change in this reagent over the years of the study. Animal plasma Simplastin times were short in the rabbit and the carnivores, but long in the echidna, the marsupials, the guinea pig, and the cow. Actually, in mammals, the animal prothrombin times were longer than human in 9 samples, shorter than human in 13 samples, and almost exactly the same (within 1 sec) in 11 samples.

The animal brain times were usually shorter on the homologous plasma than on human plasma. Of the 21 tested, 19 were shorter on homologous plasma. RVV was an excellent thromboplastin reagent for almost all the mammals as well as human plasma.

APTTs were close to the human range with three exceptions: The APTT was very long in porpoises (XII-deficient) and bats (XI-deficient) and moderately long in horses.

Simplastin times were longer than 20 sec in all the premammalian vertebrates except necturus (17.8 sec) and three of the fishes: porgy (15.4 sec), sculpin (18.9 sec) and tautog (15.8 sec). Prothrombin times using homologous tissue were tested whenever possible. The animal tissue had no thromboplastic effects on human, but were often very potent clot accelerators in the species from which they had been prepared. RVV was active in all species except Chondrichthyes. The diverse APTT reagents used gave different results in the normal humans (23.8–51.4 sec). Most of the premammalian species tested gave very long plasma APTTs (>120 sec).

1.4. Thrombin Time

The Clot T of plasma on addition of a standard thrombin is called the *thrombin time* (Thromb T). Other reagents that clot fibrinogen directly, without added Ca, such as the snake venoms, CHH from *Crotalus horridus horridus* or Atroxin/Reptilase from *Bothrops atrox,* may be used. The strength of the thrombic agent is adjusted to give a thrombin time (0.2 ml plasma + 0.1 ml agent) of 15 ± 3 sec on human plasma.

In human plasma, the thrombin time may be long for a number of reasons: deficiency or dysfunction (D/D) of factor I (fibrinogen); presence of an inhibitor: heparin, fibrinogen split products (FgSP), or other inhibitor; or liver disease.

It was possible to make some differentiation among these diagnoses by the simple tests shown in Table 27.4. The Thrombin time was long in all patient groups. The Atroxin/Reptilase time (Atrox/Rept T) was normal in the presence of heparin or a heparinlike inhibitor, but long in the others. If Ca was added to the thrombin, the thrombin time of liver disease patients was normalized. The thrombin time mix (0.1 ml normal plasma + 0.1 ml patient plasma + 0.1 ml thrombin) was prolonged in the presence of an inhibitor in the patient's plasma. Treatment of the inhibitor plasma with barium removed heparin; gentle heating for 5 min at 56°C destroyed fibrinogen or its larger split products.

Table 27.5 lists the bovine thrombin times of 173 animals from 33 mammalian species. Bovine thrombin times on control human plasma and on the animals are shown. The thrombin times on human plasma fell between 12.0 and 17.8 sec, which falls within the allowable range of 15 ± 3 sec. The thrombin times of most of the carnivores and the sheep were within the human range. Most of the other animal plasmas gave long times. CHH and Atroxin/Reptilase (A/R) times were usually shorter than thrombin times, suggesting heparin.

Table 27.5 also shows the bovine thrombin times for 107 animals from 31 premammalian species. The thrombin times are quite variable, ranging from 8.4 to more than 120 sec. Most of the animal plasmas gave long CHH and Atroxin times.

1.5. Thromboplastin Generation Test

The old-fashioned, somewhat tedious thromboplastin generation test (TGT) is almost obsolete in human studies but still important in comparative animal research because it requires no heterologous reagents; thus, it is essentially independent of the effects of species specificity. It involves preparing reagents [see the Appendix (Section 2)]; diluting them with I/S in the proportions serum = 1:10, Ba Pl = 1:5, and platelets = 1:10; and adding equal volumes, usually 0.3 ml, to a generating mixture (mother) tube at 37°C. The mother stopwatch is started as 0.3 ml 0.025 M $CaCl_2$ is added and mixed. Then, 0.1 ml of this generating mixture is added to a substrate (daughter) tube at precisely 2, 4, 6, 8, 10, and 12. The substrate tube contains 0.1 ml platelet-poor plasma to which 0.1 ml warm 0.025 M $CaCl_2$ is added just before the 0.1 ml generating mixture. For normal humans, the daughter clotting times can be expected to become shorter and level off after 8–10 min incubation, giving a curve somewhat like this: 40+, 25, 14.5, 9.9, 9.9, 9.9 sec.

TABLE 27.3
Prothrombin Times and Activated Partial Thromboplastin Times

		Prothrombin time (sec)							
		Simplastin[a]		Homolog Br[a]		RVV[a]		APTT (sec)[a]	
Mammal	*N*	H	A	H	A	H	A	H	A
Echidna	3	13.6	32.9	58.5	29.4	9.5	15.1	47.4	38.1
Opossum	6	11.2	30.9	18.7	12.3	15.5	16.5	32.8	23.2
Quokka	1	13.8	20.0	13.4	9.5	9.5	12.5	42.0	19.5
Wallaby	1	13.2	22.9	17.9	9.6	13.1	21.8	52.8	22.5
Hedgehog	2	12.2	12.9	19.8	10.3	20.4	15.5	49.0	19.6
Bat	5	11.4	10.5	21.8	17.0	17.4	16.4	48.0	233
Baboon	6	12.0	12.1	—	—	15.8	17.6	32.4	33.1
Chimpanzee	5	11.4	10.6	—	—	14.1	15.9	27.4	22.0
Monkey	10	11.5	11.5	—	—	15.8	21.6	33.8	35.5
Armadillo	6	12.3	15.9	28.4	10.5	15.1	10.8	45.4	29.2
Rabbit	6	12.1	7.5	15.9	14.6	15.4	14.3	33.9	32.8
Guinea pig	11	11.4	26.0	[b]	19.7	17.4	14.4	32.8	28.7
Hamster	15	12.1	9.0	73.6	19.3	15.3	15.2	27.8	22.2
Rat									
S-D	6	10.7	10.1	[b]	18.5	16.2	13.8	30.0	18.8
Wistar	15	12.3	15.3	53.5	28.1	15.8	10.6	28.9	22.6
Porpoise	12	12.3	13.1	—	—	18.8	25.6	44.1	328
Cat	8	12.2	8.8	23.0	17.2	15.8	17.7	55.0	17.3
Dog	6	11.2	7.6	33.0	8.0	18.4	14.4	24.3	12.3
Ferret	6	11.1	10.3	—	—	14.1	14.1	25.5	18.4
Fox	2	11.5	10.2	24.0	9.2	—	—	43.1	18.2
Mink	3	10.9	6.9	32.8	12.9	17.4	13.3	54.2	15.8
Raccoon	3	11.7	9.6	9.8	13.2	15.8	15.5	50.0	26.2
Seal	5	11.5	11.0	—	—	14.7	12.7	41.4	41.2
Elephant	6	11.8	10.5	—	—	15.7	15.8	28.5	22.0
Manatee	6	11.5	9.8	—	—	17.1	14.3	36.1	14.5
Horse	6	10.7	9.0	16.8	16.6	16.4	15.9	30.2	69.9
Pig	6	12.5	13.5	24.8	14.9	15.4	15.9	27.2	23.2
Camel	2	12.0	9.6	—	—	15.0	14.4	48.2	50.0
Guanaco	1	11.4	10.2	—	—	15.5	19.2	48.0	36.4
Llama	1	12.0	9.1	—	—	19.4	14.0	50.4	26.4
Cow	6	12.1	22.6	26.8	32.3	16.1	24.7	48.0	59.6
Goat	6	11.1	11.7	—	—	16.1	18.5	33.9	32.4
Sheep	10	11.5	13.6	20.9	20.2	14.2	21.7	33.9	26.3

(*continued*)

TABLE 27.3 *(Continued)*

Premammal	N	Simplastin[a] H	Simplastin[a] A	Homologous[c] Br	Homologous[c] Gill/lung	Homologous[c] Skin	RVV[a] H	RVV[a] A	APTT[a] H	APTT[a] A
Chicken	10	13.5	[b]	13.9	—	—	17.8	15.7	36.8	[b]
Duck	4	13.5	40+	12.2	—	—	17.8	19.3	36.8	50+
Goose	6	13.5	[b]	12.9	—	—	17.8	14.5	36.8	90+
Pigeon	2	13.5	46.8	22.5	—	—	17.8	12.0	36.8	[b]
Turkey	8	13.5	60+	13.6	—	—	17.8	18.2	36.8	40+
Turtles										
Green	1	12.6	22.0	24.2	11.4	[b]	17.4	13.4	24.8	[b]
Loggerhead	2	12.4	93.0	32.0	16.5	[b]	22.0	33.1	34.0	[b]
PSE	3	11.8	28.8	12.4	18.4	—	17.2	16.8	32.0	80.7
Snake	4	12.4	[b]	26.2	27.2	—	17.2	24.8	27.8	[b]
Alligator	4	11.2	68.0	38.0	—	—	16.2	14.1	28.1	71.0
Amphiuma	3	13.1	29.0	26.5	17.4	16.3	14.8	15.4	42.8	87.0
Cryptobranchus	3	10.8	38.0	28.2	15.2	21.6	14.5	14.0	47.4	[b]
Necturus	3	13.0	17.8	21.0	16.6	9.8	15.0	14.1	28.8	[b]
Salamander	1	12.4	37.8	51.8	31.6	23.8	13.8	13.4	34.8	68.0
Frog	5	12.9	[b]	15.7	17.3	12.9	17.8	18.0	47.4	[b]
Bullfrog	6	12.8	[b]	42.9	17.4	12.8	14.8	16.4	34.2	66.6
Toad	3	12.1	24.8	35.5	11.3	14.2	15.2	15.6	45.5	33.4
Giant toad	5	12.4	24.6	32.1	15.4	15.6	17.4	33.4	35.6	[b]
Amberjack	1	12.8	60.8	—	—	—	—	—	34.2	[b]
Cobia	1	12.1	57.4	20.4	[b]	[b]	15.1	45.3	—	—
Flounder	4	12.4	87.0	68.8	14.8	18.8	15.1	37.4	—	—
Goosefish	4	11.2	88.9	21.1	22.0	13.3	14.5	[b]	33.2	[b]
Porgy	1	12.1	15.4	8.1	7.6	13.6	15.0	17.8	—	—
Sculpin	4	12.2	18.9	19.2	22.8	14.6	14.6	31.0	23.8	[b]
Sea bass	1	12.2	91.2	18.0	15.6	21.0	16.7	31.8	—	—
Sturgeon	5	11.8	[b]	—	—	—	15.2	[b]	27.0	[b]
Tautog	1	12.1	15.8	15.0	10.2	16.5	14.6	13.8	—	—
Toadfish	1	12.1	31.5	10.9	18.0	54.0	15.1	[b]	—	—
Triggerfish	1	12.1	[b]	13.4	18.0	14.5	—	—	—	—
Dogfish	7	12.0	57.7	35.0	10.3	31.4	15.0	[b]	23.8	[b]
Shark										
Black-tipped	1	12.1	[b]	38.4	32.4	51.7	15.0	115	36.6	[b]
Blue shark	2	12.1	[b]	27.8	25.4	57.2	14.8	118	34.8	[b]
Skate	3	12.1	[b]	44.0	68.0	39.0	17.4	118	36.6	[b]
Guitarfish	1	12.1	65.0	—	—	—	15.0	58.7	36.6	[b]
Lamprey	3	12.7	[b]	—	—	30.2	14.2	21.7	51.4	[b]

[a]Times for human (H) or animal (A) plasmas.
[b]More than 120 sec.
[c]Homologous tissue (brain, gill/lung, or skin) tested on the same animal plasma.

TABLE 27.4
Differential Diagnosis of a Long Thrombin Time in Humans[a]

Test	Fibrinogen D/D	Heparin	FgSP	Liver disease
Thromb T	L	L	L	L
Atrox/Rept T	L	N	L	L
Thromb + Ca	—	—	—	N
Thromb mix	N	L	L	N
56° C-labile %	—	No	Yes	—
Ba-labile %	—	Yes	No	—

[a]Key: Clotting time: (L) long, (N) normal; (No) inhibitor effect not diminished, (Yes) inhibitor effect diminished or eliminated.

Table 27.6 shows the results of TGTs on many different animals. The table lists the animals by their common names in phylogenetic order, the shortest ClotTs of the control human generating mixture on human or on animal substrate, and the results of animal generating mixtures on human or animal substrate. Mammals were tested at 37°C. All the human mixtures on human substrate tests gave fast Clot Ts: 8.2–11.1 sec (mean: 9.9 sec). When the human mixture was clotted on animal plasma substrate, clotting times were close to those on human plasma with the exception of the Australian mammals. The latter times were over 20 sec, which could be due to an abnormality in the animal substrate such as D/D of fibrinogen or prothrombin or to an inhibitor or a problem with species specificity. Generating mixtures from all animal ingredients gave more variable clotting times, but were mostly faster than 15 sec. Again, echidna, quokka, and wallaby were long. Porpoise was over 40 sec on both substrates. Horse was slow and manatee and goat borderline.

The lower vertebrates were often tested at both 37°C and ambient temperatures, and the shortest Clot Ts were recorded in Table 27.6. Again, the human-on-human clotting times were consistently rapid (8.1–13.0 sec). The activated human mixtures clotted Aves (13.8–21.6 sec) and Reptilia plasmas (13.0–38.4 sec) fairly rapidly. Plasmas from Amphibia, Osteichthyes, and Chondrichthyes clotted very slowly with the human generating mixtures. The premammalian generating mixtures had little or no plasma thromboplastin activity on human or homologous substrates.

1.6. Coagulation Factors

As far as possible, each animal plasma was tested in each coagulation factor assay usually used for humans. Most of the mammalian results fell within a wide

TABLE 27.5
Thrombin Times with Bovine Thrombin, *Crotalus horridus horridus* Venom, or Reptilase/Atroxin[a]

Mammal	N	Thromb T (sec)			Premammal	N	Thromb T (sec)		
		Bov Thr	CHH	A/R			Bov Thr	CHH	A/R
Echidna	3	[b]	—	—	Chicken	9	[b]	[b]	[b]
Opossum	10	60+	—	—	Duck	4	24.9	—	—
Quokka	1	48.0	—	—	Goose	6	19.2	—	—
Wallaby	1	115.0	—	—	Pigeon	2	[b]	—	—
Hedgehog	2	61.4	—	—	Turkey	6	[b]	[b]	[b]
Bat	5	57.7	—	—	Turtle				
Baboon	9	24.5	25.6	14.6	Green	1	9.8	—	—
Chimpanzee	5	24.5	18.1	11.8	Loggerhead	2	10.4	—	—
Monkey	6	22.1	—	—	PSE	8	69.6	64.4	78.4
Armadillo	6	67.0	—	16.1	Snake	6	[b]	—	—
Rabbit	6	22.9	—	15.3	Alligator	6	[b]	[b]	[b]
Guinea Pig	6	24.5	26.2	13.6	Congo eel	3	28.2	—	—
Hamster	7	[b]	[b]	14.7	Hellbender	4	63.8	—	—
Rat					Mud puppy	3	37.3	—	—
S-D	6	66.0	—	14.6	Frog	5	53.0	54.8	105.0
Wistar	6	[b]	[b]	15.3	Bullfrog	8	[b]	[b]	[b]
Porpoise	12	17.9	—	—	Toad	3	55.0	—	—
Cat	8	15.8	—	13.8	Giant toad	4	[b]	—	[b]
Dog	6	16.0	14.9	13.8	Amberjack	1	83.2	—	53.0
Ferret	6	28.8	[b]	14.1	Cobia	1	[b]	—	40.0
Fox	2	17.6	—	—	Flounder	4	[b]	58.8	21.0
Mink	3	14.1	—	—	Goosefish	4	59.4	[b]	[b]
Raccoon	3	16.9	—	—	Porgy	1	12.3	—	—
Seal	5	13.4	—	—	Sculpin	4	28.0	48.9	24.6
Elephant	6	37.3	—	—	Sea bass	1	13.8	—	—
Manatee	4	39.8	24.0	16.7	Sturgeon	4	[b]	[b]	[b]
Horse	6	60+	—	15.4	Tautog	1	8.4	—	—
Pig	12	50.0	33.3	14.6	Toadfish	1	9.4	—	—
Camel	2	19.6	—	—	Triggerfish	1	[b]	—	—
Guanaco	1	24.8	—	—	Dogfish	2	57.4	[b]	[b]
Llama	1	25.9	—	—	Shark				
Cow	6	21.8	—	—	Black-tipped	1	33.8	—	—
Goat	6	27.0	—	—	Blue	1	29.8	—	97.4
Sheep	5	15.3	13.0	15.1					

[a]Range on normal human plasma for Bov Thr, CHH, and A/R: 12–18 sec.
[b]More than 120 sec.

TABLE 27.6
Thromboplastin Generation Tests

	Substrate clotting time (sec)								
	Mammals					Premammals			
	Generating mixture					Generating mixture			
	Human		Animal			Human		Animal	
	Substrate					Substrate			
	H	A	H	A		H	A	H	A
Mammal					Premammal				
Echidna	11.1	21.3	18.1	19.9	Chicken	10.4	21.6	40+	40+
Opossum	10.1	9.8	10.4	10.8	Duck	8.6	18.1	40+	40+
Quokka	11.1	23.6	19.3	26.5	Goose	9.6	14.2	40+	40+
Wallaby	11.1	21.3	21.5	18.1	Pigeon	9.2	17.8	30.2	40+
Hedgehog	10.4	10.8	13.2	11.0	Turkey	8.2	13.8	40+	40+
Baboon	9.9	9.8	10.4	11.1	Turtle				
Monkey	9.9	10.2	10.7	11.3	Green	10.2	13.0	40+	40+
Armadillo	9.9	10.7	13.3	14.6	Loggerhead	13.0	26.0	40+	40+
Rabbit	9.9	9.0	12.5	10.8	PSE	8.2	14.8	40+	40+
Guinea pig	10.5	11.1	11.2	11.7	Snake	10.5	—	40+	40+
Rat					Alligator	10.6	38.4	40+	40+
Wistar	8.2	11.8	13.3	13.6	Amphiuma	10.5	40+	40+	37.0
S-D	9.0	14.6	8.6	9.8	Cryptobranchus	10.2	40+	40+	40+
Porpoise	9.3	10.6	40+	40+	Necturus	10.7	24.6	40+	24.6
Cat	9.9	14.9	9.0	12.8	Salamander	10.5	40+	40+	40+
Dog	9.8	9.0	8.5	7.4	Bullfrog	10.1	40+	40+	40+
Fox	9.8	7.2	10.0	7.6	Frog	10.5	40+	40+	40+
Mink	10.4	6.8	10.6	10.0	Giant toad	9.9	40+	40+	40+
Raccoon	9.8	8.8	9.1	9.1	Toad	10.6	40+	40+	40+
Seal					Cobia	9.9	—	40+	27.0
Harp	9.9	10.2	14.4	14.2	Flounder	10.8	40+	26.2	25.4
Harbor	9.6	10.6	12.6	12.6	Orange filefish	10.8	40+	40+	40+
Elephant	9.8	9.2	12.6	10.8	Porgy	8.1	—	18.6	18.0
Manatee	9.8	12.5	15.2	14.0	Blue shark	9.4	40+	40+	40+
Horse	11.1	10.1	23.2	22.4	Skate	10.0	40+	40+	40+
Pig	8.6	12.7	8.7	10.4	Dogfish	8.8	40+	40+	40+
Camel	9.2	10.2	10.8	10.2					
Guanaco	9.2	10.4	9.8	9.8					
Goat	9.4	10.1	16.1	18.2					
Sheep	9.8	11.4	11.5	11.4					

human range of 0.50–3.0 U/ml. For each species, the mean for the number of animals tested is presented. Many animal factors were not assayed in sufficient dilutions to give results greater than 2.0+ U/ml. Premammalian coagulation factor assays were very much lower than mammalian. In calculating the mean value for each species, a value of less than 0.01 was counted as 0 and 2.0+ as 2.0.

The first part of Table 27.7 shows the results in mammals. Coagulation factors present in zero to trace levels are indicated by *a*. These levels appear in factors VII for wallaby, X for opossum and wallaby, XI for bat, and XII for porpoise. These latter two appear to be species characteristics because all animals of the species gave the same results.

In the other mammals the levels of fibrinogen (factor I) and XIII fell within or close to the human range. The other factor levels were very variable and, frequently, well above human. Unfortunately, factors Fletcher (Flet) and Fitzgerald (Fitz) were done on fewer than half the animals.

The second part of Table 27.7 shows the results found in the premammals. Only one value, factor V in the *Pseudemys scripta elegans* (PSE) turtles, was high, 3.47 U/ml, and one wonders whether this could be an aberration.

In the premammals, all fibrinogen (factor I) levels were within the human range except that of the bullfrogs, which was 93 mg/dl. The factor II levels varied from less than 0.01 to 0.30 U/ml and factor V levels from less than 0.01 to 0.34 U/ml. Only a few animals had measurable levels of factors VII and X. Factor VIII levels were moderate in Aves, 0.18–0.57 U/ml, but low (<0.01 to 0.05 U/ml) in the premammals of other classes. Factors IX, XI, and XII were low in most of the animals. Flet and Fitz factors were rarely assayed and were almost always very low. Some antithrombin III (ATIII) was found in all the premammals who were tested. Factor XIII (R) was usually low.

Table 27.8 shows a great variableness in the few assays of recently recognized coagulation factors. Fletcher factor assay results are widely variable as compared to findings in human blood. Guinea pig and pig are high; baboon, chimpanzee, hamster, rat, ferret, and manatee fall within the human range. Rabbit, cat, dog, and horse assayed in the low range, and the two premammals (alligator and bullfrog) show less than 0.01 U/ml. Fitzgerald factor is high in manatee and pig and normal in baboon, chimpanzee, guinea pig, and ferret. It is low in rabbit, hamster, rat, and dog and less than 0.01 U/ml in alligator and bullfrog. Protein C activity was high in baboon and normal in chimpanzee, guinea pig, ferret, pig, and sculpin (a fish!). It was low in hamster, dog, alligator, and dogfish and less than 0.01 U/ml in the Wistar rat. Protein S by immunological assay, which may be human specific, was low in all animals tested except the chimpanzee. Plasmogen (plgn) also assayed in the low range in all the animals tested, suggesting that the activation system is not satisfactory. Antiplasmin was in the human range in all the mammals and in the bullfrog and sturgeon. It was moderately low in

TABLE 27.7
Coagulation Factors

Animal	N	I (mg/dl)	II (U/ml)	V (U/ml)	VII (U/ml)	X (U/ml)	VIII (U/ml)	IX (U/ml)	XI (U/ml)	XII (U/ml)	Flet (U/ml)	Fitz (U/ml)	ATIII (U/ml)	XIII (R)
Echidna	3	474	0.55	1.00	0.02	0.02	1.13	0.37	0.40	1.0+	—	—	—	4
Opossum	10	288	0.10	2.0+	0.03	[a]	2.0+	1.20	1.80	1.5+	1.10	—	1.10	10
Quokka	1	208	0.46	1.28	0.05	0.05	0.90	0.08	1.0+	1.0+	—	—	—	4
Wallaby	1	105	0.26	1.40	[a]	[a]	0.75	0.10	1.0+	1.0+	—	—	—	8
Hedgehog	2	288	0.21	2.18	0.01	0.03	2.03	2.13	1.63	1.50	—	—	2.0+	16
Bat	5	370	0.46	5.11	0.28	1.44	1.70	0.81	[a]	0.21	—	—	0.86	14
Baboon	9	196	0.93	1.09	3.63	0.90	0.78	0.44	0.40	2.80	0.93	0.61	1.17	18
Chimpanzee	5	288	1.12	1.06	1.32	1.07	1.32	1.44	2.24	2.86	1.17	1.37	1.26	—
Man														
Low	100+	150	0.70	0.65	0.50	0.75	0.75	0.65	0.75	0.50	0.50	0.50	0.80	4
High	100+	450	1.30	1.45	1.30	1.25	1.40	1.70	1.40	1.45	1.50	1.50	1.20	16
Monkey	10	305	1.14	1.05	1.85	0.80	1.09	1.01	1.00	0.93	—	—	0.79	34
Armadillo	6	608	0.77	2.0+	0.58	0.53	2.0+	2.0+	2.0+	2.0+	—	—	0.90	22
Rabbit	6	193	2.40	100+	3.00	1.80	3.20	1.20	2.50	0.80	0.03	0.05	0.80	27
Guinea Pig	6	262	0.42	6.10	0.09	0.12	4.50	0.97	1.71	7.00	1.74	1.00	1.25	58
Hamster	9	141	0.68	5.30	9.45	3.43	1.03	0.78	0.79	1.14	1.33	0.22	0.86	—
Rat														
Wistar	6	164	0.45	2.82	3.91	0.30	1.74	0.77	0.63	1.28	1.26	0.30	1.00	—
S-D	7	268	0.70	1.42	1.63	0.36	4.00	0.41	1.28	1.77	1.44	—	2.38	10
Porpoise	12	403	1.29	1.05	0.95	1.25	10+	1.12	1.08	[a]	—	—	—	23

(continued)

TABLE 27.7 *(Continued)*

Animal	N	I (mg/dl)	II (U/ml)	V (U/ml)	VII (U/ml)	X (U/ml)	VIII (U/ml)	IX (U/ml)	XI (U/ml)	XII (U/ml)	Flet (U/ml)	Fitz (U/ml)	ATIII (U/ml)	XIII (R)
Cat	8	477	1.33	11.9	4.82	3.16	18.9	1.84	2.84	2.22	0.09	—	0.93	14
Dog	6	215	1.43	6.02	4.70	2.18	4.49	1.12	2.79	1.06	0.32	0.30	0.87	10
Ferret	5	188	0.81	5.41	1.59	1.68	2.50	1.14	0.98	6.23	0.92	1.50	1.18	16
Fox	2	688	1.50	1.90	2.15	2.14	2.0+	2.0+	1.50	1.10	—	—	—	8
Mink	3	155	2.30	2.85	2.45	1.57	2.0+	1.70	1.70	1.60	—	—	—	43
Raccoon	3	410	2.04	3.47	2.40	1.69	2.0+	1.50	1.50	1.42	—	—	1.20	16
Seal	5	375	1.51	1.93	1.90	2.40	3.60	1.80	1.80	1.64	—	—	—	13
Elephant	6	403	1.79	2.03	1.33	1.77	1.99	1.64	1.44	1.80	—	—	0.08	36
Manatee	6	187	0.76	2.53	4.10	2.15	8.75	4.61	1.57	1.40	1.40	1.50	—	64
Horse	6	273	1.34	2.67	6.90	2.10	0.98	0.61	0.31	0.78	0.17	—	1.02	16
Pig	14	309	0.64	4.73	2.09	1.89	6.13	4.18	1.43	4.95	1.67	3.30	1.11	21
Camel	2	259	1.25	1.54	1.12	0.50	2.0+	0.58	0.75	1.38	—	—	1.05	16
Guanaco	1	318	1.10	2.0+	1.30	1.37	2.0+	1.50	1.25	1.25	—	—	0.60	8
Llama	1	237	1.03	1.70	0.87	0.89	7.10	2.50	1.35	1.75	—	—	1.28	16
Cow	6	180	0.42	2.0+	0.19	1.09	2.0+	0.76	1.01	0.97	—	—	0.93	9
Goat	6	336	0.62	6.45	1.63	1.75	12+	3.33	1.48	2.47	—	—	0.91	24
Sheep	10	272	0.62	2.0+	0.36	0.44	2.0+	2.0+	0.98	0.98	—	—	1.05	19
Chicken	8	371	0.14	0.06	[a]	[a]	0.31	[a]	[a]	[a]	—	—	—	—
Duck	4	358	0.08	0.30	[a]	[a]	0.33	[a]	[a]	[a]	—	—	—	—
Goose	6	223	0.08	0.16	[a]	[a]	0.18	0.06	0.02	[a]	—	—	—	—
Pigeon	2	243	0.21	0.07	0.03	[a]	0.55	0.03	0.08	0.03	—	—	—	—
Turkey	8	441	0.30	0.19	[a]	[a]	0.57	[a]	[a]	[a]	—	—	—	29
Turtle														
Green	1	267	[a]	0.20	[a]	[a]	[a]	[a]	0.10	0.05	—	—	—	2

Loggerhead	2	388	0.05	0.22	[a]	[a]	[a]	[a]	[a]	[a]	—	—	—	4
PSE	6	329	0.12	3.47	0.04	0.24	[a]	[a]	0.03	0.03	[a]	—	[a]	2
Iguana	1	281	[a]	[a]	[a]	[a]	0.05	[a]	0.02	0.05	—	—	—	—
Snake	4	162	[a]	[a]	[a]	[a]	[a]	[a]	[a]	[a]	—	—	—	—
Alligator	6	178	0.21	0.02	[a]	[a]	[a]	0.02	0.02	0.02	[a]	[a]	1.05	—
Congo eel	3	213	0.03	0.02	[a]	[a]	0.04	0.18	0.27	0.15	[a]	—	—	2
Hellbender	4	257	0.06	0.06	[a]	[a]	0.04	[a]	0.48	0.19	—	—	—	2
Mud puppy	4	284	0.06	0.13	0.02	[a]	0.01	0.02	0.02	0.20	[a]	—	1.13	10
Salamander	1	363	0.11	0.20	[a]	[a]	[a]	[a]	[a]	[a]	—	—	—	2
Frog	5	337	0.04	0.04	[a]	[a]	0.04	0.16	[a]	0.18	—	—	—	2
Bullfrog	7–14	93	0.02	0.11	0.04	0.02	[a]	[a]	[a]	0.04	[a]	[a]	0.21	12
Toad	2–3	152	0.17	0.03	[a]	[a]	0.04	0.02	0.02	0.5	—	—	—	2
Giant toad	4	179	0.06	0.18	[a]	0.03	[a]	[a]	[a]	[a]	[a]	0.02	—	2
Goosefish[b]	4	204	[a]	[a]	[a]	[a]	[a]	[a]	[a]	[a]	—	—	0.44	9
Sculpin	4	135	0.05	0.34	[a]	0.04	0.02	0.02	0.01	0.03	—	—	0.38	4
Sturgeon	5	212	[a]	0.04	0.03	[a]	[a]	[a]	[a]	[a]	—	—	—	—
Dogfish	7	—	[a]	[a]	[a]	[a]	[a]	[a]	[a]	[a]	—	—	0.16	—
Shark														
Black-tipped	1	150	[a]	[a]	[a]	[a]	[a]	[a]	[a]	[a]	—	—	—	—
Blue	2	260	[a]	[a]	[a]	[a]	[a]	[a]	0.15	0.02	—	—	—	—
Mako	1	190	[a]	[a]	[a]	[a]	[a]	[a]	—	0.05	—	—	—	—
Skate	2	160	[a]	[a]	[a]	[a]	[a]	[a]	0.05	0.05	—	—	—	—
Guitarfish	1	212	[a]	[a]	[a]	[a]	[a]	[a]	[a]	0.40	—	—	—	—
Lamprey	3	175	[a]	[a]	[a]	[a]	0.05	0.07	0.05	[a]	—	—	—	—

[a]Less than 0.01 U/ml. [b]See also Table 6.4.

TABLE 27.8
Other Coagulation Factors

Animal	Flet (U/ml)	Fitz (U/ml)	Prot C (A) (U/ml)	Prot S (I) (U/ml)	Plgn (U/ml)	Antipl (U/ml)	RCF (U/ml)
Baboon	0.93	0.61	1.62	0.44	0.09	1.17	2.68
Chimpanzee	1.17	1.37	0.77	0.94	0.38	1.00	0.87
Rabbit	0.03	0.05	—	—	—	—	—
Guinea pig	1.74	1.00	0.72	—	0.02	0.97	[a]
Hamster	1.33	0.22	0.51	—	0.15	0.72	—
Rat							
Wistar	1.26	0.30	[a]	0.04	0.13	1.18	—
S-D	1.44	—	—	—	—	—	0.49
Cat	0.09	—	—	—	—	—	0.48
Dog	0.32	0.30	0.15	0.08	0.25	1.10	0.64
Ferret	0.92	1.50	0.77	—	0.44	1.45	—
Manatee	1.40	1.51	[a]	—	0.18	1.85	0.75
Horse	0.17	—	—	—	—	—	0.86
Pig	1.67	3.30	0.71	0.08	0.08	1.25	—
Alligator	[a]	[a]	0.08	—	0.06	0.42	[a]
Bullfrog	[a]	[a]	[a]	—	0.04	1.21	—
Sculpin	—	—	1.32	—	—	0.38	—
Sturgeon	—	—	—	—	0.21	1.01	—
Dogfish	—	—	0.19	0.07	—	0.32	—

[a]Less than 0.01 U/ml.

alligator, sculpin, and dogfish. Ristocetin cofactor, assayed on fixed human platelets, was high in baboon; normal in chimpanzee, dog, manatee, and horse; low in rat and cat; and less than 0.01 U/ml in guinea pig and alligator.

1.7. Fibrinolytic Enzyme System

Animal experimentation is vastly important in the study of the cause and treatment of thromboembolic disease, which accounts for morbidity and mortality in 40 million human beings each year. Most of this sickness and death occurs in middle-aged people in the Western hemisphere. Prevention is the first concern and effective treatment is the second. In choosing an animal model, some knowledge of its reaction to known fibrinolytic activators may be helpful. Only meager studies were done on these animals, but they are reviewed here for the sake of completeness. Methods used were variable; results in each test run were qualitatively compared to human controls.

Figure 27.1 presents a simplified view of the author's concept of the human fibrinolytic system. Plasminogen (Plgn) is the precursor of the active enzyme

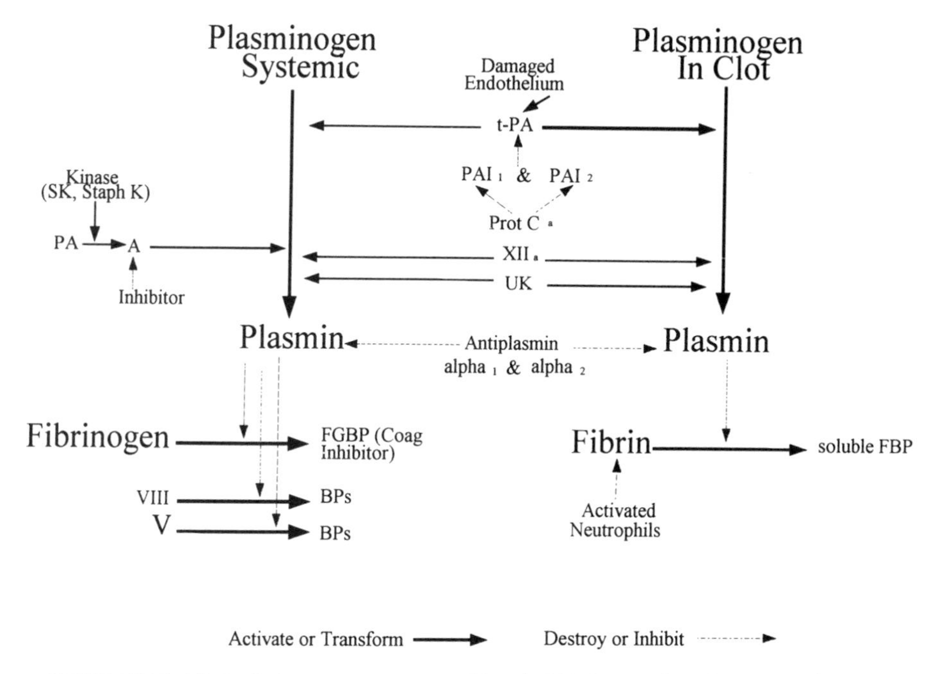

FIGURE 27.1. Fibrinolytic enzyme system. Key: (t-PA) tissue plasminogen activator; (PAI_1, PAI_2) plasma activator inhibitors; (SK) streptokinase; (Staph K) staphylokinase; (Prot Ca) activated protein C; (PA) plasma activator; (A) Activator; (UK) Urokinase; (FGBP) fibrinogen breakdown products; (FBP) fibrin breakdown products; (BP) breakdown products.

plasmin also known as fibrinolysin. We may consider Plgn in two circumstances as shown in the figure. First is systemic, whereby it is distributed essentially evenly throughout the circulation, and second is when it is confined within a pathological clot (thrombosis). In either case, it must be converted to the active serpin before it can attack the structural substance of the clot, fibrin. Natural activation occurs through the release of tissue plasminogen activator (t-PA) from the damaged endothelium. Tissue PA is moderately effective at the site of the thrombosis, but has little systemic effect because it is so highly diluted in the circulation and inhibitors are present. The same is true of plasmin itself. Tissue PA is inhibited by two plasma activator inhibitors (PAI_1 and PAI_2). Interestingly, both these inhibitors can be inhibited by activated protein C (Prot C^a), thus allowing protein C to have a modulating effect by inhibiting inhibitors.

There are a number of other ways by which Plgn may be activated. Factor XIIa formed at the site of thrombosis has a slow and incomplete activating effect. Urokinase (UK) is present in plasma in low concentration.

Therapeutic activation is usually attempted through the infusion, often aimed directly toward the clot, of a kinase such as streptokinase (SK), UK, or t-PA. Here, a delicate balance between systemic and intraclot activation is necessary. Systemic formation of plasmin in the circulating blood leads to breakdown of fibrinogen and factors V and VIII and formation of fibrinogen breakdown products (FgBPs) that are clot-inhibitory, all of which may cause severe bleeding.

It is known that activated neutrophils are attracted to and accumulate within the fibrin clot. These cells can ingest and digest fibrin, but have not been explored as a therapeutic fibrinolytic agent.

Table 27.9 shows spontaneous Clot Lys found in the mammals studied. Only two whole blood clots lysed, and these were from chimpanzees that had received ketamine. Only one Recal clot lysed, that from an armadillo. All euglobulin lysis times were longer than 3 hr. Many of the animals had less reactivity to SK, UK, or staphylokinase (Staph K) than did humans. Reaction to SK was low in opossum, baboon, the rodents, and the Artiodactyla.

Table 27.9 also shows the fibrinolytic activity in the premammals. Whole blood and Recal Clot Lys was not seen in Aves or Reptilia, but was common in Amphibia and Chondrichthyes. Whole blood lysis was seen in four of the seven sturgeons. SK, UK, and Staph K had no effects on conversion of plasminogen to plasmin in premammals. Antiplasmin (Antipl) was low or absent in birds and fish, but variable, and sometimes high, in the other premammals.

Table 27.10 summarizes the effects of SK, UK, and Staph K on fibrinolytic activity in the animals studied.

1.8. Platelet Count, Platelet Glass Bead Retention, and Platelet Aggregation in Mammals

For hemostasis in normal human beings, platelets are essential and play at least three major roles.

Platelet plug or white thrombus formation: Following an injury, circulating platelets quickly adhere to exposed subendothelial collagen fibers, forming a temporary plug that acts as the nidus for the more permanent RBC-fibrin-containing thrombus. ADP released from the initial aggregated platelets or vessel wall, or both, and von Willebrand factor cause more platelets to aggregate. Finally, thrombin formed *in situ* causes permanent platelet agglutination.

There are a number of tests to determine the efficiency of this phase. The bleeding time gives an overall estimate. The adhesiveness of platelets is measured by the glass bead retention index. Aggregation of platelets is measured by adding the reagent to platelet-rich plasma (PRP) and recording the clearing of the hazy PRP as the platelets aggregate into clumps, allowing light to penetrate.

Clot formation: The second major role of platelets in hemostasis is the

TABLE 27.9
Fibrinolytic Activities

Animal plasma	Spontaneous clot lysis			Activated lysis[a]			
	Blood	Plasma	Euglobulin	SK	UK	Staph K	Antipl
Echidna	0	0	>300	+	+	—	++++
Opossum	0	0	>240	0	0	0	++++
Quokka	0	0	>300	++	+	—	++
Wallaby	0	0	>300	+++	++	—	++++
Hedgehog	0	0	210	+++	+++	—	—
Bat	0	0	—	+++	+++	—	—
Baboon	0	0	>180	0–+	++	++	+++
Chimpanzee	2/5	0	—	+++	+++	—	++++
Man	0	0	>300	++++	++++	++++	++++
Monkey	0	0	—	++++	++++	+++	+++
Armadillo	0	1/6	>120	++	++	++	++
Rabbit	0	0	>240	++	++	++	+++
Guinea pig	0	0	>240	0–+	++	++	++
Hamster	0	0	>240	0–+	+	+	—
Rat							
S-D	0	0	>240	0–+	+	0–+	++
Wistar	0	0	>240	0	+	0	++
Porpoise	0	0	—	+++	++++	++++	+++
Cat	0	0	>180	++++	+	+	++++
Dog	0	0	>180	++++	+++	+++	+++
Ferret	0	0	—	—	—	—	—
Fox	0	0	—	—	—	—	—
Mink	0	0	>180	+++	+	++	+++
Raccoon	0	0	>180	—	—	—	—
Seal	0	0	>180	+++	++	++	+
Elephant	0	0	—	+	+	—	—
Manatee	0	0	—	++++	+	+	++
Horse	0	0	—	0–+	+	0–+	++++
Pig	0	0	>240	0–+	+	0	++++
Camel	0	0	—	0	—	—	—
Guanaco	0	0	—	0	—	—	—
Llama	0	0	—	0	—	—	—
Cow	0	0	>300	0	++	0–+	++++
Goat	0	0	>300	0	++	++++	+++
Sheep	0	0	>300	0	++	++++	++++
Chicken	0	0	>300	0	0	0	+
Buck	0	0	>300	0	—	—	0–+
Goose	0	0	>300	0–+	0	0	++
Pigeon	0	0	—	—	—	—	0
Turkey	0	0	—	—	—	—	—

(*continued*)

TABLE 27.9 *(Continued)*

Animal plasma	Spontaneous clot lysis			Activated lysis[a]			
	Blood	Plasma	Euglobulin	SK	UK	Staph K	Antipl
Turtle							
Green	0	0	>300	0	0	0	++
Loggerhead	0	0	>300	0	0	0	+++
PSE	0	0	—	+	—	—	+
Iguana	0	0	>240	—	—	—	—
Snake	0	0	—	0	—	—	++++
Alligator	0	0	>240	0	0	0	++++
Amphiuma	3/3	3/3	—	0	0–+	0	+
Cryptobranchus	1/4	1/3	—	0	0–+	0	+
Necturus	0/3	1/3	—	—	—	—	—
Salamander	0/1	—	—	—	—	—	—
Frog	5/5	2/5	—	0	—	—	++
Bullfrog	6/6	6/6	—	0	—	—	++++
Toad	0/3	0/3	—	0	—	—	++++
Giant toad	4/5	4/5	—	—	—	—	—
Cobia	0	0	—	0	0	0	+
Goosefish	0	0	>240	0	0	0	+
Grouper	0	0	—	0–+	0–+	—	0
Toadfish	0	0	—	0–+	—	—	+++
Sturgeon	4/7	0	—	0	0	0	—
Dogfish	6/6	7/7	—	0–+	0	0	++
Shark, black-tipped	1/1	1/1	—	0	0	0	0
Lamprey	0	0	—	—	—	—	—

[a]Concentrations: streptokinase (SK), 50–1000 U/ml; urokinase (UK), 50–550 U/ml; staphylokinase (Staph K), 2%.

release of platelet phospholipid. This phospholipid forms complexes with the activated cofactors, factors Va and VIIIa, that are essential to the intrinsic and common coagulation pathways.

Clot retraction: The third role of platelets in hemostasis is to cause the clot to retract or shrink to about half its original size. Serum is squeezed out. Without platelets, Clot Retr does not occur. In some platelet diseases, retraction is also deficient. The RBC volume affects the degree of retraction.

Table 27.11 shows the platelet counts (Plat Cs) of the mammals. The range of normal human Plat Cs is 150–450 $\times$ $10^3/mm^3$. Of the 32 mammals, 16 had higher counts, 3 lower, and 13 about the same as human. Only 4 mammals—echidna, hedgehog, elephant, and manatee—showed less Clot Retr than the normal human range of 3+ to 4+. The platelet glass retention test on all the mammals on which this test was performed gave retention indices of greater than 20%, which is the same range as that found in normal humans.

TABLE 27.10
Summary of Activation by Streptokinase, Urokinase, and Staphylokinase

Activated by streptokinase, urokinase, and staphylokinase	
Armadillo	Manatee
Cat	Mink
Chimpanzee	Monkey
Dog	Porpoise
Man	Seal
Activated by urokinase and staphylokinase (but not streptokinase)	
Goat	Hamster
Guinea pig	Sheep
Activated by urokinase (but not streptokinase or staphylokinase)	
Cow	Pig
Horse	Rat (Sprague-Dawley and Wistar)
No activation or only trace activation	
Opossum	All premammalian vertebrates

Table 27.11 shows that not all mammals were tested with all of the aggregating agents. In general, animal platelets did not aggregate to as great an extent as did human platelets. Baboon, chimpanzee, and pig showed biphasic curves with ADP. Animal platelets that aggregated to 10% or less were as follows: **collagen:** guinea pig, Wistar rat, sheep; **ristocetin:** bat, rabbit, Wistar rat, cat, dog, ferret, pig, cow; **thrombin:** guinea pig, Wistar rat, cat, manatee, pig, camel; **arachidonic acid:** chimpanzee, pig; **pig or bovine plasma:** rabbit, Wistar rat, manatee, pig, llama.

3. HEREDITARY HEMORRHAGIC/COAGULATION DISEASES IN MAMMALS

3.1. Definition

A hereditary hemorrhagic disease is defined as a genetically controlled deficiency or dysfunction (D/D) of any components of normal hemostasis. In humans, a great number of such components derived from plasma, platelets, or blood vessel walls have been recognized and their roles in various aspects of hemostasis defined. D/Ds of a portion of these have been identified as causes for bleeding or thrombotic disease in man.

Similar bleeding disorders have been described in a number of mammals, particularly those that are (1) frequently observed (pets and farm and laboratory

TABLE 27.11
Platelet Counts, Clot Retraction, Platelet Glass-Retention, and Aggregation Tests

Animal	N	Plat C ($\times 10^3$/ mm^3)	Clot Retr	Plat Glass Ret Ind. (%)	Aggregation (%)[a] ADP	Col	Risto	Thr	AA	Pig/Bov
Echidna	3	549	2–2	—	—	—	—	—	—	—
Opossum	4–8	498	3–4	52	95	22	—	95	—	—
Quokka	1	1180	4–4	—	—	—	—	—	—	—
Wallaby	1	390	3–3	—	—	—	—	—	—	—
Hedgehog	2	113	1–1	—	100	100	—	—	—	—
Bat	3	819	3–4	—	100	100	0	—	—	—
Baboon	6	299	4–4	—	60[b]	89	97	—	67	66
Chimpanzee	5	406	2–4	—	48[b]	61	90	22	3	51
Man										
Low	100+	150	3	21	70	80	80	60	60	60
High	100+	450	4	100	100	100	100	100	100	100
Monkey	5	510	3–4	63	98	95	100	—	—	—
Armadillo	3–6	356	2–4	—	100	100	—	—	—	—
Rabbit	4–10	411	3–4	90	47	52	7	97	63	5
Guinea pig	3–7	434	3–4	59	71	8	34	4	59	68
Hamster	6	368	4–4	—	—	—	—	—	—	—
Rat										
Wistar	8–10	656	3–4	66	78	7	0	10	43	0
S-D	12	770	3–4	—	—	—	—	—	—	—
Porpoise	6	136	3–4	31	93	36	—	—	—	—
Cat	8	382	3–4	52	94	95	0	0	—	13
Dog	6	450	3–4	52	73	89	8	88	22	16
Ferret	5	497	1–3	—	67	74	++	—	—	17
Fox	1	109	3–4	—	—	—	—	—	—	—
Mink	3	773	5–5	50	67	69	—	39	—	—
Raccoon	3	824	2–3	72	100	100	—	—	—	—
Seal	5	651	4–4	69	98	100	—	—	—	—
Elephant	4–6	620	2–2	26	63	69	—	—	—	—
Manatee	4–7	347	1–2	—	69	91	45	0	45[b]	9
Horse	6	235	2–4	64	99	100	13	100	—	—
Pig	11–14	465	4–4	32	49[b]	71	<5	<5	<5	<5
Camel	1–2	260	4–4	26	80	70	40	0	—	—
Llama	1	680	4–4	68	100	50	100	—	—	0
Cow	4–6	350	3–4	56	87	55	0	—	—	—
Goat	4–7	551	4–4	64	98	80	12	100	—	—
Sheep	4–9	479	4–4	46	77	10	43	67	49	11

[a]Reagents: (ADP) adenosine 5-diphosphate; (Col) bovine soluble collagen; (Risto) Ristocetin; (Thr) bovine thrombin; (AA) arachidonic acid; (Pig/Bov) pig or bovine plasma.
[b]Biphasic curve, showing aggregation followed by disaggregation.

animals) or (2) highly inbred. Special breeding to emphasize desirable characteristics and produce recognizable, reproducible strains or breeds increases the risk of producing genetic disease.

Hereditary bleeding in mammals (other than man) has been recognized in association with D/Ds of plasma factors I, II, VII, X, VIII, IX, XI, and von Willebrand. In addition, bleeding may be associated with platelet disorders that occur in an inherited pattern.

3.2. Coagulation Disorders without Bleeding in Man

Rarely, D/Ds of factors XI, XII, Flet, and Fitz are found in humans. About half the factor XI and all the others have no bleeding symptoms at all. They are often found in other members of the same families and are called *hereditary coagulation disorders*.

3.3. Hemorrhagic/Coagulation Diseases in Mammals

Table 27.12 lists the D/Ds of coagulation factors or platelets in humans that have also been found in other mammals. Bleeding has been associated with all except factor XII D/D and about half of factor XI D/D. In general, all these disorders are rare, and the relative frequency indicated in the table is only an impression from the numbers of reported cases (see References).

TABLE 27.12
Recognized Hereditary Coagulation Factor Deficiencies or Dysfunctions in Mammals

Factor D/D	Estimate relative frequency from rare (+) to common (++++)								
	Man	Dog	Cat	Cow	Pig	Goat	Horse	Rabbit	Rat
I	+	+				+			
II	+	+							
V	+								
VII	+	+							
X	+	+							
VIII	++++	++++	+				+		
IX	++	+							
XI	++	+	+						
XII	+		+						
vW	++++	+	+		+			+	
Platelet	+	+	+	+	+				+

Factor I: Afibrinogenemia was one of the first congenital bleeding disorders recognized in animals. Breukind *et al.* (1972) identified this disorder in goats and established an important colony that exists to this day. Afibrinogenemia in dogs has been recognized in Dodds's laboratory. In humans, hypofibrinogenemia and dysfibrinogenemia also occur, but their counterparts have not been defined in other mammals.

Factor II: A true congenital deficiency of prothrombin (factor II) has not been described and must be extremely rare or not compatible with life, but dysprothrombinemias have been described in man and dogs.

Factor V: D/D of factor V has been seen only in man.

Factors VII and X: D/Ds of these two vitamin-K-dependent factors are rare in man and have been reported in dogs.

Factor VIII: D/D of factor VIII, called *hemophilia A,* is the most frequent hemorrhagic disease in man, occurring in about 1 in 10,000 males. It has been described in dogs, cats, and horses. It occurs almost entirely in males and is inherited in a sex-linked pattern. Hemophilia A has been identified in many different strains or breeds of dogs: Arctic sled dogs, Labradors, Pomeranian, poodles, retrievers, St. Bernard, Samoyed, and vizsla. The gene mutation that causes hemophilia A must have occurred in the dim past or been replicated many times.

Factor IX: D/D of factor IX in humans is called *hemophilia B* and appears to relate to many different molecular changes in the structure of this vitamin-K-dependent protein. D/D of factor IX has been reported in dogs. Both human and dog diseases are inherited in sex-linked patterns. It will be fascinating to see whether dog factor IX D/D has multiple etiological molecular patterns.

Factor XI: D/D of factor XI is quite frequent in man and occurs predominantly in families derived from the Mediterranean area. Bleeding symptoms are seen in only about half the families. Factor XI D/D has also been seen in dogs and cats.

Factor XII: D/D of factor XII without bleeding symptoms has been found in occasional humans and about 5% of alley cats.

vW factors: Many variants of von Willebrand's disease serve as the most frequent causes of mild to severe bleeding in man. Similar diseases have been diagnosed in dogs, cats, pigs, and rabbits. One pig cohort has been studied extensively by Dr. Walter Bowie and the Mayo Clinic Group and appears to have a disease similar to human type III von Willebrand's. In this von Willebrand's pig colony, a new disease apparently due to a platelet-related problem is occurring (Thiele *et al.,* 1986).

Platelets or platelet factors: Hereditary platelet disorders in man are relatively rare and may be quantitative or qualitative. Similar platelet-related disorders have been described in dogs, cats, cattle, rats, and now pigs.

3.4. A Hemophiliac Dog Colony

Our hemophiliac dog colony (J. H. Lewis *et al.*, 1983) was built following the gift of a male hemophiliac Pomeranian by Dr. Paul Didisheim (Didisheim and Bunting, 1964). In order to increase the dog size, this small lapdog was mated with three normal beagle bitches. From the first litters, two females, Nos. 1050 and 1056 (carriers by definition), were mated first with large normal male beagles. The five possible types of matings are shown in Fig. 27.2. Type I occurred when the original hemophiliac dog was paired with normal beagles. Type II occurred three times when the two second-generation carriers (1050 and 1056) were mated with large beagles. Type III occurred eight times. Type IV is illustrated in Fig. 24.4 and Type V in Fig. 24.3. A total of 29 hemophiliac dogs (17 male and 12 female) were born and lived at least 6 months. At 6 months, the puppies were tested, and normals (XY or XX) or carriers (xX) were given away.

The offspring of the five possible patterns of mating (see Fig. 27.2) were as follows:

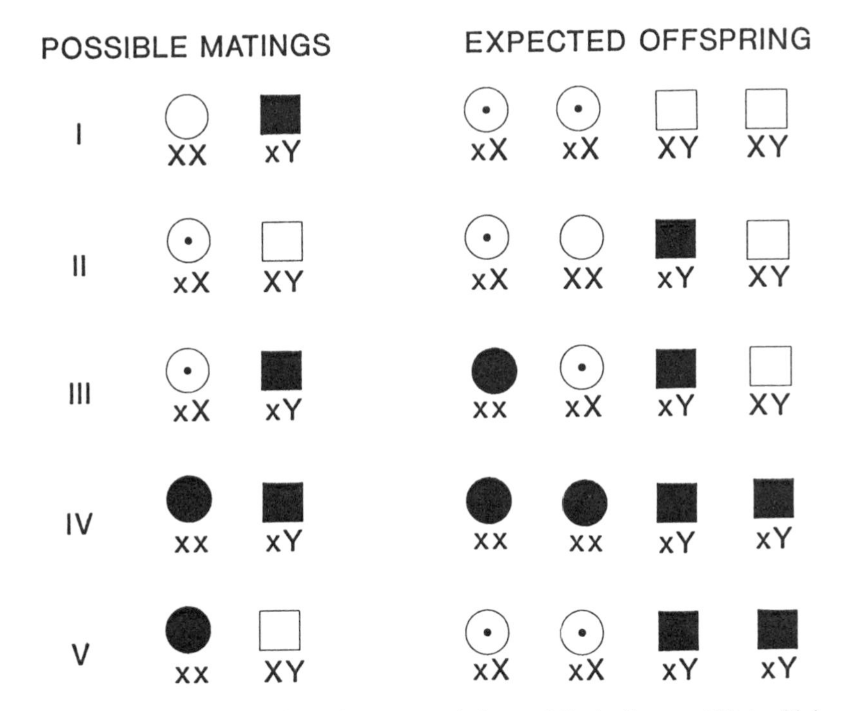

FIGURE 27.2. The five possible mating patterns in hemophilia A: (I) normal bitch with hemophiliac male; (II) carrier bitch with normal male; (III) carrier bitch with hemophiliac male; (IV) hemophiliac bitch with hemophiliac male; (V) hemophiliac bitch with normal male.

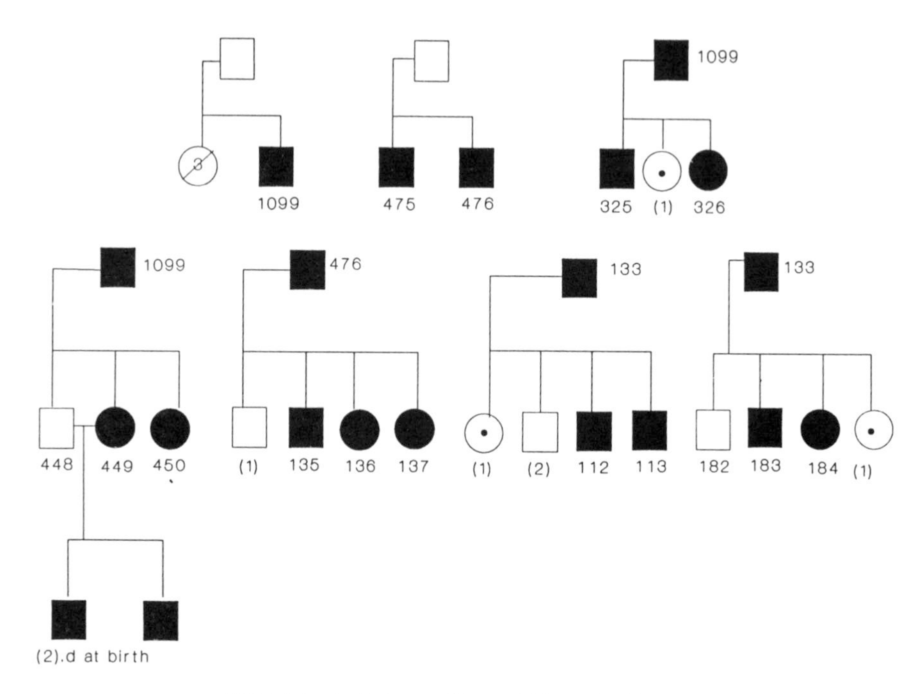

FIGURE 27.3. Matings of factor VIII carrier No. 1050.

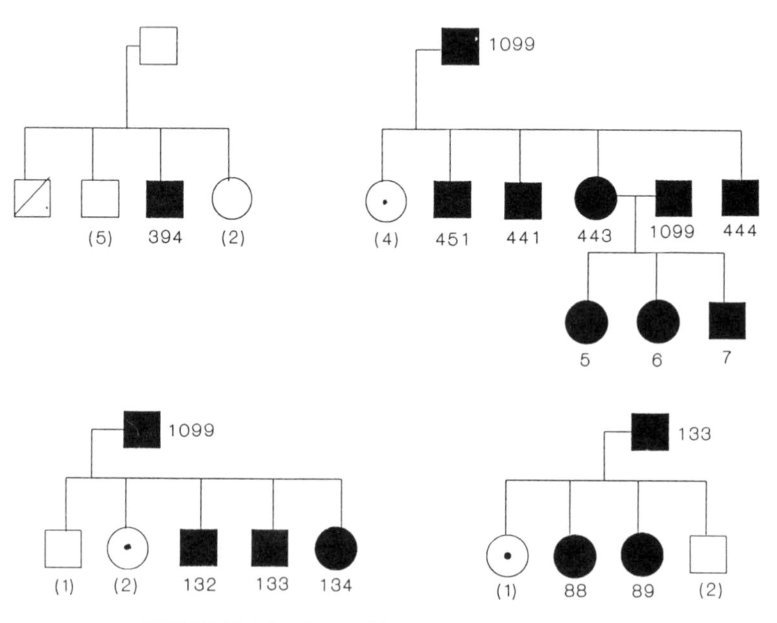

FIGURE 27.4. Matings of factor VIII carrier No. 1056.

I. Normal bitch (XX) with hemophiliac male (xY): The original Pomeranian mating illustrated the expected offspring: 50% xX, 50% XY.

II. Carrier (xX) with normal male (XY): 25% xX, 25% XX, 25% xY, 25% XY.

III. Carrier (xX) with hemophiliac male (xY): This type of mating should produce equal numbers of xx (female hemophiliacs), xX, xY, and XY. Type III matings occurred eight times in carriers No. 1050 and 1056. There were a total of 38 offspring: 10 xx, 10 xX, 10 xY, 8 XY. The gene distribution was almost as predicted.

IV. This type of mating, in which both parents are hemophiliacs (xx and xY), should produce all hemophiliac offspring: 50% female and 50% male. The only example of a type IV mating (No. 443 to No. 1099 in Fig. 27.4) did produce only hemophiliacs: two female and one male.

V. This type of mating, hemophiliac bitch (xx) with normal male (XY), was inadvertent and occurred between No. 449 and a male sibling (see Fig. 27.3). The mother, No. 449, and two male offspring, probably hemophiliac, died during delivery. The other male hemophiliac survived and prospered as a pet.

The hemophiliac dogs were frisky and healthy without visible deformities of the joints. Each dog was observed in the morning and late afternoon and treated with canine plasma cryoprecipitate when indicated. In the 29 hemophiliac dogs who lived at least 6 months, there was no evidence of hemarthrosis. Only 22 nonfatal treatable hemorrhages were seen in 6 years. These hemorrhages included subcutaneous and scalp hematomata, epistaxis, hematuria, melena, and one episode of postpartum bleeding. Response was prompt following infusion of 1–6 U canine cryoprecipitate. Two postcoital hemorrhages pointed the way for later precoital treatment with cryoprecipitate. Each unit was prepared from 200 ml of plasma obtained by plasmapheresis from a donor dog that was pheresed at regular intervals.

There were 21 deaths due to hemorrhage in these hemophiliac animals. In each case, the dog was observed and thought to be well on an afternoon check, then found dead the following morning. Autopsy examinations were made on all deceased dogs. Characteristically, the hemorrhages were into loose tissues or cavities where a great deal of blood could be lost before the back-pressure would slow the flow. The most common site was intraabdominal. One other dog died in anaphylactic shock following treatment with human factor VIII.

3.5. Summary

Many mammals, particularly those frequently observed and often inbred, have been found to suffer from bleeding or coagulation diseases similar to those encountered in humans. It is very important that these animals be maintained and bred in special colonies in which genetic patterns and disease characteristics can be studied.

REFERENCES

Breukind, H. J., Hart, H. C., Arkel, C., Velden, N. A., and Watering, C. C., 1972, Congenital afibrinogenemia in goats, *Zentralbl. Veterinaermed. Reihe A* **19:**661.78.

Didisheim, P., and Bunting, D. L., 1964, Canine hemophilia, *Thromb. Diath. Haemorrh.* **12:**377.

Lewis, J. H., Spero, J., and Hasiba, U., 1983, A hemophiliac dog colony: Genetic studies and coagulation findings in hemophiliac and normal dogs, *Comp. Biochem. Physiol.* **75:**147.

Mielka, C. H., Kaneshiro, M. M., Maher, I. A., Weiner, J. M., Rapaport, S. I., 1969, The standardized normal ivy bleeding time and its prolongation by aspirin, *Blood* **34:**294.

Thiele, G. L., Rempel, W. E., Fass, D. N., Bowie, E. J. W., Stewart, M., and Zoecklein, L., 1986, Inheritance of a new bleeding disease in a herd of swine with Willebrand's disease, *J. Hered.* **77:**179.

SUGGESTED READINGS

Fibrinolytic Enzyme System (Section 1.7)

Baillie, A. J., and Sim, A. K., 1971, Activation of the fibrinolytic enzyme system in laboratory animals and man—a comparative study, *Thromb. Diath. Haemorrh.* **25:**499.

Cliffton, E. E., and Cannamela, D. A., 1951, Variations in proteolytic activity of serum of animals including man, *Proc. Soc. Exp. Biol. Med.* **77:**305.

Hedlin, A. M., Monkhouse, F. C., and Milojevic, S. M., 1972, A comparative study of fibrinolytic activity in human, rat, rabbit and dog blood, *Can. J. Physiol. Pharmacol.* **50:**11.

Irfan, M., 1968, Fibrinolytic activity in animals of different species, *Q. J. Exp. Physiol.* **53:**374.

Kowalski, E., Budzynski, A. Z., Kopec, M., Latallo, Z. S., Lipinski, B., and Wegrzynowicz, Z., 1964, Studies on the molecular pathology and pathogenesis of bleeding in severe fibrinolytic states in dogs, *Thromb. Diath. Haemorr.* **12:**69.

Lewis, J. H., and Wilson, J. H., 1964, Fibrinogen breakdown products, *Am. J. Physiol.* **207:**1053.

Lewis, J. H., Szeto, I. F., Bayer, W. L., and Curiel, B. C., 1972, Leukofibrinolysis, *Blood* **40:**844.

Mason, R. G., and Read, M. S., 1971, Some species differences in fibrinolysis and blood coagulation, *J. Biomed. Mater. Res.* **5:**121.

Todd, A. S., 1964, Localization of fibrinolytic activity in tissues, *Br. Med. Bull.* **20:**210.

Tsapogas, M. J., and Flute, P. J., 1964, Experimental thrombolysis with streptokinase and urokinase, *Br. Med. Bull.* **20:**223.

Hereditary Hemorrhagic/Coagulation Diseases in Animals (Section 3)

Archer, R. K., 1961, True haemophilia (haemophilia A) in a thoroughbred foal, *Vet. Rec.* **73:**383.

Archer, R. K., and Allen, B. V., 1972, True haemophilia in horses, *Vet. Rec.* **85:**655.

Archer, R. K., and Bowden, R. S., 1959, A case of true haemophilia in a Labrador dog, *Vet. Rec.* **71:**560.

Aufderheid, W. M., Skinner, S. F., and Kaneko, J. J., 1975, Clearance of cryoprecipitated factor VIII in canine hemophilia A, *Am. J. Vet. Res.* **36:**367.

Bellars, A. R. M., 1969, Hereditary disease in British Antarctic sledge dogs, *Vet. Res.* **85:**600.

Benn, D. M., 1978, Classic hemophilia A in a family of collies, *Can. Vet. J.* **19:**221.

Benson, R. E., Bouma, B. N., and Dodds, W. J., 1975, Factor VIII–related antigen in canine hemophilia and von Willebrand's disease: Variation when measured with different agaroses, *Thromb. Res.* **7:**383.

Benson, R. E., Catalfamo, J. L., and Dodds, W. J., 1992, A multispecies enzyme-linked immunosorbent assay for von Willebrand's factor, *J. Lab. Clin. Med.* **119:**420.

Bentino-Smith, J., Roberts, S. J., and Katz, E. M., 1960, A bleeding disease of newborn calves, *Cornell Vet.* **50:**15.

Bird, R. M., Hampton, J. W., Lain, K. C., Brock, W. E., and Buckner, R. G., 1964, Mild canine hemophilia: Clinical and cytogenetic studies in females, *Trans. Am. Clin. Climatol. Assoc.* **76:**131.

Bogart, R., and Muhrer, M. E., 1942, The inheritance of a hemophilia-like condition in swine, *J. Hered.* **33:**59.

Bowie, E. J. W., Owen, C. A., Jr., Zollman, P. E., Thompson, J. H., Jr., and Fass, D. N., 1973, Tests of hemostasis in swine: Normal values in pigs affected with von Willebrand's disease, *Am. J. Vet. Res.* **34:**1405.

Bowie, E. J. W., Woods, J. E., Fass, D. N., Zollman, P. E., and Owen, C. A., Jr., 1975, Liver transplantation in pigs with von Willebrand's disease, *Br. J. Haematol.* **31:**37.

Bowie, E. J., Fass, D. N., and Owen, C. A., 1980, Hemostatic effect of transfused Willebrand factor in porcine von Willebrand's disease: Similarities to the human disease, *Haemostasis* **9:**352.

Bowie, E. J. W., Solberg, L. A., Jr., Fass, D. N., Johnson, C. M., Knutson, G. J., Stewart, M. L., and Zoecklein, L. J., 1986, Transplantation of normal bone marrow into a pig with severe von Willebrand's disease, *J. Clin. Invest.* **78:**26.

Brinkhous, K. M., and Graham, J. B., 1950, Hemophilia in the female dog, *Science* **111:**723.

Brinkhous, K. M., Morrison, F. C., Jr., and Muhrer, M. E., 1952, Comparative study of clotting defects in human, canine, and porcine hemophilia, *Fed. Proc. Fed. Am. Soc. Exp. Biol.* **11:**409.

Brinkhous, K. M., Davis, P. D., Graham, J. B., and Dodds, W. J., 1973, Expression and linkage of genes for X-linked hemophilias A and B, *Blood* **41:**577.

Brock, W. E., Buchner, R. G., Hampton, J. W., Bird, R. M., and Wulz, C. E., 1963, Canine hemophilia: Establishment of a new colony, *Arch. Pathol.* **76:**464.

Brown, R. C., Swanton, M. C., and Brinkhous, K. M., 1963, Canine hemophilia and male pseudohermaphroditism, *Lab. Invest.* **12:**961.

Brown, R. C., Cole, K. W., Cornell, C. N., and Brinkhous, K. M., 1968, Haemophilia-like disease in swine: Cytogenetic studies, *Am. J. Vet. Res.* **29:**1491.

Buckner, R. G., Hampton, J. W., Bird, R. M., and Brock, W. E., 1967, Hemophilia in the vizsla, *J. Small Anim. Pract.* **8:**511.

Capel-Edwards, K., and Hall, D. E., 1968, Factor VII deficiency in the beagle dog, *Lab. Anim.* **2:**105.

Cook, W. R., 1974, Epistaxis in the racehorse, *Equine Vet. J.* **6:**45.

Cornell, C. N., Cooper, R. G., Muhrer, M. E., and Garb, S., 1972, Splenectomy and factor VIII response in bleeder swine, *Am. J. Physiol.* **22:**1610.

Cornell, C. N., Cooper, R. G., Muhrer, M. E., and Garb, S., 1972, Effect of plasma and platelet concentrate infusion on factor VIII, platelet adhesiveness, and bleeding time in bleeder swine, *Thromb. Diath. Haemorrh.* **28:**431.

Cotter, S. M., Brenner, R. M., and Dodds, W. J., 1978, Hemophilia A in three unrelated cats, *J. Am. Vet. Med. Assoc.* **172:**166.

Dam, H., 1934, Haemorrhages in chicks reared on artificial diets: A new deficiency disease, *Nature (London)* **133:**909.

Dodds, W. J., 1966, Familial canine thrombocytopathy, *Blood* **28:**1013.

Dodds, W. J., 1967, Familial canine thrombocytopathy: Platelets: Their role in hemostasis and thrombosis, *Thromb. Diath. Haemorrh. Suppl.* **26:**241.

Dodds, W. J., 1968, Current concepts of hereditary coagulation disorders in dogs, *Exp. Anim.* **1:**243.

Dodds, W. J., 1970, Congenital thrombopathies and related coagulation disorders in dogs, in: *Proceedings of "Les mutants pathologiques chez l'animal, leur intérèt dans la recherche bio-médicale"* (M. Sabourdy, ed.), pp. 317–326, CNRS Press, Paris.

Dodds, W. J., 1970, Canine von Willebrand's disease, *J. Lab. Clin. Med.* **76:**713.

Dodds, W. J., 1971, Hemorrhagic disorders, in: *Current Veterinary Therapy,* 4th ed. (R. W. Kird, ed.), p. 247, W. B. Saunders, Philadelphia.

Dodds, W. J., 1973, Canine factor X (Stuart-Prower factor) deficiency, *J. Lab. Clin. Med.* **82:**560.

Dodds, W. J., 1974, Hereditary and acquired hemorrhagic disorders in animals, in: *Progress in Hemostasis and Thrombosis,* Vol. 2 (T. H. Spaet, ed.), pp. 215–247, Grune & Stratton, New York.

Dodds, W. J., 1975, Further studies in canine von Willebrand's disease, *Blood* **45:**221.

Dodds, W. J., 1975, Inherited hemorrhagic disorders, *J. Am. Anim. Hosp. Assoc.* **11:**366.

Dodds, W. J., and Blaisdell, F. S., 1972, Platelet and fibrinogen survivals in canine von Willebrand's disease, Program of the 15th Meeting of the American Society of Hematology, Hollywood, FL.

Dodds, W. J., and Kull, J. E., 1971, Canine factor XI (plasma thromboplastin antecedent) deficiency, *J. Lab. Clin. Med.* **78:**746.

Dodds, W. J., Packham, M. A., Rowsell, H. C., and Mustard, J. F., 1967, Factor VII survival and turnover in dogs, *Am. J. Physiol.* **213:**36.

Dodds, W. J., Webster, W. P., Brinkhous, K. M., Owen, C. A., Jr., and Bowie, E. J. W., 1975, Porcine and canine von Willebrand's disease, in: *Handbook of Hemophilia* (K. M. Brinkhous and H. C. Hemker, eds.), pp. 140–147, Excerpta Medica, Amsterdam.

Fass, D. N., Didisheim, P., Lewis, J. C., and Grabowski, E. F., 1976, Adhesion of porcine von Willebrand (VWD) platelets, *Circ. Suppl.* **54:**116 (abstract).

Fass, D. N., Knutson, G. J., and Bowie, E. J. W., 1978, Porcine Willebrand factor: A population of multimers, *J. Lab. Clin. Med.* **91:**307.

Fass, D. N., Bowie, E. J. W., Owen, C. A., Jr., and Zollman, P. E., 1979, Inheritance of porcine von Willebrand's disease: Study of a kindred of over 700 pigs, *Blood* **53:**712.

Fay, P. J., Cavallaro, C., Marder, V. J., Pancham, N., Fournel, M. A., and Schroeder, D. D., 1986, Comparison with the in vivo survival of human factor VIII with and without von Willebrand factor in the hemophilic dog, *Thromb. Res.* **37:**425.

Feldman, B. F., 1981, Coagulopathies in small animals, *J. Am. Vet. Med. Assoc.* **179:**559.

Feldman, B. F., and Giacopuzzi, R. L., 1982, Hemophilia A (factor VIII deficiency) in a colt, *Equine Pract.* **4:**24.

Field, R. A., Rickard, C. G., and Hutt, F. B., 1946, Hemophilia in a family of dogs, *Cornell Vet.* **36:**283.

Flute, P. T., and Howard, A. N., 1959, Blood coagulation in scorbutic guinea-pigs: A defect in activation by glass contact, *Br. J. Haematol.* **5:**421.

Fuster, V., Bowie, E. J. W., Lewis, J. C., Fass, D. N., Owen, C. A., and Brown, A. L., 1978, Resistance to arteriosclerosis in pigs with von Willebrand's disease, *J. Clin. Invest.* **61:**722.

Garner, R., Hermoso-Perez, C., and Conning, D. M., 1967, Factor VII deficiency in beagle dog plasma and its use in the assay of human factor VII, *Nature (London)* **216:**1130.

Gentry, P. A., 1977, Diagnosis of classic hemophilia A in a standard poodle, *Can. Vet. J.* **18:**79.

Gentry, P. A., Crane, S., and Lotz, F., 1975, Factor XI (plasma thromboplastin antecedent) deficiency in cattle, *J. Can. Vet. Med. Assoc.* **16:**160.

Giles, A. R., Tinlin, S., and Greenwood, R., 1982, A canine model of hemophilic (factor VIII:C deficiency) bleeding, *Blood* **60:**727.

Graham, J. B., and Barrow, E. M., 1957, The pathogenesis of hemophilia, *J. Exp. Med.* **106:**273.

Graham, J. B., Buckwalter, J. A., Hartley, L. J., and Brinkhous, K. M., 1949, Canine hemophilia: Observations on the course, the clotting anomaly, and the effect of blood transfusions, *J. Exp. Med.* **90:**97.

Graham, J. B., Collins, D. L., Jr., Godwin, I. D., and Brinkhous, K. M., 1951, Assay of plasma antihemophilic activity in normal heterozygous (hemophilia) and prothrombopenic dogs, *Proc. Soc. Exp. Biol. Med.* **77:**294.

Graham, J. B., Brinkhous, K. M., and Dodds, W. J., 1975, Canine and equine hemophilia, in: *Handbook of Hemophilia* (K. M. Brinkhous and H. C. Hemker, eds.), pp. 119–139, Excerpta Medica, Amsterdam.

Green, R. A., and White, F., 1977, Feline factor XII (Hageman) deficiency, *Am. J. Vet. Res.* **38:**893.

Greenlee, J. C., and Carper, H. A., 1968, Blood coagulation. III. Hemophilia A, *Am. J. Vet. Clin. Pathol.* **2:**27.

Griggs, T. R., Webster, W. P., Cooper, H. A., Wager, R. W., and Brinkhous, K. M., 1974, Von Willebrand factor: Gene dosage relationships and transfusion response in bleeder swine—a new bioassay, *Proc. Natl. Acad. Sci. U.S.A.* **71:**2087.

Groth, C. G., Hathaway, W. E., Custafsson, A., Geis, W. P., Putnam, C. W., Bjorken, C., Porter, K. A., and Starzl, T. E., 1974, Correction of coagulation in the hemophilic dog by transplantation of lymphatic tissue, *Surgery* **75:**725.

Hall, D. E., 1972, *Blood Coagulation and Its Disorders in the Dog,* Williams & Wilkins, Baltimore.

Hampton, J. W., Buckner, R. G., Gunn, C. G., Miller, L. R., and Mayes, J. W., 1973, Canine hemophilia in beagles: Genetics, site of factor VIII synthesis and attempts at experimental therapy, in: *Proceedings of the VII Congress of the World Federation of Haemophilia,* (F. Ala and K. W. E. Denson, eds.), p. 26, Excerpta Medica, Amsterdam.

Hathaway, W. E., Mull, M. M., Githens, J. H., Groth, C. G., Marchioro, T. L., and Starzl, T. E., 1969, Attempted spleen transplant in classical hemophilia, *Transplantation* **7:**73.

Hegreberg, G. A., 1975, Animal model: Ehlers–Danlos syndrome in dogs and mink, canine cutaneous asthenia, *Am. J. Pathol.,* **79:**383.

Hegreberg, G. A., Padgett, G. A., and Henson, J. B., 1970, Connective tissue disease of dogs and mink resembling the Ehlers–Danlos syndrome of man, *Arch. Pathol.* **90:**159.

Hinton, M., and Jones, D. R. E., 1977, A clotting defect in an Arab colt foal, *Equine Vet. J.* **9:**1.

Hogan, A. G., Muhrer, M. E., and Bogart, R., 1941, A hemophilia-like disease in swine, *Proc. Soc. Exp. Biol. Med.* **48:**217.

Hovig, T., Rowsell, H. C., Dodds, W. J., Jorgensen, L., and Mustard, J. F., 1967, Experimental hemostasis in normal dogs and dogs with congenital disorders of blood coagulation, *Blood* **30:**636.

Hovig, T., Dodds, W. J., Rowsell, H. C., and Mustard, J. F., 1968, The transformation of hemostatic platelet plugs in normal and factor IX deficient dogs, *Am. J. Pathol.* **53:**355.

Howard, A. N., and Flute, P. T., 1959, Defective blood coagulation in scorbutic guinea-pigs, *Proc. Nutr. Soc.* **18:**32.

Howell, J. M., and Lambert, P. S., 1964, A case of haemophilia A in the dog, *Vet. Rec.* **76:**1103.

Hutchins, D. R., Lepherd, E. E., and Crook, I. G., 1967, A case of equine hemophilia, *Aust. Vet. J.* **43:**83.

Hutt, F. B., Rickard, C. G., and Field, R. A., 1948, Sex-linked haemophilia in dogs, *J. Hered.* **39:**3.

Jones, D. R. E., and Hill, F. W. G., 1974, Suspected thrombasthenia in a Shetland sheepdog, *Vet. Rec.* **94:**558.

Kammerman, B., Gmur, J., and Stunzi, H., 1971, Afibrinogenamie beim Hund, *Zentralbl. Veterinaermed. Reihe A* **18:**192.

Kaneko, J. J., Cordy, D. R., and Carson, J., 1967, Canine hemophilia resembling classic hemophilia A, *J. Am. Vet. Med. Assoc.* **150:**15.

Kier, A. B., Bresnahan, J. F., White, F. J., and Wagner, J. E., 1980, The inheritance pattern of factor XII (Hageman) deficiency in domestic cats, *Can. J. Comp. Med.* **44:**309.

Kociba, G. J., Ratnoff, O. D., Loeb, W. F., Wall, R. L., and Heider, L. E., 1969, Bovine plasma thromboplastin antecedent (factor XI) deficiency, *J. Lab. Clin. Med.* **74:**37.

Langdell, R. D., Wagner, R. H., and Brinkhous, K. M., 1955, Antihemophilic factor (AHF) levels following transfusions of blood, plasma, and plasma fractions, *Proc. Soc. Exp. Biol. Med.* **88:**212.

Lewis, E. F., and Holman, H. H., 1951, Haemophilia in a St. Bernard dog, *Vet. Rec.* **63:**666.

Lewis, J. C., and Bowie, F. J. W., 1978, Ultrastructural studies of platelets of von Willebrand and normal swine, *Mayo Clin. Proc.* **53:**179.

McKinna, W. R., 1936, Haemophilia, *Vet. J.* **92:**370.

Merkens, J., 1938, Haemophilie bij honden, *Ned.-Ind. Bladen Diergeneeskd. Dierenteelt* **50:**149.

Mertz, E. T., 1942, The anomaly of a normal Duke's and a very prolonged saline bleeding time in swine suffering from an inherited bleeding disease, *Am. J. Physiol.* **136:**360.

Muhrer, M. E., Hogan, A. G., and Bogart, R., 1942, A defect in the coagulation mechanism of swine blood, *Am. J. Physiol.* **136:**355.

Muhrer, M. E., Lechler, E., Cornell, C. N., and Kirkland, J. L., 1965, Antihemophilic factor levels in bleeder swine following infusions of plasma and serum, *Am. J. Physiol.* **208:**508.

Mustard, J. F., Rowsell, H. C., Robinson, G. A., Hoeksema, T. D., and Downie, H. G., 1960, Canine haemophilia B (Christmas disease), *Br. J. Haematol.* **6:**259.

Mustard, J. F., Basser, W., Hedgardt, G., Secord, D., Rowsell, H. C., and Downie, H. G., 1962, A comparison of the effect of serum and plasma transfusions on the clotting defect in canine haemophilia B, *Br. J. Haematol.* **8:**36.

Mustard, J. F., Secord, D., Hoeksema, T. D., Downie, H. G., and Rowsell, H. C., 1962, Canine factor VII deficiency, *Br. J. Haematol.* **8:**43.

Myers, L. J., Pierce, K. R., Gowing, G. M., and Leonpacker, R. J., 1972, Hemorrhagic diathesis resembling pseudohaemophilia in a dog, *J. Am. Vet. Med. Assoc.* **161:**1028.

Nichols, T. C., Bellinger, D. A., Reddick, R. L., Read, M. S., Koch, G. G., Brinkhous, K. M., and Griggs, T. R., 1991, Role of von Willebrand factor in arterial thrombosis: Studies in normal and von Willebrand disease pigs, *Circulation* **83:**IV–56.

Norman, J. C., Covelli, V. H., and Sise, H. S., 1968, Transplantation of the spleen: Experimental cure of hemophilia, *Surgery* **64:**1.

Nossel, H., Archer, R. K., and MacFarlaine, R. G., 1962, Equine haemophilia: Report of a case and its response to multiple transfusions of hetrospecific AHG, *Br. J. Haematol.* **8:**355.

Osbaldiston, G. W., 1974, Hemophilia in the cat, *Bull. Am. Soc. Vet. Clin. Pathol.* **3:**64.

Osweiler, G. D., Prankratz, D. C., Prasse, K. W., Stahr, H. M., and Buck, W. B., 1970, Porcine hemorrhagic disease, *Mod. Vet. Pract.* **51:**35.

Owen, C. A., Jr., Bowie, E. J. W., Zollman, P. E., Fass, D. N., and Gordon, H., 1974, Carrier of porcine von Willebrand's disease, *Am. J. Vet. Res.* **35:**245.

Parks, B. J., Brinkhous, K. M., Harris, P. F., and Penick, G. D., 1969, Laboratory detection of female carriers of canine hemophilia, *Thromb. Diath. Haemorrh.* **12:**368.

Poller, L., Thomson, J. M., Sear, C. H. J., and Thomas, W., 1971, Identification of a congenital defect of factor VII in a colony of beagle dogs: The clinical use of the plasma, *J. Clin. Pathol.* **24:**626.

Reddick, R. L., Griggs, T. R., Lamb, M. A., and Brinkhous, K. M., 1982, Platelet adhesion to damaged coronary arteries: Comparison of normal and von Willebrand diseased swine, *Proc. Natl. Acad. Sci. U.S.A.* **79:**5076.

Rowsell, H. C., 1963, Hemorrhagic disorders in dogs: Their recognition, treatment and importance, in: *The Newer Knowledge about Dogs,* Twelfth Gaines Veterinary Symposium, p. 9.

Rowsell, H. C., Downie, H. G., Mustard, J. F., Leeson, J. E., and Archibald, J. A., 1960, A disorder resembling hemophilia B (Christmas disease) in dogs, *J. Am. Vet. Med. Assoc.* **137:**347.

Sanger, V. L., Mairs, R. E., and Trapp, A. L., 1964, Haemophilia in a foal, *J. Am. Vet. Med. Assoc.* **144:**259.

Saunders, C. N., Kinch, D. A., and Imlah, P., 1966, Thrombocytopenic purpura in pigs, *Vet. Rec.* **79:**549.

Sawada, Y., Fass, D. N., Katzmann, J. A., Bahn, R. C., and Bowie, E. J. W., 1986, Hemostatic plug formation in normal and von Willegrand pigs: The effect of the administration of cryoprecipitate and a monoclonal antibody to Willebrand factor, *Blood* **67:**1229.

Schofield, F. W., 1924, Damaged sweet clover: The cause of a new disease in cattle simulating haemorrhagic septicaemia and blackleg, *J. Am. Vet. Med. Assoc.* **64:**553.

Sharp, A. A., and Dike, G. W. R., 1963, Haemophilia in the dog: Treatment with heterologous antihaemophilic globulin, *Thromb. Diath. Haemorrh.* **10:**494.

Sherwood, L., Schmidt, D. A., Brett, I. J., and Howard, D. L., 1966, Canine hemophilia due to antihemophilic factor deficiency: A case report, *Mich. State Univ. Vet.* **26:**52.

Slappendahl, R. J., 1975, Hemophilia A and B in a family of French bulldogs, *Tijdschr. Diergeneeskd.* **100:**1075.

Spurling, N. W., Burton, L. K., Peacock, R., and Pilling, T., 1972, Hereditary factor VII deficiency in the beagle, *Br. J. Haematol.* **23:**59.

Spurling, N. W., Burton, L. K., and Pilling, T., 1974, Canine factor-VII deficiency: Experience with a modified thrombotest method in distinguishing between the genotypes, *Res. Vet. Sci.* **16:**228.

Spurling, N. W., Peacock, R., and Pilling, T., 1974, The clinical aspects of canine factor-VII deficiency including some case histories, *J. Small Anim. Pract.* **15:**229.

Storb, R., Marchioro, T. L., Graham, M. W., Hougie, C., and Thomas, E. D., 1972, Canine hemophilia and hemopoietic grafting, *Blood* **40:**234.

Stormorken, H., Egeberg, O., and Austad, R., 1965, Haemophilia A in Samoyed dog, *Scand. J. Haematol.* **2:**174.

Stuart, M. J., 1975, Inherited defects of platelet function, *Semin. Hematol.* **12:**233.

Swanton, M. C., 1959, Hemophilic arthropathy in dogs, *Lab. Invest.* **8:**1269.

Taskin, J., 1935, Un cas grave d'hémophylie chez une chienne, *Bull. Acad. Vet. Fr.* **8:**595.

Tschopp, T. B., and Zucker, M. B., 1972, Hereditary defect in platelet function in rats, *Blood* **40:**217.

Veltkamp, J. J., Asfaou, E., van de Torren, K., van der Does, J. A., van Tilburg, N. H., and Pauwels, E. K. J., 1974, Extrahepatic factor VIII synthesis: Lung transplants in hemophilic dogs, *Transplantation* **18:**56.

Wagner, R. H., Langdell, R. D., Richardson, B. A., Farrell, R. A., and Brinkhous, K. M., 1957, Antihemophilic factor (AHF): Plasma levels after administration of AHF preparations to hemophilic dogs, *Proc. Soc. Exp. Biol. Med.* **96:**152.

Weaver, R. A., Price, R. E., and Langdell, R. D., 1964, Antihemophilic factor in cross-circulated normal and hemophilic dogs, *Am. J. Physiol.* **206:**335.

Webster, W. P., Penick, G. D., Peacock, E. E., and Brinkhous, K. M., 1967, Allotransplantation of spleen in hemophilia, *N. C. Med. J.* **28:**505.

Webster, W. P., Mandel, S. R., Reddick, R. L., Wagner, J. L., and Penick, G. D., 1974, Orthotopic liver transplantation in canine hemophilia B, *Am. J. Physiol.* **226:**496.

Webster, W. P., Dodds, W. J., Mandel, S. R., and Penick, G. D., 1975, Biosynthesis of factors VIII and IX: Organ transplantation and perfusion studies, in: *Handbook of Hemophilia* (K. M. Brinkhous and H. C. Hemker, eds.), pp. 149–164, Excerpta Medica, Amsterdam.

White, J. M., and Holm, G. C., 1965, Colony of hemophilic dogs, *Science* **150:**1766.

Wurzel, H. A., and Lawrence, W. C., 1961, Canine hemophilia, *Thromb. Diath. Haemorrh.* **6:**98.

CHAPTER 28

COMPARATIVE HEMATOLOGY

1. ERYTHROCYTE SIZE AND SHAPE

1.1. Introduction

Erythrocyte size and shape vary widely. For example, there are striking differences between the small, round, nonnucleated cell of the mouse and the large, flat-oval, nucleated cells of the tailed amphibians: amphiuma (Congo eel) and necturus (mud puppy). These differences in erythrocyte size and shape could possibly be used for additional criteria to define the seven classes of Chordata (Cyclostomata, Chondrichthyes, Osteichthyes, Amphibia, Reptilia, Aves, and Mammalia), for which it is customary to use such other characteristics as skeletal structure, skin features (hair, feathers, scales), the site of oxygen exchange, and the modes of reproduction.

The first person to study animal erythrocytes or red blood cells (RBCs) was Antonie van Leeuwenhoek, a Dutch merchant who invented the microscope (Dobell, 1932; Schierbeek, 1959). In the decade 1673–1683, he built a number of single- and compound-lens microscopes that he used to examine different fluids, including frog blood. About 200 years later, Gulliver (1875), employing the excellent microscopes of his time, studied stained blood smears and published an extensive account of vertebrate red cell measurements. Since then, sporadic reports have appeared (Kisch, 1949; Wintrobe, 1961; Lucas and Jamroz, 1961; Hartman and Lessler, 1963, 1964). Figure 28.1 was prepared by a student in our laboratory to illustrate the relative sizes and shapes of erythrocytes and their nuclei in the major vertebrate classes.

1.2. Methods

Cell and nucleus measurements were performed on Wright's stained blood films with a carefully calibrated eyepiece grid at a magnification of 970×. For

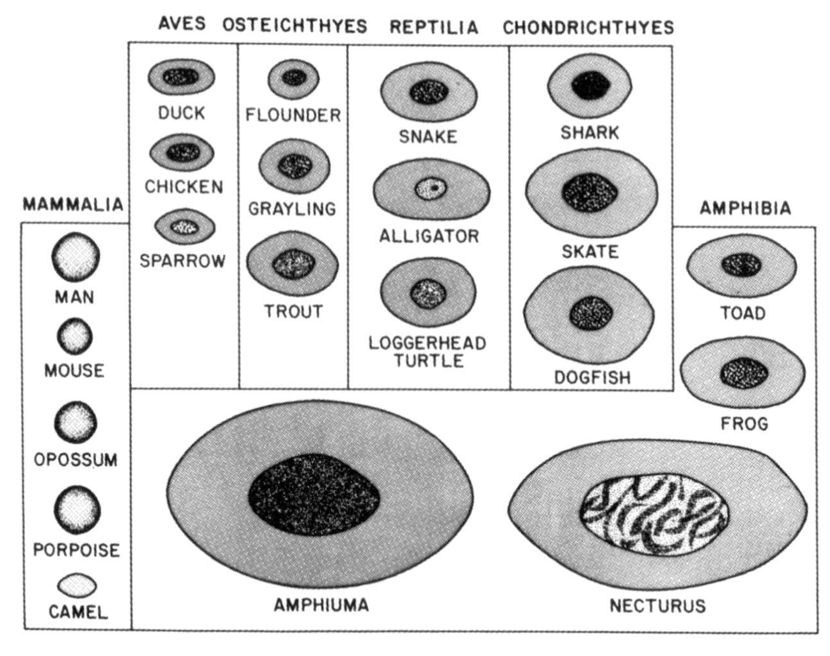

FIGURE 28.1. Relative sizes and shapes of vertebrate erythrocytes and their nuclei.

TABLE 28.1
Erythrocyte Measurements in Mammalia

Common name	Length (μM)	Width (μM)	L/W
Camel	6.6	3.9	1.69
	Diameter (μM)		
Armadillo	8.1	Human	7.3
Bat	5.7	Monkey	7.2
Cat	5.0	Mouse	5.8
Cow	5.5	Opossum	6.8
Dog	6.3	Pig	5.8
Elephant	9.2	Porpoise	8.1
Goat	3.2	Rabbit	6.1
Guinea Pig	7.1	Rat	6.0
Hamster	6.6	Seal	7.2
Horse	6.0	Sheep	4.8

each species, 20–50 RBCs were measured. The mean values are presented in Tables 28.1 and 28.2. Also included in these Tables are the hematocrit (HCT), hemoglobin (HGB), red blood cell (RBC) count, and the calculated index mean corpuscular volume (MCV).

1.3. Mammalia

Table 28.1 lists the average values for HCT, HGB, and RBCs of 34 mammals, the diameter or length × width (μm), and the MCV. The animals are listed in order of MCV from the one with the largest erythrocyte (manatee) to the one with the smallest (goat). MCV was used rather than diameter because it was available for all mammals. On smear, all the mammalian RBCs appeared as round disks except those of Camelidae, which were elliptical. The MCV values corresponded fairly well to the measured sizes. Differences may reflect cell thicknesses.

Table 28.1 also lists mean corpuscular hemoglobin (MCH), mean corpuscular hemoglobin concentration (MCHC), and white blood cell (WBC) parameters. There was a wide range of WBC counts: 4.0–17.5 × $10^3/cm^3$. In many of the animals, the lymphocyte was the predominant cell. Platelet counts (Plat Cs) are also listed. They varied from 109 to 1180 without relationship to other parameters.

1.4. Premammalian Vertebrates

The RBCs of the premammalian vertebrates are oval and nucleated, as shown in Table 28.2. Within each class, the species are listed in order of decreasing erythrocyte length. The tailed Amphibia, the Urodela, clearly had the largest erythrocytes. Cytosome lengths of the smaller Anura overlapped those of the largest Chondrichthyes. In turn, the lengths of the Chondrichthyes overlapped those of Reptilia and a few fish. The latters' lengths were similar to those of Aves, but the cells and nuclei were rounder.

1.5. Cell and Nucleus Dimensions

The length/width ratios for cytosome and nucleus were calculated and listed in Table 28.2. Roundness is indicated by ratios of 1.0–1.6, flatness by ratios greater than 1.6. These findings are summarized in Table 28.3.

Mammalian RBCs were round (ratio 1.0), with the exception of Camelidae

TABLE 28.2
Erythrocyte Values and Dimensions in Premammalian Blood[a]

Class and common name	HCT (%)	HGB (g/dl)	RBCs (10 10^6/mm^3)	MCV (fl)	Cytosome Length (μM)	Cytosome Width (μM)	Cytosome L/W ratio	Nucleus Length (μM	Nucleus Width (μM)	Nucleus L/W ratio
Chondrichthyes										
Skate	23	3.4	0.21	1095	23.6	16.4	1.44	6.4	5.1	1.25
Spiny dogfish	30	5.1	0.40	750	23.1	15.4	1.50	8.7	5.7	1.53
Blue shark	20	5.4	0.25	800	21.6	15.4	1.40	7.9	6.8	1.16
Black-tipped shark	34	4.7	0.46	739	15.1	10.4	1.45	5.8	4.3	1.35
Osteichthyes										
Toadfish	16	2.5	0.20	800	17.1	11.5	1.49	5.9	4.3	1.37
Artic trout	—	—	—	—	16.0	11.3	1.42	6.5	4.4	1.48
Artic grayling	—	—	—	—	12.2	9.1	1.34	5.4	3.8	1.42
Sea bass	33	5.9	1.90	173	11.8	8.2	1.44	4.7	3.6	1.31
Goosefish	21	4.5	0.96	215	11.6	9.2	1.26	5.6	4.2	1.33
Tautog	49	9.5	4.8	102	10.8	7.8	1.38	4.4	3.5	1.26
Grouper	24	—	2.10	114	10.1	8.0	1.26	4.1	3.7	1.11
Red snapper	22	3.4	1.04	211	9.7	5.8	1.67	3.8	2.7	1.40
Trigger	47	4.3	3.69	127	9.2	6.6	1.39	4.0	2.7	1.48
Porgy	37	6.1	1.76	210	8.8	7.5	1.17	4.0	3.0	1.33
Flounder	24	8.3	2.00	120	8.8	6.3	1.40	3.8	2.6	1.46
Orange filefish	35	4.4	1.93	181	8.2	7.2	1.14	3.9	2.9	1.34
Cobia	35	—	1.35	259	8.0	6.0	1.33	3.7	2.7	1.37
Amphibia										
Congo eel	41	14.8	0.056	7391	54.2	32.5	1.67	24.3	15.3	1.59
Mud puppy	27	8.2	0.038	7325	54.2	26.3	1.98	25.2	13.8	1.83
Hellblender	34	8.6	0.091	3938	39.7	25.4	1.56	21.4	10.0	2.14

TABLE 28.2 *(Continued)*

Class and common name	HCT (%)	HGB (g/dl)	RBCs (10 $10^6/mm^3$)	MCV (fl)	Cytosome Length (μM)	Cytosome Width (μM)	Cytosome L/W ratio	Nucleus Length (μM	Nucleus Width (μM)	Nucleus L/W ratio
Tiger salamander	33	8.8	0.124	2698	28.1	15.1	1.86	11.7	7.0	1.66
Bullfrog	30	9.2	0.44	714	25.0	12.4	2.02	9.9	5.7	1.74
Grass frog	31	7.0	0.50	654	21.5	12.2	1.76	8.8	5.2	1.69
Toad	24	6.8	0.33	755	19.0	10.3	1.84	7.7	3.9	1.97
Reptilia										
Alligator	33	13.8	0.77	433	20.1	10.4	1.93	4.8	3.8	1.26
Soft-shelled turtle	—	—	—	—	17.4	10.5	1.66	5.3	3.8	1.39
Painted turtle	—	—	—	—	17.3	11.2	1.54	6.2	4.5	1.38
Pond turtle	21	5.6	0.63	333	17.3	8.4	2.06	5.1	3.9	1.30
Loggerhead turtle	34	8.5	0.46	739	17.0	12.6	1.35	5.7	4.5	1.27
Box turtle	—	—	—	—	17.0	9.5	1.79	5.5	3.9	1.41
Snake	31	7.9	1.22	254	16.9	10.0	1.69	6.0	3.9	1.54
Green turtle	40	5.9	0.84	476	16.8	10.1	1.66	4.4	3.3	1.33
Musk turtle	—	—	—	—	16.3	9.3	1.75	5.2	4.0	1.30
Iguana	36	6.3	1.31	275	14.3	7.8	1.83	5.4	4.0	1.35
Aves										
Goose	45	14.5	2.9	153	14.3	8.2	1.74	7.6	2.5	3.04
Duck	40	13.2	2.6	158	11.6	6.1	1.90	5.4	2.1	2.57
Pigeon	30	13.5	3.1	95	11.4	6.5	1.75	5.8	2.3	2.54
Chicken	37	11.6	2.7	135	10.7	6.3	1.70	4.5	2.7	1.67

[a]Listed in order of decreasing cell length within each class.

TABLE 28.3
Vertebrate Class Characteristics:
Erythrocyte Length and Shapes of Cell and Nucleus

Name	Cell length (μM)	Cytosome shape	Nucleus shape
Mammalia	3.2–9.2	Round	None
Camelidae	6.6–7.5	Flat–oval	None
Osteichthyes	8–18	Round–oval	Round–oval
Aves	10–14	Flat–oval	Flat–oval
Reptilia	14–20	Flat–oval	Round–oval
Chondrichthyes	15–24	Round–oval	Round–oval
Amphibia	19–55	Flat–oval	Flat–oval

erythrocytes, which were slightly flat–oval (L/W ratio 1.6–1.7). They were without nuclei.

Avian erythrocytes were flat–oval (L/W ratio >1.6) in shape with flat–oval nuclei and were small, varying in overall length from 10.7 to 14.3 μm. They were easily distinguished from fish erythrocytes, which may have the same overall length but have almost round cytosomes and nuclei.

Reptilian RBCs varied in length from 14+ to 20+ μm, and often the cytosome was flat–oval in shape and the nucleus much more round. Notable exceptions were most snake erythrocytes, which have been described by Hartman and Lessler (1964) as having round–oval cytosomes with long nuclei.

Amphibian erythrocytes were usually very large, 19–54 μM in length and flat–oval in shape cells with moderately flat nuclei. Notably, the RBCs of the urodeles were much larger than those of frogs and toads.

Osteichthyes and Chondrichthyes were difficult to distinguish on the basis of erythrocyte shape. Both had roundish cells with correspondingly shaped nuclei. As a general rule, cells from elasmobranchs were larger than those from the bony fish.

1.6. Summary

Our observations are in good agreement with those in the published literature. The extensive studies of Hartman and Lessler (1964) on 124 avian species describe all as small long–oval cells with long–oval nuclei. Of particular interest are the handsome colored photographs in the *Atlas of Avian Hematology* of Lucas and Jamroz (1961). These authors showed that erythrocytes from chick embryos were quite round with round nuclei, in contrast to the elongated oval adult cells. The studies reported in this volume have confirmed and extended previous works by others. They emphasize cell/nucleus dimensions as vertebrate class characteristics.

2. OSMOTIC FRAGILITY OF VERTEBRATE ERYTHROCYTES

2.1. Introduction

The measurement of osmotic fragility assesses the strength of the RBC membrane to withstand internal expansion following exposure to hypotonic solutions (usually NaCl). In order to equalize the internal and external ionic strength, water enters the cell through its semipermeable surface membrane, expanding the cell's internal volume. When the surface membrane has stretched to its full capacity, it ruptures, emitting the cell contents—hemoglobin—which can be assayed. Cells that lyse at high saline concentrations, 0.7% or above, are said to be *fragile,* while those that do not rupture until low concentrations, 0.5% or below, are *resistant.*

The osmotic fragility of erythrocytes from man and domestic and laboratory mammals has been studied extensively in both healthy and disease states (Wintrobe, 1956; Altman, 1961; Schalm, 1961). Nucleated red cells of premammalian vertebrates have received little attention. Jacobs (1931) measured the rate of hemolysis in distilled water. Suboski and Colavita (1964) measured the osmotic fragility of erythrocytes from a few premammalian vertebrates and suggested that resistance to osmotic stress may be dependent on red cell size. Frei and Perk (1964) studied the osmotic behavior of chicken and frog erythrocytes by measuring the swelling in hypotonic solution. Laying hens and cocks differed from each other.

In the studies described below, the amount of hemolysis that occurred after mixture of blood with salt solutions of increasing concentrations was measured, and an attempt was made to relate this amount to the plasma osmolarity and RBC size.

2.2. Source

Whenever possible, a tube of fresh heparinized blood was obtained from the animals listed in the classification (Table 1.1). Sufficient plasma to measure the osmolarity was available from the animals listed in Table 28.4. Some of the results shown here have been previously published (Lewis and Ferguson, 1966) and are reprinted with the kind permission of the editor of *Comparative Biochemistry and Physiology.*

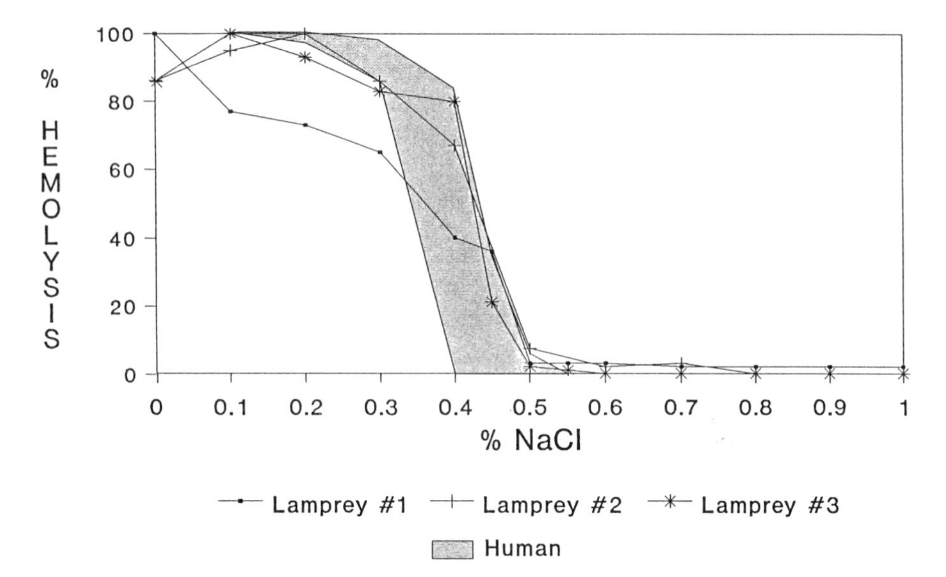

FIGURE 28.2. Osmotic fragility of three lampreys.

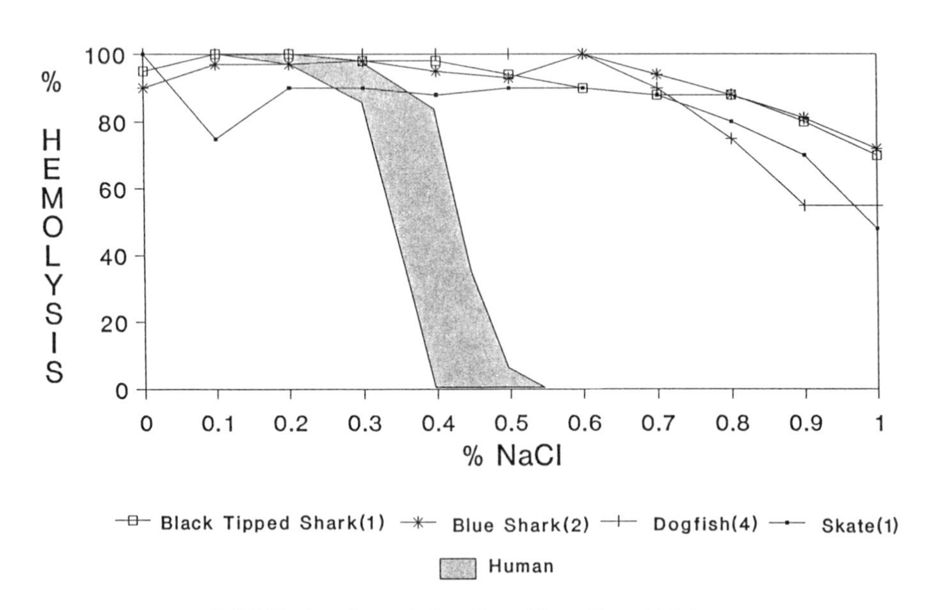

FIGURE 28.3. Osmotic fragility of four Chondrichthyes.

2.3. Cyclostomata

The osmotic fragility of the heparinized blood samples from three lampreys is shown in Fig. 28.2. Results resembled human or birds, with 50% lysis falling between 0.4 and 0.45% NaCl.

2.4. Chondrichthyes

The osmotic fragility of black-tipped reef shark, blue shark, dogfish, and skate is shown in Fig. 28.3. The erythrocytes were very fragile and lysed even in 1.0% NaCl. At 2.5% NaCl, the RBCs were stable.

2.5. Osteichthyes

Figure 28.4 shows the osmotic fragility curves for bony fish. All fish RBCs were more fragile than human, and there was a good deal of variation from species to species. No differences between freshwater (Arctic grayling and trout) and marine fishes were noted.

2.6. Amphibia

Figure 28.5 illustrates the curves obtained on several species of Amphibia. The RBCs from both urodeles and anurans were highly resistant to osmotic lysis and did not lyse until concentrations between 0.3% and 0.1% NaCl were reached. At higher concentrations of NaCl, small amounts of hemolysis, up to 10%, were often found.

2.7. Reptilia

Figure 28.6 demonstrates that reptilian erythrocytes were more resistant to osmotic lysis than human cells. Most of the curves fell sharply between 0.2% and 0.4% NaCl. Once no hemolysis was seen, the curves did not bounce up and down as did those of the amphibians.

2.8. Aves

Figure 28.7 shows that the osmotic fragility of bird RBCs closely resembled that of human cells.

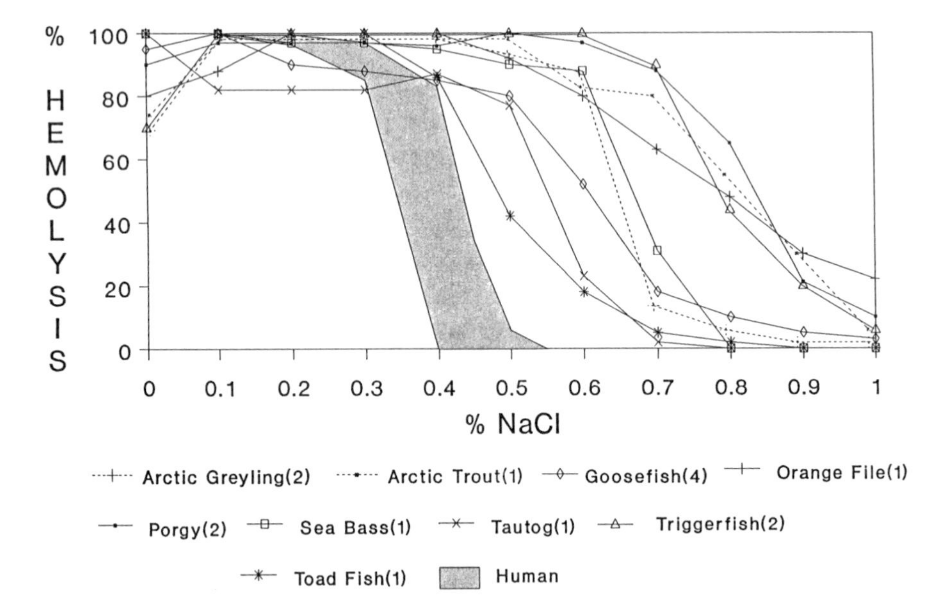

FIGURE 28.4. Osmotic fragility of nine Osteichthyes.

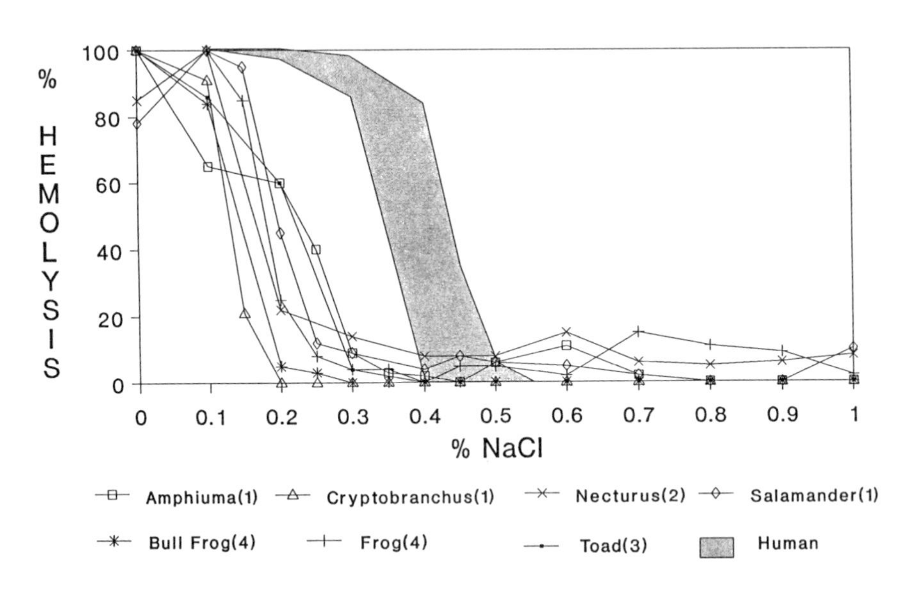

FIGURE 28.5. Osmotic fragility of seven Amphibia.

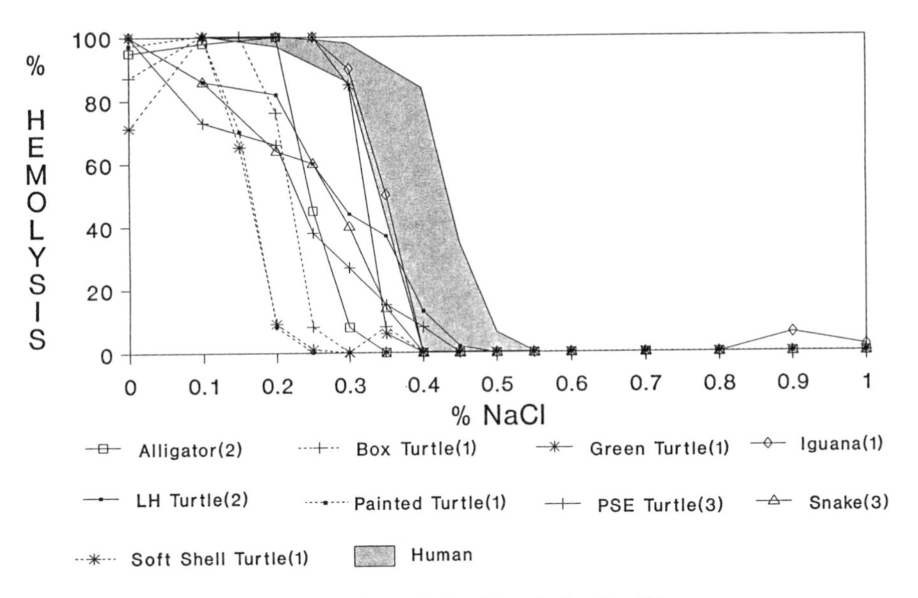

FIGURE 28.6 Osmotic fragility of nine Reptilia.

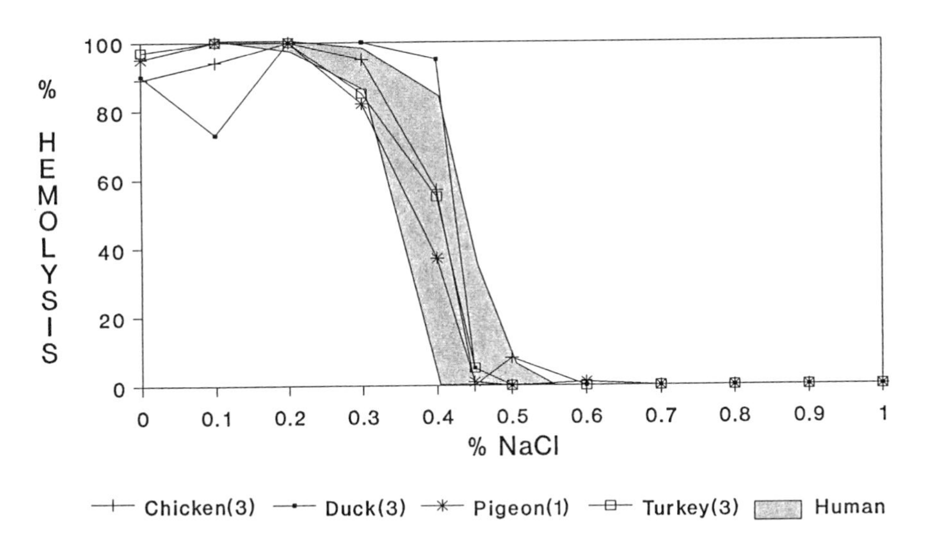

FIGURE 28.7. Osmotic fragility of four Aves.

2.9. Mammalia

Figures 28.8 and 28.9 show the 15 mammals that were studied. Many of the fragility curves fell within the human pattern or were slightly more fragile. The elliptical camel erythrocytes were more resistant, and seal and the small goat and sheep cells were more fragile, being partly lysed at 0.5% NaCl.

2.10. Comparison of Plasma Osmolarity and Erythrocyte Fragility

Table 28.4 illustrates the plasma osmolarity in those animals from which sufficient plasma was available to measure the freezing point. This was compared to the NaCl concentration that resulted in 50% lysis. The close correlation is apparent.

2.11. Osmotic Fragility versus Cell Size

Table 28.5 shows the lack of correlation between cell size and osmotic fragility.

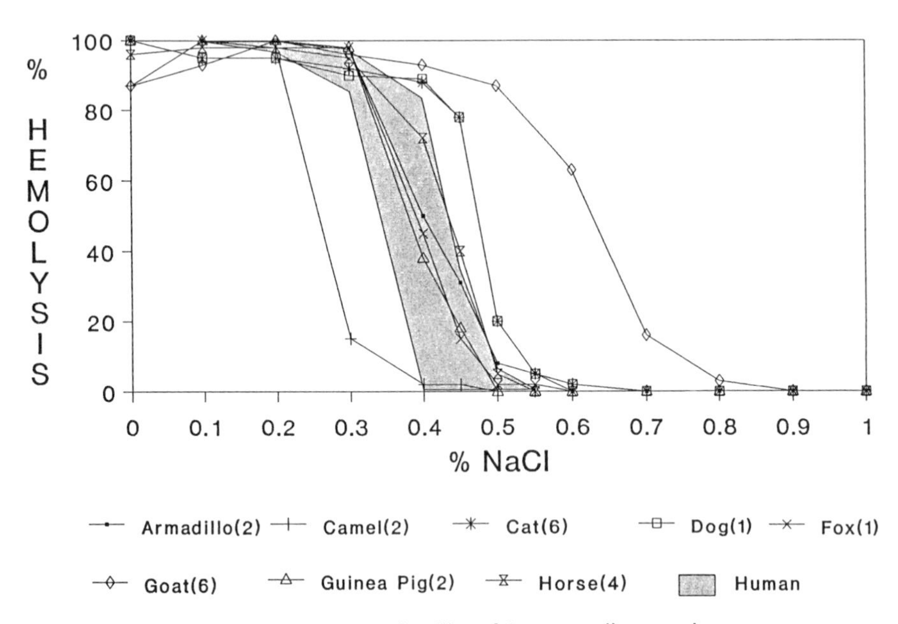

FIGURE 28.8. Osmotic fragility of 8 mammalian species.

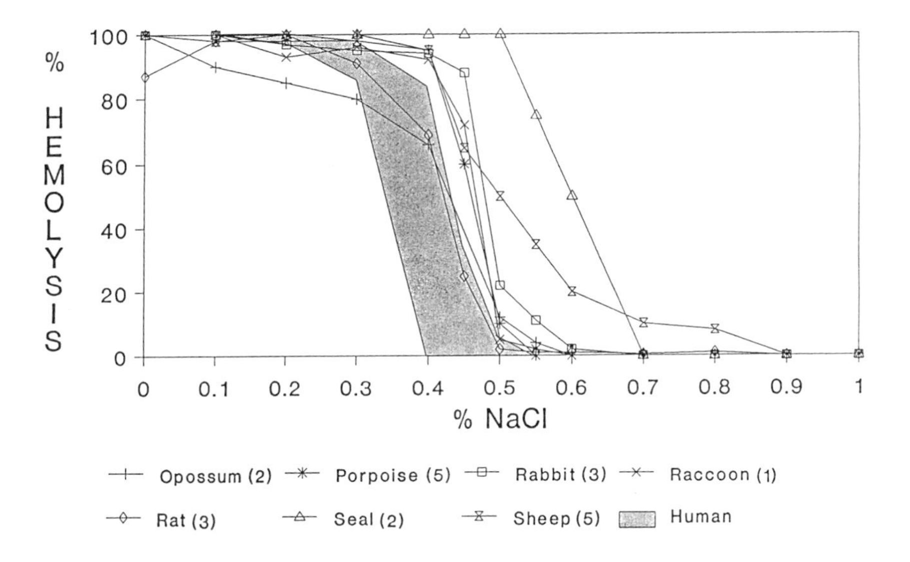

FIGURE 28.9. Osmotic fragilty of 7 mammalian species.

TABLE 28.4
Comparison of Plasma Osmolarity and Sodium Chloride Concentration That Resulted in 50% Lysis of Erythrocytes

Animal	Osmolarity (mosmoles/kg)	NaCl conc. (%) resulting in 50% lysis
Dogfish	1050	>1.00
Shark		
No. 2	990	>1.00
No. 3	848	>1.00
Grayling	403	0.81
Tautog	345	0.57
Chicken	345	0.41
Mammals (except camel and seal)	294–302	0.4–0.5
Green turtle	340	0.35
Alligator	320	0.22
Bullfrog	218	0.13

TABLE 28.5
Vertebrate Classes Listed in Order of Decreasing Osmotic Fragility and Increasing Erythrocyte Size

Fragility	Size (cell length)
Chondrichthyes	Mammalia
Osteichthyes	Osteichthyes
Aves	Aves
Mammalia	Reptilia
Reptilia	Chondrichthyes
Amphibia	Amphibia

2.12. Summary

The degree of hemolysis in graded NaCl solutions (the osmotic fragility) was determined on blood from 36 premammalian vertebrates of different species and from 15 mammalian species. Amphibian cells were the toughest—that is, the most resistant to hemolysis—lysing only in very dilute saline. Reptilian cells were the next most resistant; measurable hemolysis started between 0.4% and 0.25% saline. In birds, lampreys, and the majority of mammals, erythrocyte lysis started at 0.55% NaCl and was complete by 0.35%. Fish and elasmobranch cells were the most fragile, being partially lysed even in 1.0% saline. Despite these wide variations in fragility, the cells from each class were well adapted to their surrounding media, as shown by the good correlation between plasma osmolarity and NaCl concentration necessary to produce 50% lysis. Osmotic fragility did not correlate with cell size.

REFERENCES

Altman, P. L., 1961, *Biological Handbooks: Blood and Other Body Fluids*, p. 540, Federation of the American Society for Experimental Biology, Washington, DC.

Dobell, C., 1932, *Antonie van Leeuwenhoek and His "Little Animals,"* Harcourt, Brace, New York.

Frei, Y. F., and Perk, K., 1964, Osmotic hemolysis of nucleated erythrocytes, *Exp. Cell Res.* **35**:230.

Gulliver, G., 1875, Observations on the size and shape of the red corpuscles, *Proc. Zool. Soc. London* **1875**:474.

Hartman, F. A., and Lessler, M. A., 1963, Erythrocyte measurements in the bird, *Auk* **80**:467.

Hartman, F. A., and Lessler, M. A., 1964, Erythrocyte measurements in fishes, amphibia, and reptiles, *Biol. Bull.* **126**:83.

Jacobs, M. H., 1931, Osmotic hemolysis and zoological classification, *Proc. Am. Philos. Soc.* **70**:363.

Kisch, B., 1949, Hemoglobin content, size, and amount of erythrocytes in fishes, *Exp. Med. Surg.* **7:**118.
Lewis, J. H., and Ferguson, E. E., 1966, Osmotic fragility of pre-mammalian erythrocytes, *Comp. Biochem. Physiol.* **18:**589.
Lucas, A. M., and Jamroz, C., 1961, *Atlas of Avian Hematology,* U.S. Department of Agriculture, Washington, DC.
Schierbeek, A., 1959, *Measuring the Invisible World: The Life and Works of Antonie van Leeuwenhoek,* Abelard-Schuman, London and New York.
Schalm, O. W., 1961, *Veterinary Hematology,* Lea and Febiger, Philadelphia.
Suboski, M. D., and Colavita, F. B., 1964, Comparative osmotic hemolysis of some representative vertebrates, *Life Sci.* **3:**519.
Wintrobe, M. M., 1956, *Clinical Hematology,* 4th ed. revised, Lea and Febiger, Philadelphia.
Wintrobe, M. M., 1961, *Clinical Hematology,* Lea and Febiger, Philadelphia.

SUGGESTED READINGS

Erythrocyte Size and Shape

Holman, H. H., 1952, A negative correlation between size and number of the erythrocytes of cows, sheep, goats and horses, *J. Pathol. Bacteriol.* **64:**379.
Jain, N. C., 1972, Morphology of blood cells in three dimensions, *Calif. Vet.* **26:**16.
Mattern, C. F. T., Brackett, F. S., and Olson, B. J., 1957, Determination of number and size of particles by electrical gating: Blood cells, *J. Appl. Physiol.* **10:**55.
Wintrobe, M. M., 1933, Variations in the size and haemoglobin content of erythrocytes in the blood of various vertebrates, *Folia Haematol.* **51:**32.
Wintrobe, M. M., Shumacker, H. B., and Schmidt, W. J., 1936, Values for number, size, and haemoglobin content of normal dog, rabbits and rats, *Am. J. Physiol.* **114:**502.

Osmotic Fragility of Vertebrate Erythrocytes

Castle, W. B., and Daland, G. A., 1967, Susceptibility of mammalian species to hemolysis with hypotonic solutions, *Arch. Intern. Med.* **60:**949.
Coldman, M. F., Gent, M., and Good, W., 1969, The osmotic fragility of mammalian erythrocytes in hypotonic solutions of sodium chloride, *Comp. Biochem. Physiol.* **31:**605.
Coldman, M. F., Gent, M., and Good, W., 1970, Relationships between osmotic fragility and other species-specific variables of mammalian erythrocytes, *Comp. Biochem. Physiol.* **34:**759.
Creed, E. F. F., 1938, The estimation of the fragility of red blood corpuscles, *J. Pathol. Bacteriol.* **46:**331.
Fogel, B. J., Shields, C. E., and von Doenhoff, A. E., Jr., 1966, The osmotic fragility of erythrocytes in experimental malaria, *Am. J. Trop. Med. Hyg.* **15:**269.
Fox, M. W., 1964, Osmotic resistance of erythrocytes, *J. Hered.* **55:**146.
Hall, D. E., and Follett, A. J., 1972, Red cell osmotic fragility in the beagle dog: Normal values and diagnostic application, *Vet. Rec.* **91:**263.
Hudson, A. E. A., 1955, Fragility of erythrocytes in blood from swine of two age groups, *Am. J. Vet. Res.* **16:**120.

Jain, N. C., 1973, Osmotic fragility of erythrocytes of dogs and cats in health and in certain hematologic disorders, *Cornell Vet.* **63:**411.

Perk, K., Frei, Y. F., and Herz, A., 1964, Osmotic fragility of red blood cells of young and mature domestic and laboratory animals, *Am. J. Vet. Res.* **25:**1241.

Ponder, E., 1935, The measurement of red cell volume. VI. The different "fragility" of the red cells of various mammals, *J. Physiol.* **83:**352.

Schalm, O. W., 1965, *Veterinary Hematology,* 2nd ed., Lea and Febiger, Philadelphia.

Soliman, M. K., and El Amrousi, S., 1966, Erythrocyte fragility of healthy fowl, dog, sheep, cattle, buffalo, horse, and camel blood, *Vet. Rec.* **78:**429.

CHAPTER 29

COMPARATIVE SEROLOGY

1. SERUM PROTEINS AND CROSS-IMMUNOREACTIVITY

1.1. Introduction

A number of investigators have suggested a relationship between the evolutionary advancement of a vertebrate and the concentration and complexity of its serum proteins. This study measured total proteins and protein electrophoretic patterns in plasmas for 50 animal species representing the six major vertebrate classes. Paper electrophoresis employing sodium barbital buffer at pH 8.6 and ionic strength 0.05 was done on all the animals except the hedgehogs. Gel electrophoresis (Panagel) and immunoelectrophoresis (IEP) were also done on many of the animal plasmas and sera. Antisera to some of the animal plasmas were prepared in rabbits and used for the IEPs and for cross-immunoreactivity tests in agar gel.

1.2. Human Results

On electrophoresis at pH 8.6, normal human albumin migrated toward the anode and formed a heavy band. About 1% of human specimens showed a prealbumin band that migrated slightly more rapidly than albumin and was visualized just anodal to it. There were three or more globulin bands ending in a wide band migrating closest to the cathode and called γ-*globulin*. Fibrinogen was a narrow, sharp band just anodal to the point of application.

1.3. Mammalian Results

Electrophoretic patterns of 24 mammalian species showed many variations, but all were alike in visualizing a heavy albumin band (+++ to ++++) and at

TABLE 29.1
Total Protein, Albumin, and γ-Globulin in Premammalian Plasma or Serum

Animal	TP (g/dl)	Albumin	γ-Globulin	Animal	TP (g/dl)	Albumin	γ-Globulin
Chicken	3.6	++++	+	Arctic grayling	4.6	0	+
Duck	3.5	++++	++	Arctic trout	5.0	0	++
Goose	3.1	++++	0	Cobia	4.5	0	+
Turkey	4.5	++++	+++	Codfish	3.5	0	++
Turtle				Flounder	2.9	0	0
Green	3.1	0	+++	Goosefish	1.7	0	0
Loggerhead	4.5	0	+++	Grouper	3.9	0	0
PSE	2.4	0	++	Red snapper	3.4	0	+
Iguana	4.4	++++	+	Sculpin	3.4	0	0
Snake	6.3	0	+	Sea bass	2.6	0	+
Alligator	5.2	+++	++++	Sea robin	2.8	0	0
Amphiuma	0.9	0	++	Shark, black-tipped	2.5	0	+
Necturus	1.9	0	++	Dogfish	1.9	0	+
Salamander	1.8	0	++	Guitarfish	2.4	0	++
Frog	2.9	++	++	Skate	2.3	0	+
Toad	2.8	++	++	Mammal (high)	8.9	++++	++++
Mammal (low)	4.8	+++	+++				

least three globulin peaks with a γ-globulin of +++ to ++++. Total protein (TP) of mammals ranged from 4.8 to 8.9 g/dl.

1.4. Premammalian Results

Table 29.1 gives the TP, albumin, and γ-globulin of 30 premammalian species. Avian plasmas showed many mammalian characteristics. Approximately 50% of the total protein was albumin and prealbumin was common. γ-Globulin was minimal or absent in chicken and goose.

Reptilian proteins varied greatly from species to species. The three turtles were alike in being analbuminemic, a finding that supports the publication of Crenshaw (1962). Snake was also analbuminem with low γ-globulin. Iguana and alligator plasmas showed heavy albuminlike bands with a mobility somewhat greater than human. γ-Globulin was heavy in alligator but low in iguana.

Two types of protein patterns were found in Amphibia. The primitive urodeles were analbuminemic, but the more "advanced" anurans, the frogs and toads, showed albuminlike components. These components are said to appear after metamorphosis (Horner and Frieden, 1960).

A wide variety of plasma protein concentrations and electrophoretic patterns was seen in 11 Osteichthyes (teleosts). None showed typical heavy mammalian albumin bands, although some of the fish showed a light band with the same mobility. γ-Globulin was absent or low in all these fish.

TP of Chrondichthyes was low, albumin was absent, and γ-globulin was low but present. Lack of albumin in elasmobranch plasma has been described by Irisawa and Irisawa (1954) and Drilhon *et al.* (1956).

1.5. Cross-Immunoreactivities

Antisera were prepared against the plasmas of all the mammals listed horizontally in Tables 29.2 and 29.3 by repeated injections of the plasma plus adjuvant into rabbit foot pads.

Mammals were tested against all antisera, and the strength of precipitin reactions after double diffusion in agarose is indicated on a scale of 1 to 4 in Table 29.2. All antisera reacted to inciting antigens. Antisera to opossum, armadillo, guinea pig, and Sprague-Dawley rat reacted only with their own antigens. Antisera to baboon, cat, goat, and sheep reacted to the inciting plasma and other plasmas from animals of the same order. Antisera to man, monkey, dog, seal, horse, pig, and cow cross-reacted with some nonrelated mammalian plasmas.

Antisera reactions with premammalian plasmas are shown in Table 29.3.

TABLE 29.2
Cross-Immunoreactivities in Mammalia

Antigen	Antiserum to[a,b]														
	Op	Bb	Man	Mo	Ar	Gp	Rat	Cat	Dog	Se	Ho	Pig	Cow	Go	Sh
Opossum (Op)	**3**	0	0	0	0	0	0	0	0	0	0	0	0	0	0
Baboon (Bb)	0	**2**	3	4	0	0	0	0	0	2	1	0	1	0	0
Man	0	1	**4**	4	0	0	0	0	1	2	1	0	1	0	0
Monkey (Rhesus) (Mo)	0	1	3	**4**	0	0	0	0	1	2	0	0	0	0	0
Armadillo (Ar)	0	0	1	0	**3**	0	0	0	0	1	1	1	0	0	0
Guinea pig (Gp)	0	0	0	0	0	**4**	0	0	0	1	1	0	0	0	0
Rat (S-D)	0	0	1	0	0	0	**1**	0	0	0	0	0	0	0	0
Cat	0	0	1	0	0	0	0	**3**	2	2	1	0	0	0	0
Dog	0	0	1	1	0	0	0	1	**4**	4	0	0	1	0	0
Seal (Se)	0	0	1	0	0	0	0	0	0	**4**	1	0	0	0	0
Horse (Ho)	0	0	0	0	0	0	0	0	0	1	**4**	0	0	0	0
Pig	0	0	0	0	0	0	0	0	1	1	0	**4**	0	0	0
Cow	0	0	0	0	0	0	0	0	2	1	0	1	**4**	2	3
Goat (Go)	0	0	0	0	0	0	0	0	1	2	0	2	3	**3**	4
Sheep (Sh)	0	0	0	0	0	0	0	0	1	1	0	1	3	3	**4**
Premammals	0	0	0	0	0	0	0	0	0	0	0	0	0	0	0

[a](0) No reactivity.
[b]Boldface = antigen-corresponding antiserum reactions.

There were more antigens than antibodies, as indicated; if no test was performed, the table so indicates. Human plasma was nonreactive with all antisera except specific antihuman, which did not react to any premammalian plasma. The only extraclass cross-reactions were those of antiduck antisera, which reacted (1+) with loggerhead (LH) and *Pseudemys scripta elegans* (PSE) turtles, and anti-LH turtle antisera,which reacted (1+) with chicken and duck. Antichicken and anti-duck cross-reacted with other Aves. Anti–green turtle reacted with other turtles, slightly (1+) with alligator, but not with snake or iguana. Antisnake reacted only with snake. Anti-LH turtle and antialligator showed some cross-reactivity with other reptiles. Antifrog and antitoad reacted with each other, but not with the urodeles. Anticobia reacted with the other Osteichthyes, with the exception of goosefish. Antidogfish reacted only with dogfish, not with other Chondrichthyes.

No mammal reacted with a nonmammalian antiserum or vice versa. These studies indicate only a rough correlation between the number, mobility, and antigenicity of plasma proteins. The concentration or presence of serum albumin

TABLE 29.3
Cross-Immunoreactivities in Premammalian Vertebrates

Antigen	Antisera to[a,c]											
	Ch	Du	GT	LhT	Sn	Al	Fr	Td	At	Co	Df	Mammals
Chicken[b] (Ch)	**4**	3	0	1	0	0	0	0	0	0	0	0
Duck[b] (Du)	2	**4**	0	1	0	0	0	0	0	0	0	0
Goose	2	4	0	0	0	0	0	0	0	0	0	0
Pigeon	1	1	0	0	0	0	0	0	0	0	0	0
Turtle												
Green[b] (GT)	0	0	**4**	4	0	0	0	0	0	0	0	0
Loggerhead[b] (LhT)	0	1	4	**4**	0	0	0	0	0	0	0	0
PSE	0	1	3	4	0	1	0	0	0	0	0	0
Iguana	0	0	0	0	0	0	0	0	0	0	0	0
Snake[b] (Sn)	0	0	0	0	**4**	0	0	0	0	0	0	0
Alligator[b] (Al)	0	0	1	1	0	**4**	0	0	0	0	0	0
Amphiuma	0	0	0	0	0	0	0	0	0	0	0	0
Necturus	—	—	—	—	—	—	0	0	0	0	—	—
Salamander	—	—	—	—	—	—	0	0	0	0	—	—
Frog[b] (Fr)	0	0	0	0	0	0	4	1	0	0	0	0
Bullfrog	0	0	0	0	0	0	**4**	1	0	0	0	0
Toad[b] (Td)	0	0	0	0	0	0	1	**4**	0	0	0	0
Arctic grayling	0	0	0	0	0	0	0	0	3	1	0	0
Arctic trout[b] (At)	0	0	0	0	0	0	0	0	**3**	1	0	0
Cobia[b] (Co)	0	0	0	0	0	0	0	0	0	**3**	0	0
Goosefish	—	—	—	—	—	—	—	—	0	0	0	—
Grouper	0	0	0	0	0	0	0	0	0	1	0	0
Trigger	0	0	0	0	0	0	0	0	0	2	0	0
Dogfish[b] (Df)	0	0	0	0	0	0	0	0	0	0	**3**	0
Guitarfish	0	0	0	0	0	0	0	0	0	0	0	0
Shark, black-tipped	0	0	0	0	0	0	0	0	0	0	0	0
Skate	—	—	—	—	—	—	—	—	0	0	0	—
Mammals	0	0	0	0	0	0	0	0	0	0	0	3–4

[a](0) No reactivity; (—) not tested. [b]Antigen.
[c]Boldface = antigen-corresponding antiserum reactions.

in the adult animal would seem to be the best indicator of evolutionary advancement.

2. ANIMAL PLASMA AGGREGATION OF HUMAN PLATELETS

2.1. Introduction

The aggregation of human platelets by bovine or porcine antihemophilic factor (AHF) (factor VIII) has been described *in vitro* and *in vivo* and is dose-dependent (Sharp and Bidwell, 1957). After treatment of a hemophilic horse with bovine AHF, Nossel *et al.* (1962) found transient thrombocytopenia. Forbes *et al.* (1972), in an extensive *in vitro* study, showed that human platelets were aggregated with the animal plasmas indicated in Table 29.4. They also showed that this reaction was not inhibited by heparin. Brinkhous *et al.* (1977) compared this animal plasma activity, which they called *platelet aggregation factor* (PAF), to animal von Willebrand factor. They found PAF activity using fixed human platelets, but not with horse, cow, pig, or dog fixed platelets. Their results are also listed in Table 29.4.

2.2. Method

Platelet-rich plasma (PRP) was prepared from 5 normal human donors, and the platelet counts were adjusted to 250,000/mm^3. Then 0.05 ml of the animal citrated platelet-poor plasma (PPP) was added to 0.4 ml fresh human PRP, and aggregation was recorded in a platelet aggregometer.

2.3. Results and Summary

Table 29.4 lists the animal plasmas tested, observed results on PRP from five individuals, summary (Lewis), and comparison with the results of Forbes *et al.* (1972) and Brinkhous *et al.* (1977). All studies agree that PAF is found in plasma from cow, goat, pig, and sheep. In addition, we found the activity in porpoise plasma and to a very small extent in rat plasma. Figure 29.1 shows an experiment that compared the effects of rat, hamster, manatee, and pig plasmas on PRP from donor No. 3. The rat mixture caused a small degree of aggregation that appeared neglible compared to that seen with pig plasma. Small and seemingly insignificant activity was found in many of the mammalian plasmas. At least one example of each of the major premammalian classes was without PAF activity.

3. CLUMPING OF STAPHYLOCOCCI BY ANIMAL PLASMAS

3.1. Introduction

Many strains of staphylococci clump in the presence of human fibrinogen or its high-molecular-weight derivatives. This clumping was first reported by Much (1908), a German bacteriologist. More recently, Lipinski *et al.* (1967) and Allington (1967) noted that this reaction would also occur with fibrin monomers and large fibrin breakdown products (FBP) and their complexes. It was sug-

TABLE 29.4
Aggregation (%) of Human Platelets by Animal Plasmas[a]

Animal	Human PRP donor No. 1	2	3	4	5	Lewis	Forbes et al. (1972)	Brinkhous et al. (1977)
Opossum	0	—	—	—	—	0	0	0
Baboon	—	0	13	6	—	0	Vsl	—
Chimpanzee	—	—	12	9	4	0	Vsl	0
Man	0	—	4	8	—	0	—	0
Monkey	0	0	—	—	—	0	Vsl	0
Rabbit	0	0	11	—	0	0	0	0
Guinea pig	0	0	8	5	0	0	0	Vsl
Hamster	—	—	6	9	—	0	0	—
Rat	—	—	17	27	13	Vsl	0	Vsl
Porpoise	80	90	—	—	—	+	—	—
Cat	0	0	—	—	—	0	0	—
Dog	0	0	9	6	—	0	0	0
Fox	0	—	—	—	—	0	—	—
Raccoon	0	0	—	—	—	0	—	—
Seal	0	—	—	—	—	0	—	—
Elephant	0	0	—	—	—	0	—	—
Manatee	—	—	13	12	—	0	—	—
Horse	0	0	—	—	—	0	+[b]	0
Pig	100	100	92	80	100	+	+	+
Llama	0	0	—	—	—	0	—	—
Cow	100	100	—	—	—	+	+	+
Goat	100	90	—	—	—	+	—	+
Sheep	100	100	—	—	100	+	+	+
Chicken	0	0	0	8	0	0	0	—
Duck	0	0	—	—	—	0	—	—
Goose	0	0	—	—	—	0	—	—
Turkey	—	0	6	8	4	0	0	—
Alligator	—	—	0	0	0	0	—	—
Bullfrog	—	—	—	—	4	0	—	—
Sturgeon	—	—	0	3	0	0	0[c]	—
Dogfish	—	—	4	11	0	0	—	—

[a]Aggregation: (Vsl) very slight; (0) none; (—) not tested.
[b]A review of the data suggests Vsl. [c]Paddlefish.

gested by Hawiger *et al.* (1970) that the staphylococcal clumping test might be a useful standarized test for fibrinogen/fibrin split products in man and in experimental animals. For this reason, we have tested animal plasmas to see whether they clump staphylococci that contain surface clumping factor. Results from a previous publication (Lewis and Wilson, 1973) are reprinted here with the kind permission of the editor of *Thrombosis Research*.

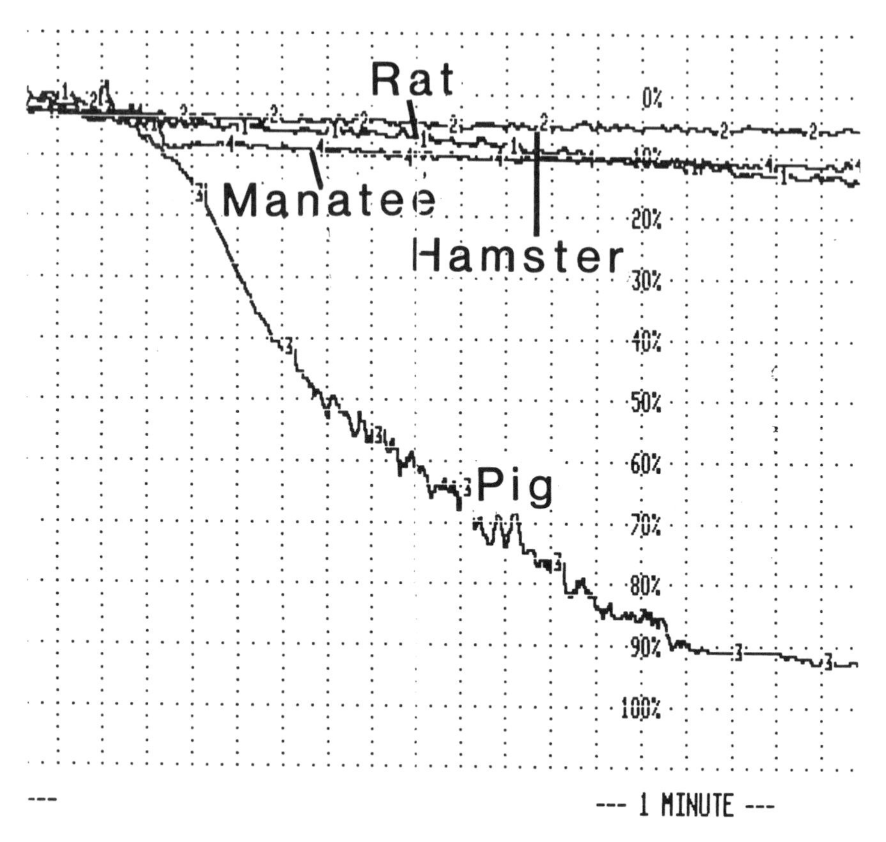

FIGURE 29.1. Effects of animal plasmas in aggregation of normal human plasma.

3.2. Importance

The study of thrombosis and thrombolysis in animal models is essential in mapping desirable therapy for human disease. Because this simple test is useful in such studies, it is important to know whether the fibrinogen from the animal model clumps staphylococci.

3.3. Technique

Until recently, the staphylococcal preparations used were cultures of the Newman D_2C strain of *Staphylococcus aureus,* grown in our laboratory, heat-killed, washed, and suspended in buffered saline. This strain did not produce staphylocoagulase. In the last few years, a commercially prepared suspension (**SI**) has been used.

The plasma dilutions used were serial doubling dilutions (1st = 1:1, 2nd = 1:2, 3rd = 1:4, 4th = 1:8th . . . 12th = 1:2048). Normal human plasma caused visible clumping up to the 12th dilution.

3.4. Results

Table 29.5 lists the animals studied and the number tested, the presence or absence of clumping, and the titer (last dilution) of plasma that was positive. The animals are listed in phylogenetic order, and it is apparent that premammalian vertebrates had little or no staphylococcal clumping activity. Testing amphibian plasma was unsatisfactory because of the active fibrinolysis. Most of the mammals were very active. Only opossum, guinea pig, elephant, goat, and sheep did not clump the standard preparation at any dilution.

TABLE 29.5
Animals Tested for Staphylococcal Clumping, Presence of Clumping, and Titer

Animal (*N*)	Clump	Titer[a]	Animal (*N*)	Clump	Titer	Animal (*N*)	Clump	Titer[a]
Lamprey (3)	0	—	Echidna (3)	+	12	Cat (2)	+	12
Dogfish (3)	0	—	Wallaby (1)	+	8	Dog (4)	+	12
Shark			Opossum (10)	0	—	Fox (1)	+	12
Blue (2)	0	—	Hedgehog (2)	+	12	Mink (3)	+	9
Mako (1)	0	—	Bat (2)	+	12	Raccoon (1)	+	12
Skate (2)	0	—	Baboon (9)	+	10	Seal (3)	+	12
Amberjack (1)	0	—	Chimpanzee (3)	+	6	Elephant (8)	0	—
Goosefish (3)	0	—	Monkey (3)	+	11	Manatee (3)	+	10
Sculpin (4)	0	—	Armadillo (6)	+	9	Horse (2)	+	7
Sturgeon (3)	0	—	Rabbit (4)	+	11	Pig (2)	+	12
Turtle (PSE) (2)	+	2	Guinea pig (9)	0	—	Camel (1)	+	9
Alligator (2)	+	4	Hamster (1)	+	7	Guanaco (1)	+	9
Chicken (4)	0	—	Rat (2)	+	12	Llama (1)	+	9
Goose (2)	+	1	Mouse (2)	+	12	Cow (2)	+	9
Pigeon (1)	+	2	Porpoise (8)	+	12	Goat (2)	0	—
Turkey (2)	+	1				Sheep (3)	0	—

[a](—) No dilution clumped the staphylococci.

3.5. Miscellaneous Tests

In human plasma, the staphylococcal clumping reaction was positive to the 10th doubling dilution or beyond. Thus, the test was sensitive to less than 1 μg/ml. Plasmas from two siblings suffering from afibrinogenemia, not under treatment for 3 weeks, did not cause staphylococcal clumping in any dilution.

Plasmas from only five mammals failed to clump the staphylococcal preparation: opossum, guinea pig, elephant, goat, and sheep. In each case, one or more fresh samples were obtained from different animals, and repeat testing confirmed the original results.

The possibility that the failure to clump was due to a lack of cofactor was studied by adding human afibrinogenemic plasma to the plasmas that did not clump the staphylococci. No clumping was observed.

The presence of an inhibitor was sought by preparing dilutions of human plasma using animal (goat) plasma as diluent instead of saline. The staphylococcal clumping titer for the human plasma was not changed by more than one dilution.

Crude fibrinogen fractions were prepared from human and goat plasmas by ammonium sulfate fractionation (at 29% saturation) or as the euglobulin fraction. These fractions were incubated with various concentrations of trypsin (**DI**). The human materials retained staphylococcal clumping activity even when thrombin clottability disappeared. Goat fibrinogen fractions with or without trypsin did not clump staphylococci.

3.6. Summary

Plasma from premammalian vertebrates clumped staphylococci poorly or not at all. Among the mammals, only opossum, guinea pig, elephant, sheep, and goat fibrinogens did not have the ability to clump staphylococci. Presumably this was due to structural differences from other mammalian fibrinogens. No inhibitor could be demonstrated in goat plasma. Afibrinogenemic human patient plasma did not supply a cofactor. Attempts to split the goat fibrinogen molecule with trypsin did not uncover a staphyloccal clumping site.

REFERENCES

Allington, M. J., 1967, Fibrinogen and fibrin degradation products and the clumping of staphylococci by serum, *Br. J. Haematol.* **13:**550.

Brinkhous, K. M., Thomas, B. D., Ibrahim, S. A., and Read, M. S., 1977, Plasma levels of platelet aggregation factor/von Willebrand factor in various species, *Thromb. Res.* **11:**345.

Crenshaw, J. W., Jr., 1962, Variation in the serum albumin and other blood proteins of turtles of the Kinosternidae, *Physiol. Zool.* **35:**157.

Drilhon, A., Fine, F. M., Vriel, J., and Lerourdelles, F., 1956, Analysis by immuno-electrophoresis of the constituents of serum, *Comp. Rendu* **243:**1802.

Forbes, C. D., Barr, R. D., McNicol, G. P., and Douglas, A. S., 1972, Aggregation of human platelets by commercial preparations of bovine and porcine antihemophilic globulin, *J. Clin. Pathol.* **25:**210.

Hawiger, J., Niewiarowski, S., and Thomas, D. P., 1970, Measurement of fibrinogen and fibrin degradation products in serum by staphylococcal clumping test, *J. Lab. Clin. Med.* **75:**93.

Horner, A. E., and Frieden, E., 1960, Changes in serum proteins during spontaneous and induced metamorphasis, *J. Biol. Chem.* **235:**2851.

Irisawa, H., and Irisawa, A. F., 1954, Blood serum protein of the marine Elasmobranchii, *Science* **120:**849.

Lewis, J. H., and Wilson, J. H., 1973, Variations in abilities of animal fibrinogens to clump staphylococci, *Thromb. Res.* **3:**419.

Lipinski, B., Hawiger, J., and Jeljaszewicz, J., 1967, Staphylococcal clumping with soluble fibrin monomer complexes, *J. Exp. Med.* **126:**979.

Much, H., 1908, Über eine Vorstufe des Fibrinfermentes in Kulturen von *Staphylokokkus aureus, Biochem. Z.* **14:**143.

Nossel, H. L., Archer, R. K., and MacFarlane, R. G., 1962, Equine haemophilia: Report of a case and its response to multiple infusions of heterospecific AHG, *Br. J. Haematol.* **8:**335.

Sharp, A. A, and Bidwell, E., 1957, The toxicity and fate of injected animal antihemophilic globulin, *Lancet* **1:**359.

SUGGESTED READINGS

Clumping of Staphylococci by Animal Plasmas

Hawiger, J., Hawiger, A., and Koenig, M. G., 1971, Staphylococcal clumping and fibrinogen and fibrin degradation products in inflammatory exudate, *Proc. Soc. Exp. Biol. Med.* **136:**132.

APPENDIX

1. MANUFACTURERS AND SUPPLIERS

Following is a list of the manufacturers and suppliers of the reagents, tests, and instruments used in these studies. Those mentioned in the text are identified by the abbreviations in the first column.

AB	Abbott Diagnostics A Division of Abbott Laboratories 1 Abbott Park Road Abbott Park, IL 60064
AC	American Cyanamid Company One Cyanamid Plaza Wayne, NJ 07407
ACL	Automated Coagulation Laboratory Instrumentation Laboratory Company Lexington, MA 02173–3190
AD	American Dade American Hospital Supply Company Miami, FL 33152-0672
AL	American Labor Corp. 8801 Midway West Road Raleigh, NC 27613

AR	Armour Pharmaceutical Company Kankakee, IL 60901
BA	Baker Instrument Company 100 Cascade Drive PO Box 2168 Allentown, PA 18001
BD	Bio/Data Corporation 3615 Davisville Road PO Box 250 Hatboro, PA 19040–0250
BDL	Becton Dickinson Labware 2 Bridgewater Lane Lincoln Park, NJ 07035
BE	Behring Diagnostics Division of American Hoechst Corporation 10933 North Torrey Pines Road La Jolla, CA 92037
BI	Beckman Instruments, Inc. Spinco Division Palo Alto, CA 94304
BR	Bio-Rad Laboratories, Inc. 3726 East Miratoma Avenue Anaheim, CA 92806
BW	Burroughs Wellcome 3030 Cornwallis Road Research Triangle Park, NC 27709
CB	Calbiochem PO Box 12087 San Diego, CA 92112–9939
CH	Chromogenix Chromogenix AB Taljegardsgatan 3 S-431 53 Molndal, Sweden
CI	Coleman Instruments 42 Madison Street Maywood, IL 60153

CL	Coagulation Laboratory Reference Laboratory Alliance 3636 Boulevard of the Allies Pittsburgh, PA 15213
CM	Curtin Matheson Scientific, Inc. 9999 Veterans Memorial Drive Houston, TX 77251
CO	Coulter 601 West 20th Street Hialeah, FL 33012
DA	Damon/IEC Needham Heights, MA 02194
DI	Difco Laboratories PO Box 331058 Detroit, MI 48232
DS	Diagnostica Stago 9, rue des Frères Chausson 92600 Asnieres-sur-Siene France
DU	E. I. du Pont de Nemours & Co. Clinical & Instrument Systems Division Concord Plaza Wilmington, DE 19898
EM	Electron Microscopy Sciences Box 251 Fort Washington, PA 19034
FI	Fisher Scientific Company 585 Alpha Drive Pittsburgh, PA 15238
GD	General Diagnostics (see OT)
GK	George King Bio-Medical, Inc. 11771 West 112th Street Overland Park, KS 66210
HE	Helena Laboratories 1530 Lindbergh Drive PO Box 752 Beaumont, TX 77704

HY	Hyland Laboratories (Baxter Healthcare Company) 444 West Glen Oaks Boulevard Glendale, CA 91202–2936
IL	Instrumentation Laboratory Company 113 Hartwell Avenue Lexington, MA 02173
JO	JOUAN Inc. USA PO Box 2716 Winchester, VA 22604-9879
LE	Lederle Laboratories A Division of American Cyanamid Company One Cyanamid Plaza Wayne, NJ 07407
LI	Eli Lilly Company Indianapolis, IN 46285
LS	Lab Systems and Flow Laboratories NTD Laboratories 393 Old Country Road Carle Place, NY 11514–2127
MI	Miami Serpentarium Laboratories Miami, FL 33156
MIL	Millipore Biomedica 80 Ashby Road Bedford, MA 01730
MLA	Medical Laboratory Automation, Inc. 270 Marble Avenue Pleasantville, NY 10570–2982
MM	Minnesota Mining Company (3M Company) St. Paul, MN 55144–1000
OR	Ortho Diagnostic Systems, Inc. Route 202 Raritan, NJ 08869
OT	Organon Teknika (USA) 100 Alzo Avenue Durham, NC 27704
PD	Parke Davis Division of Warner-Lambert Company 201 Tabor Road Morris Plains, NJ 07950

PH	Pharmacia Diagnostics AB F35–3 S25182 Uppsala, Sweden
PL	Pharmacia LKB Nuclear, Inc. 9319 Gaither Road Gaithersburg, MD 20877
SB	Serono-Baker Diagnostics, Inc. 100 Cascade Drive Allentown, PA 18103–9562 (see also BA)
SH	Shimadzu Spectronics 7102 Riverside Drive Columbia, MD 21046
SI	Sigma Diagnostics 545 South Ewing Avenue PO Box 14508 St. Louis, MO 63103
SQ	Sequoia Turner 755 Ravendale Drive Mountain View, CA 94043
UP	The Upjohn Company Kalamazoo, MI 49001
WL	WA Lemberger 2500 Waukau Avenue Oshkosh, WI 54903–2482
WO	Worthington Biochemical Corporation Halls Mill Road Freehold, NJ 07728

2. REAGENTS AND DEVICES

Reagents and devices used in these studies are listed alphabetically below by abbreviation and full name followed by the manufacturer's or supplier's identification in parentheses. For commercial preparations, no details are given because they are fully described in the package inserts. Reagents developed or modified in this laboratory are identified by the notation (CL).

ADP	Adenosine 5′ diphosphate (SI)
Arach A	Arachidonic acid (BD)

Atroxin	*Bothrops atrox* venom (SI)
Ba Pl	Barium-treated plasma (CL) 1. 1 ml oxalated pl + 50 mg Ba_2SO_4 2. 1 ml Cit pl + 8 mg $BaCl_2$. Stir 10 min, centrifuge, decant
Barb buff	Barbital buffers, ph 7.3 or 8.6 (FI)
Biuret	Biuret reagent (FI)
BR	Brain—dried, powdered (CL) Whole fresh animal brain picked free of surface membranes and blood vessels is dehydrated by repeated grinding with 10 changes of acetone. Thoroughly dried powders can be stored in a desiccator or freezer for years. The suspensions employed in the tests are prepared by gently mixing 300 mg brain powder with 10 ml I/S and allowing the mixture to stand for 10 min at 37°C. The top portion is decanted. Before use, it is tested for thrombin contaminant by adding it to human plasma without calcium and observing for clot formation. If no clot forms in 1 hr, the suspension is used in the animal study.
Br fresh	*See* Tissue F
Ca	0.025 M $CaCl_2$ (CL)
Casein	Gel-stabilized casein plates: RADP 36 N 663 (WO)
CHH	*Crotalus horridus horridus* venom (MI, SI)
Cit	0.13 M (3.8%) trisodium citrate (CL, FI)
Cit-I/S	Citrated I/S; 1 part Cit + 9 parts I/S
Cit plasma	Citrated plasma (CL) 1 part Cit + 9 parts whole blood, mix, centrifuge at 1500 g for 15 min., decant pl
Collagen	Collagen 1. Lyophilized soluble calf skin collagen (BD) 2. Animal tendon, chopped, frozen, and ground (CL)
Control	Control human plasma 1. Cit plasma is obtained from 6 females and 6 males previously shown to test as normal in all clotting tests. It is pooled, divided into 1 ml aliquots, and stored frozen at −75°C (CL) 2. Verify® (OT) 3. Assess™ normal control (IL) 4. Assess™ calibration plasma (IL)

DE	Dog enzyme (CL) Euglobulin from 100 ml dog plasma is dissolved in 10 ml Cit-I/S at pH 7.3. After cold refrigerator storage in a glass bottle for 3 weeks, it is tested weekly until a 1:10 dilution has the same fibrinolytic activity with or without added SK. It is then divided into 1-ml aliquots, labeled, and stored frozen at −70°C.
D/W	Distilled or deionized water
EACA	ϵ-amino caproic acid (FI)
EACA-buff	EACA-buffer (CL) 2.5 g EACA + 7.65 g NaCl + 3.8 g sodium citrate + 900 ml D/W + 30 ml Barb buff (pH 7.3) q.s. to 1 liter
EPON	EPON aradilite (EM)
E-mix	Ethanol mix (CL) 65 ml 95% ethanol + 1 ml NaOH (2.5 N) + 34 ml D/W
F sub	Factor substrates (CL, GK, SI) A. II: mixture of fresh human or bovine Ba PL and aged human serum B. V: aged human plasma C. VII, X, VIII, IX, XI, XII, Flet, Fitz: plasma from patients with severe D/Ds of these factors who are HIV-, HBB-, and HBC-antigen-free. It is phoresed from our patients or purchased.
Glass beads	Superbrite (MM)
Glut	Glutaraldehyde (EM)
I/buff	Imidazole buffer (FI)
I/S	Imidazole-buffered saline (CL) 1 part I/buff + 9 parts N/S
MCA	1% monochloroacetic acid (FI)
Nair™	Commercial depilatory (for removing hair)
NS	0.85% sodium chloride (CL)
Oxalate	0.1 M sodium oxalate (CL)
Pig Pl	Pig plasma (CL) Plasma from standard pig cit blood is centrifuged at 2000*g* for 15 min, decanted, divided into 1 ml aliquots, and frozen at −70°C. Stable 1 year +.
Plat fixed	Platelets, fixed with *p*-formaldehyde (CL)

Plat lyl	Platelets, lyophilized (BD)
PPP	Platelet-poor plasma (CL) Plasma from Cit blood centrifuged at 2000*g* for 15 min, decanted into a fresh tube and recentrifuged at 2000*g* for 15 min, decanted again to fresh tube
PRP	Platelet-rich plasma (CL) Plasma from Cit blood is centrifuged at 800*g* for 10 min. Platelet count is performed and adjusted to 250,000/μl.
Rept	Reptilase (Atroxin®) (SI).
Risto	Ristocetin (SI)
RVV	Russell's viper venom (DS)
SER	Serum (CL) Fluid is obtained from a blood clot prepared in a glass tube and allowed to clot for 2 hr at 37°C
Simpl	Simplastin®: rabbit brain thromboplastin (GD, OT)
Simplate	Simplate Bleeding Time Device (OT)
SK	Streptokinase 1. Varidase (LE) 2. Streptokinase (AC)
Staph C F	1. Cultured Newman D_2 (coagulase-free) *Staphylococcal aureus* is centrifuged, bacteria killed at 75°C for 1 hour, washed, and resuspended at 10 mg/ml in D/W (CL) 2. Commercially prepared killed staphylococci suspended at 1 mg/ml (SI)
Staph K	Staphylokinase 2% (CL)
Thromb	Bovine thrombin 1. Thrombin Topical (AR) 2. Data-Fi® Thrombin (AD)
t-PA	Tissue plasminogen activator (CH)
Tissue F	Tissue factor, fresh (CL) Animal tissue—brain, gill/lung, or skin—is picked free of visible membranes or blood vessels and minced as finely as possible. 2 gm wet weight is stirred into 10 ml I/S, then allowed to settle ½ hr at 37°C and the top decanted to a fresh tube.
Tryp	Trypsin, bovine pancreatic (SI)
UK	Urokinase (SI)

3. TEST METHODS

In this section test methods are listed alphabetically by abbreviation and are followed by the full name and the manufacturer's or supplier's identification. For commercially available tests no directions are given because such directions are in the package inserts and are followed without modification. Methods developed or modified in our coagulation laboratory are identified as (CL), and the technique is briefly described.

Anti DE Anti Tryp	Anti dog enzyme or trypsin activity (CL) Serum or dilution is incubated with standard DE or trypsin for 1 hr at 28°C. Residual fibrinolytic activity is assayed.
Anti PL	Antiplasmin Assay 1. Anti DE (CL) 2. Anti Trypsin (CL) 3. Stachrom® Anti plasmin Kit (DS) 4. Il Test™ a_2 Anti-plasmin kit (IL)
APTT	Activated partial thromboplastin time 0.1 ml plasma + 0.2 ml APTT reagent (incubate 5 min at 37°C) + 0.1 ml 0.025 M $CaCl_2$. Clot t is recorded. 1. Manual (CL) 2. Semiautomated (MLA 600, 620, 660) 3. Coag-a-mate × 2 (GD, OT) 4. Coag Screener (AL) 5. Automated (ACL)
APTT mix	Activated partial thromboplastin time mix (CL) 0.2 ml normal plasma + 0.2 ml animal or patient plasma are incubated together for 1 hr at 37°C. APTT test is done on 0.1 ml of the mixture.
ATIII	Antithrombin III Activity 1. 0.1 ml standard thrombin + 0.1 ml serum is assayed before and after 6 min incubation (CL) 2. Protopath® method (AD) 3. Stachrom® At III kit (DS) 4. Il Test™ kit (IL) immunoassay 5. Radial diffusion or Rocket EID method is done if appropriate antiserum available (HE) or (CL).
Bleed T	Bleeding time (CL) Time from making a small cut in skin or mucous membrane to cessation of bleeding. In animals, an area free of hair, scales, or feathers is sought.

When necessary, hair may be removed with the depilatory Nair. Cuts (6 × 2 mm) are made with a rounded, new Bard Parker blade or small, sharp surgical scissors. Templates (Mielke *et al.*, 1969) control the size and depth of the cut. Simplate (OT) is a widely used, sterile, disposable, convenient template. In man and some animals, a sphygmomanometer cuff is placed above the cleansed volar surface of the upper extremity, and inflated to 30 mm Hg. One or two horizontal Simplate cuts are made, and the bleeding is timed. Normal human bleeding time is 3–9 min.

Clot T,R,L

Clotting time, clot retraction, clot lysis (CL)
2 ml samples of whole blood are placed in two new glass and two siliconized 13 × 100 mm tubes (37°C) for mammals and birds, 28°C or room temperature for other animals). The first glass or siliconized tubes are gently tilted about every 30 seconds until solidified. Then the process is repeated with the second tubes and the clotting time recorded. At 4 and 24 hours after phlebotomy the tubes are observed and clot retraction recorded. The amount of serum exuded from the clot is graded as follows: 0 = none, + = 1/4 size of clot, ++ = 1/2 size of clot, +++ = 3/4 size of clot, ++++ = size of clot (dependent on hematocrit and platelet count and function). After 24 hours complete dissolution of the clot is recorded as + clot lysis.

Diff

Differential blood count (CL)
On a Wright-stained smear, a skilled technician examines 100 WBC (for mammals) or 500 to 1000 cells (for premammals)

EG

Ethanol gel test (CL)
0.2 ml plasma + 0.05 ml ethanol mix. The tube is shaken gently and placed in a melting ice bath for 15 min. Examine tube against a black background and read: 0 = liquid; 1+ = viscous liquid: 2+ = partially gelled; 3+ = almost completely gelled; 4+ = completely gelled.

ELT

Euglobulin lysis time (CL)

0.5 ml Cit plasma + 8 ml D/W + 0.2 ml 1% acetic acid. Mix, chill 10 min, centrifuge at 2000 g for 10 min. Discard supernatant. Drain tube and add 0.5 ml warm Barb buff ph 8.6. Gently mix until precipitate dissolves. Pour into a 12 × 75 mm glass tube containing 0.02 ml 0.05 M $CaCl_2$ and 0.02 ml 500 U/ml thrombin. Note clot formation time. Incubate at 37°C (mammals and birds) or room temperature (other animals). Inspect every 15 min during 5 hr. Report lysis time in hours or fractions thereof. Normal human lysis time: >5 hr.

EM

Electron microscopy (CL)

Preparation of Buffy Coat for transmission electron microscopy (TEM)

Centrifuge Cit blood for 5 min at 1200 RPM, and with a Pasteur pipette, draw off as much plasma as possible. Carefully add 0.5 ml fresh fixative (2.5% gluteraldehyde in 0.125 M sodium cacodylate buffer with 0.75 M sucrose, ph 7.4). Allow buffy coat to fix at least 30 min. Draw off and discard used fixative. Gently loosen the fixed buffy coat disk from the sides of the tube with a pointed wooden applicator stick. The fixed disk will float free of unfixed RBCs. Slide the released disk into a sample vial containing about 1 ml buffer. Rinse the disk several times by removing buffer with unfixed blood and adding fresh. Samples may be stored for up to a week in the refrigerator. Rinse disks again in 4 changes of fresh buffer. Using a dissecting microscope, cut disks into 1 mm squares so that they can be embedded to display cross sections of platelets, leukocytes, and erythrocytes. Postfix in 1% osmium tetroxide for 1 hr at room temperature. Dehydrate in ethanol and propylene oxide; embed in EPON.

F assays

Factor I fibrinogen

1. Clot-Biuret method (CL)
   0.5 ml plasma (human or animal) + 1.5 ml Ca-EACA buff + 0.1 ml bovine thrombin (1 U/ml).

Immediately place glass rod in tube and twirl gently at 37°C until fibrin strands form. In 5 min, wind all fibrin onto rod, pressing against the upper side of the tube to remove serum. Wash with a gentle stream of I/S. Then wash with D/W, continuing to squeeze against tube wall. Blot on absorbent paper. Place stirring rod with fibrin in a fresh tube, add 0.5 ml D/W + 1.0 ml NaOH. Prepare blank and fibrinogen standard. Place all tubes in simmering water bath for about 3 min, until all fibrin dissolves. Cool tubes. Add 2 ml Biuret reagent. Let stand at room temperature for 30 min. Read optical density (OD) at 540 mm in a spectrophotometer.

2. Data-Fi® (AD)
   A thrombin clotting time method

— Factors II, V, VI and X (CL)

0.1 ml F sub + 0.1 ml Cit plasma (diluted as required*) + 0.2 ml Simplastin® + $CaCl_2$. Record clotting time and convert to U/ml by comparison with a standard curve.

— Factors VIII, IX, XI, XII, Flet, Fitz (CL)

0.1 ml F sub + 0.1 ml APTT reagent, incubate 5 min at 37°C + 0.1 ml Cit plasma (diluted as required*) + 0.1 ml 0.025 M $CaCl_2$. Record clotting time and convert to U/ml by comparison with a standard curve.

Fib L Fibrinolytic Assays

1. Fibrin lysis (CL)
   200 ml of outdated human ACD plasma is diluted with 800 ml I/S and recalcified with 25 ml 1 M $CaCl_2$. After clotting in a glass container for 1 hr at 37°C the fibrin is finely minced with a blender and then resuspended and reminced in the blender 10 times with I/S and 10 times with D/W. The fibrin is collected by filtration and air-dried. Lots of 5 gms are iodinated with ^{131}I or ^{122}I, washed 10 times

*For humans use dilutions 1:10 and 1:20; for animals use 1:5, 1:10, 1:20, and 1:40.

and suspended with I/S so that each 2 ml amount contains approximately 100 mg fibrin with a count of 100,000–150,000 per tube. A flattened staple is placed in each tube and a rack of 40 counted before and after incubation with test materials. The rack is placed in a 37°C waterbath on a magnetic stirrer. The bent staple in each tube keeps the fibrin slowly whirling. The percent of fibrin dissolved is calculated from the pre- and post-incubation counts. If SK alone lysed more than 1% of the fibrin, that lot is considered contaminated with plasminogen and undergoes repeated washing or is discarded.

2. Casein-agarose plates (WO)
   Radial diffusion plates contain casein incorporated in agarose gel at pH 7.6. Each well is filled with 10 μl, and after 24 hr incubation at 37°C, the diameters of the cleared zones (mm) are read and compared to results with human plasma.
3. Clot lysis test (CL)
   Clots are prepared to contain 0.2 ml plasma (animal or human) + 0.1 ml buffer or activator + 0.1 ml Ca-Thromb. These are observed for complete clot dissolution (lysis) every few minutes for 5 hr.

HCT or PCV

Hematocrit

1. Wintrobe hematocrit tube
   A square Wintrobe tube marked with graduations to 1 ml is filled with EDTA anticoagulated blood and centrifuged at 2000 g for 10 min. The height of the total and red cell columns are read and corrected:

$$\frac{\text{height of RBCs}}{\text{height of total blood}} \times 100 = \text{HCT }\%$$

   Example:

$$\frac{0.45}{1.1} \times 100 = 40.9\ (41)\ \%$$

   Buffy coat (WBC + platelets) height and plasma color are noted.

2. Microhematocrit method
   Duplicate microhematocrit tubes are filled with EDTA anticoagulated blood and the tube end plugged with clay. The tubes are placed in the microhematocrit rotor (DA) with the plugged end outward and centrifuged for a standard time and speed. The tubes are then transferred to a microcapillary tube reader and the lengths of the total blood and red cell columns recorded. Buffy coat and plasma color are also noted.
3. Electronic Counter Method
   The hematocrit is calculated from the measured RBC and MCV

$$\frac{MCV \times RBC}{10} = HCT\ \%$$

HGB — Hemoglobulin (g/dl)

1. Eyeball comparison of standard lysed blood cell solution with color standards.
2. Photometric comparison color (wavelength 450 mμ) of lysed RBC cells (usually cyanmethemoglobin procedure).

Panagel — Panagel electrophoresis (MIL)

Plat Aggr — Platelet aggregation tests (BD)
0.45 ml PRP + 0.05 ml aggregating reagent. Record light transmission on an aggregation profiler.

Plat C — Platelet count

1. Blood is diluted and RBC lysed in a Unopette®. Platelets counted in chamber with a phase-contrast microscope.
2. Electronic particle counter
   a. Coulter Model F (C)
   b. Baker 810 or 9000 (BA)

Plat Glass Ret Ind — Platelet glass retention index (CL)
Whole fresh blood is passed slowly (1 ml/min) through a 1 g glass bead column packed in Tygon tubing, 3 drops discarded, and 1 ml collected in an EDTA tube. Platelet counts are done on the pre- and postsamples.

$$\frac{Pre - Post}{Pre} \times 100 = \%\ retention$$

Plat-MPV	Platelet-mean volume Baker 9000 (BD)
Plg	Plasminogen assay (for humans) IL Test™ Plasminogen (IL) (This test uses SK for activator and may prove useful in some mammals. For other animals, see Fib L.)
Prot C Prot S	Protein C and Protein S (for humans) 1. Il Test™ Proclot (IL) 2. Asserachrom® Protein C (DS) 3. Staclot® Protein S (DS) 4. Asserachrom® Protein S (DS)
PG	Protamine gel (CL) 0.2 ml plasma + 0.05 ml protamine in duplicate. The tubes are shaken gently and allowed to incubate at 37°C undisturbed for 15 min. The tubes are then examined against a black background and graded as follows: 0 = liquid, 1 = fine flecks, 2 = fine strands, 3 = clumps, 4 = gelation. Positive results indicate the presence of some fibrin monomer which may have been formed intravascularly or during difficult venepuncture.
PT	Prothrombin time 0.1 ml plasma + 0.1 ml thromboplastin or tissue factor + 0.1 ml 0.025 M $CaCl_2$. Record clot t in duplicate.
RBC C	Red blood cell count 1. Dilute blood with Unopette® and count cells on standard chamber 2. Electronic particle Coulter 9000 (CO)
Recal t	Recalcification time (CL) 0.2 ml Cit plasma + 0.2 ml 0.025 M $CaCl_2$. Observe without shaking, every 15 sec until clotted. Prepare in triplicate.
Recal t Lys	Recalcification time lysis (CL) Allow one clotted tube to remain at 37°C and observe every 30 min for a few hours and at 24 hours. Report as positive (+) only if all traces of clot are gone.

Recal t-MCA Lys

Recalcification time clot lysis in MCA (CL)
Loosen one clot with a sharp tap. Add 2 ml 1% MCA. Observe at 37°C for 1 hr. If clot appears absent, pour out on black surface and examine carefully before reporting MCA L as +.

RCF

Ristocetin cofactor (VIII:vW) assay (CL)

1. Manual: 0.4 ml fixed or lyophylized platelets + 0.05 ml ristocetin + 0.05 ml plasma or dilution. Record aggregation time. Compare to a standard curve and read RCF activity. Normal human range: 0.5–1.50 U/ml.
2. Semiautomated: same test run on a platelet aggregometer.

Sia T

Sia test for macroproteins (CL)
0.1 ml plasma or serum + 0.9 ml D/W. Observe for turbidity at 10 min. Record 0–4+ turbidity.

SPT

Serum prothroombin time (CL)
This test, considered "old-fashioned" by some, is very useful in detecting human coagulation abnormalities. It is not specific, but if the time is short (<20 sec), it always indicates an abnormality. Tube No. 5 [see Chapter 2 (Section 2.2)] contains serum that has been citrated after 2 hr in the 37°C bath. Ser, 0.1 ml, is withdrawn and added to a preprepared 12 × 75 mm glass tube containing 0.1 ml Ba Pl and 0.2 ml Simplastin-Ca. The tube is roll-clotted. Short times are abnormal and indicate that the blood in the original tube did not clot completely due to an abnormality in the blood or condition of clotting.

Staph C Test

Staphylococcal clumping test (SI)
Serial dilutions of plasma 1:2 to 1:2048 are prepared in 25 μl amounts in a microtiter plate, and 25 μl staphylococcal clumping factor is added to each well. The plate is covered and placed on a rotator for 20 min at room temperature, then shaken vigorously and observed for clumping on a Cooke plate recorder. The highest dilution that shows clumping is recorded as the titer.

TGT

Thromboplastin generation test (CL)

Prepare generating mixture—use all human or all animal:

- 0.3 ml Ba Pl diluted 1:5 with I/S
- 0.3 ml SPT Ser diluted 1:10 with I/S
- 0.3 ml washed fresh platelets as prepared
- 0.3 ml 0.025 M $CaCl_2$

Place generating tube at 37°C. Start stopwatch no. 1 when $CaCl_2$ is added. At 15 sec before 2, 4, 6, 8, 10, and 12 min on stopwatch No. 1, add 0.1 ml warm 0.025 M $CaCl_2$ to 0.1 ml substrate (at 37°C for mammals). On the minute, add 0.1 ml generating mixture and start stopwatch No. 2. Tilt tube in and out of 37°C bath. Record roll-clot time. With some experience, a second set of sustrate times can be run at 3, 5, 7, 9, and 11 min on stopwatch No. 1.

Both animal and animal generating mixtures are clotted on human or animal plasma.

Thromb T

Thrombin time (CL)

Thrombin Time is a rapid procedure that tests the quality and quantity of fibrinogen and the presence of such inhibitors as heparin or fibrinogen split products (FgSP).

1. Manual: 0.2 ml Cit plasma + 0.1 ml thrombin
2. Coascreener: 0.1 ml plasma + 0.05 ml thrombin

For the thrombin times, the strength of thrombin is adjusted to a clotting time of 15 ± 2 sec with normal human plasma.

4. CALCULATIONS

The calculations used in this study are simple, widely used, and need no reference.

MCH

Mean corpuscular hemoglobin (pg)

$$\frac{\text{HGB}}{\text{RBC}} \times 10$$

Example: $\frac{15}{5} \times 10 = 30$ pg

MCHC — Mean corpuscular hemoglobin concentration (g/dl)

$$\frac{\text{HBG}}{\text{HCT (\%)}} \times 10$$

Example: $\frac{15}{45} \times 10 = 33.3$ g/dl

MCV — Mean corpuscular volume (fl)

$$\frac{\text{HCT (\%)}}{\text{RBC}} \times 10$$

Example: $\frac{45}{5} \times 10 = 90$ fl

Mean — Mean $= \Sigma x/n$

SD — Standard deviation: $\sqrt{\frac{\Sigma(x - \bar{x})^2}{n - 1}}$

5. INSTRUMENTS

The animal coagulation testing described in this volume has been done over a period of 30 years, during which the instrumentation has changed profoundly, but, in reality, the actual test ingredients have changed very little. In essence, there are four steps to any plasma test or factor assay. The results have changed very little, though each decade has brought new instruments:

Step	1960–1970	1970–1980	1980–1990	1992–
Dispense plasma	Manual	Manual	Manual	Automated
Dispense reagents	Manual	Manual	Automated	Automated
Mix	Manual	Instrument	Automated	Automated
Read end point	Eyeball	Automated	Automated	Automated

For centrifugation:

- Beckman J21C (BI)
- HN-S Centrifuge GLC-25 General Laboratory Centrifuge (DA)
- Jouan CR 4.22 (JO)
- Mistral 3000i (CM)
- Sorvall Instruments GLC-2BLA (DU)

For clotting:

- Fibrometer (BDL)
- MLA 600, 620 (MLA)
- Coa Screener (ALA)
- Coag-a-mate × 2 (GD, OT)
- Coagulyzer (SH)
- Automated ACL (AC)

For counting:

- Cell-Dyn 100 (SQ)
- MK 4 IHC (BA)
- Coulter counter (CO)
- Sequoia-Turner Co (SQ)
- Baker 810 platelet analyzer (BA)
- Baker System 9000 cell counter (BA)

For platelet aggregation:

- Platelet Aggregation Profiler (PAP-4) (BD)

Spectrophotometer:

- Coleman Universal Model 14 (CI)
- Shimadzu Spectronic 210 UV (SH)
- Titertek® Multiskan (LS)

REFERENCES

Allain, J. P., Cooper, H. A., Wagner, P. H., and Brinkhous, K. M., 1975, Platelets fixed with paraformaldehyde: A new reagent for assay of von Willebrand factor and platelet aggregating factor, *J. Lab. Clin Med.* **85:**318.

Quick, A. J., Stanley Brown, M., and Bancroft, F. W., 1935, A study of the coagulation defect in hemophilia and in jaundice, *Am. J. Med. Sci.* **190:**501.

Salzman, E. W., and Ner, L. L., 1966, Adhesiveness of blood platelets in uremia, *Thromb. Diath. Haemorrh.* **15:**84.

Schalm, O. W., Jain, N. C., and Carroll, E. J., 1975, *Veterinary Hematology* (3rd edition), Lea & Febiger, Philadelphia.

BIBLIOGRAPHY

Ahmad, N., Dube, B., Agarwal, G. P., and Dube, R. K., 1979, Comparative studies of blood coagulation in hibernating and non-hibernating frogs (*Rana tigrina*), *Thrombos. Haemostas. (Stuttgaard)* **42:**959.

Alder, A., and Huber, E., 1923, Untersuchungen über Blutzellen und ellbildung bei Amphibien und Reptilien, *Folia Haematol.* **29:**1.

Andrew, W., 1965, *Comparative Hematology,* Grune and Stratton, New York.

Anstall, H. B., and Huntsman, R. G., 1960, Influence of temperature upon blood coagulation in a cold- and a warm-blooded animal, *Nature* **166:**726.

Archer, R. K., 1970, Blood coagulation in non-human vertebrates, *Symp. Zool. Soc. London* **27:**121.

Archer, R. K., 1971, Blood coagulation, in: *Physiology and Biochemistry of the Domestic Fowl* (D. J. Bell and B. M. Freeman, eds.), p. 897, Academic Press, London.

Archer, R. K., and Jeffcott, L. B., 1977, *Comparative Clinical Haematology,* Blackwell, Oxford, London, Edinburgh and Melbourne.

Astrup, T., and Buluk, K., 1963, Thromboplastic and fibrinolytic activities in blood vessels of animals, *Circ. Res.* **13:**253.

Astrup, T., and Darling, S., 1942, Blood coagulation and species specificity of fibrinogen, *Acta Physiol. Scand.* **3:**311.

Awadhiya, R. P., Vegad, J. L., and Kolte, G. N., 1980, Demonstration of the phagocytic activity of chicken thrombocytes using colloidal carbon, *Res. Vet. Sci.* **29:**120.

Bagdy, D., and Szilagyi, T., 1953, Mammalian specificity of fibrinogen, *Experientia* **9:**104.

Baitsell, G. A., 1917, A study of the clotting of the plasma of frog's blood and the transformation of the clot into a fibrous tissue, *Am. J. Physiol.* **44:**109.

Behnke, O., and Zelander, T., 1966, Substructure in negatively stained microtubules of mammalian blood platelets, *Exp. Cell Res.* **43:**236.

Belamarich, F. A., 1976, Hemostasis in animals other than mammals, in: *Progress in Hemostasis and Thrombosis,* Vol. 3 (T. H. Spaet, ed.), p. 19, Grune & Stratton, New York.

Bennett, B., and Ratnoff, W. D., 1973, Immunologic relationships of antihemophilic factor of different species detected by specific human and rabbit antibodies, *Proc. Soc. Exp. Biol. Med.* **143:**701.

Bertolini, B., and Monaco, G., 1976, The microtubule marginal band of the newt erythrocyte, observations on the isolated band, *J. Ultrastruct. Res.* **54**:59.

Blain, D., 1928, A direct method for making total white blood counts on avian blood, *Proc. Soc. Exp. Biol. Med.* **25**:594.

Blomback, B., and Doolittle, R. F., 1963, Amino acid sequence studies on fibrinopeptides from several species, *Acta Chem. Scand.* **17**:1819.

Blomback, B., and Sjoquist, J., 1960, Studies on fibrinopeptides from different species, *Acta Chem. Scand.* **14**:493.

Blomback, B., and Teger-Nilsson, A., 1965, On the thrombin–fibrinogen reaction in different species, *Acta Chem. Scand.* **19**:751.

Blomback, B., Blomback, M., and Grondahl, N. J., 1965, Studies on fibrinopeptides from mammals, *Acta Chem. Scand.* **19**:1789.

Bordet, J., 1921, The theories of blood coagulation, *Bull. Johns Hopkins Hosp.* **32**:213.

Brinkhous, K. M., 1940, Plasma prothrombin: Vitamin K, *Medicine (Baltimore)* **19**:329.

Burns, E. L., Scharles, F. H., and Aitken, L. F., 1931, The effects of mixtures of tissue extracts and blood sera of various animals on the coagulation of blood, *Am. J. Physiol.* **97**:233.

Caillard, B., Devant, J., and Klepping, J., 1962, Comparative study of blood coagulability in man and some animal species, *Can. Res. Soc. Biol.* **156**:1813.

Campbell, F., 1967, Fine structure of the bone marrow of the chicken and pigeon, *J. Morphol.* **123**:405.

Campbell, F. R., 1970, Ultrastructure of the bone marrow of the frog, *Am. J. Anat.* **129**:329.

Cartwright, T., and Kekwick, R. G. O., 1971, A comparative study of human, cow, pig and sheep fibrinogen, *Biochim. Biophys. Acta* **236**:550.

Chandrasekhar, N., and Rao, G. J. S., 1961, Species differences in prothrombins and fibrinogens, *Ann. Biochem. Exp. Med.* **21**:227.

Coates, M. L., 1975, Hemoglobin function in the vertebrates: An evolutionary model, *J. Mol. Evol.* **6**:285.

Csako, G., Gazdy, E, Csernyanszky, H., and Szilagyi, T., 1975, Species specificity of thromboplastin: A phylogenetic study, *Folia Haematol.* **102**:584.

Csako, G., Gazdy, E., Csernyanszky, H., and Szilagyi, T., 1975, Specificity of bovine thrombin and reptilase for mammalian plasmas, *Blut* **30**:283.

Dein, F. J., 1982, Avian clinical hematology, *Proc. Assoc. Avian Vet.* **1982**:5.

Delezenne, C., 1897, Sur la coagulation du sang chez les batraciens et les poissons, *C. R. Seances Soc. Biol. Paris,* **1897**:489–490.

Delisle, K., and Rose, F. L., 1973, Serum protein changes during metamorphosis in *Ambystoma tigrinum, Comp. Biochem. Physiol.* **44A**:1015.

Dent, J. H., and Schuellein, R. J., 1930, A consideration of the prothrombin times of several amphibians with notes on effects of parasitization and disease, *Physiol. Zool.* **23**:23.

Devi, A., Mitra, S. R., and Sarkar, N. K., 1955, Studies on blood coagulation. Part II. Coagulation times of the bloods of different species of animals, *Bull. Calcutta Sch. Trop. Med.* **3**:106.

Dieterich, R. A., 1970, Hematologic values of some Arctic mammals, *J. Am. Vet. Med. Assoc.* **157**:604.

DiScipio, R. L., Hermodson, M. A., Yates, S. G., and Davie, E. W., 1979, A comparison of human prothrombin, factor IX (Christmas factor), factor X (Stuart factor), and protein S, *Biochemistry* **11**:698.

Dodd, M. H. I., and Dodd, J. M., 1976, The biology of metamorphosis, in: *Physiology of the Amphibia,* Vol. III (B. Lofts, editor), p. 467, Academic Press, NY.

Dodds, W. J., 1974, Blood coagulation: Hemostasis and thrombosis, in: *Handbook of Laboratory Animal Science,* Vol. 2 (E. C. Melbey, Jr., and N. H. Altman, eds.), p. 87, CRC Press, Cleveland.

Dodds, W. J., 1978, Platelet function in animals: Species specificities, in: *Platelets: A Multidisciplinary Approach* (G. de Gaetano and S. Garattini, eds.), p. 45, Raven Press, New York.

Dodds, W. J., and Kaneko, J. J., 1971, Hemostasis and blood coagulation, in: *Clinical Biochemistry of Domestic Animals,* 2nd ed., Vol. 2 (J. J. Kaneko and C. E. Cornelius, eds.), pp. 179–206, Academic Press, New York.

Donaldson, V. H., and Breyley, R., 1979, Phylogenetic heterogeneity of plasma kininogen, *Proc. Soc. Exp. Biol. Med.* **160:**134.

Donner, L., and Houskova, J., 1972, Some properties of blood platelets in animal species, *Folia Haematol.* **98:**296.

Doolittle, R. F., and Blomback, B., 1964, Amino-acid sequence investigations of fibrinopeptides from various mammals: Evolutionary implications, *Nature (London)* **202:**147.

Dorn, A. R., and Broyles, R. H., 1982, Erythrocyte differentiation during the metamorphic hemoglobin switch of Rana catesbeiana, *Proc. Nat. Acad. Sci USA* **79:**5592.

Dorst, S. E., and Mills, C. A., 1923, Comparative studies on blood clotting in mammals, birds and reptiles, *Am. J. Physiol.* **64:**160.

Dunaway, P. B., and Lewis, L. L., 1965, Taxonomic relation of erythrocyte count, mean corpuscular volume, and body-weight in mammals, *Nature (London)* **205:**481.

Edmonds, R. H., 1968, Electron microscope studies on the hemostatic process in bird embryos. 1. The initial plug, *J. Ultrastruct. Res.* **24:**295.

Fantl, P., and Marr, A. G., 1957, Prothromboplastin activities of some mammalian plasmas and sera, *Aust. J. Biol. Med. Sci.* **10:**351.

Fass, D. N., Hewick, R. M., Knutson, G. J., Nesheim, M. E., and Mann, K. G., 1985, Internal duplication and sequence homology in factors V and VIII, *Proc. Natl. Acad. Sci. U.S.A.* **82:**1688.

Feldhoff, R. C., 1971, Quantitative changes in plasma albumin during bullfrog metamorphosis, *Comp. Biochem. Physiol.* **40B:**733.

Fey, F., 1965, Comparative hemacytology of lower vertebrates. II. Thrombocytes, *Folia Hematol.* **85:**205.

Fey, F., 1966, Comparative hemacytology of lower vertebrates. III. Granulocytes, *Folia Hematol.* **86:**1.

Forbes, C. D., 1972, Aggregation of human platelets by commercial preparation of bovine and porcine antihaemophilic factor globulin, *J. Clin. Pathol.* **25:**210.

Fossum, K., 1970, Proteolytic enzymes and biological inhibitors. II. Naturally occurring inhibitors in sera from different species and their effect upon proteolytic enzymes of various origin, *Acta Pathol. Microbiol. Scand. Sect. B* **78:**605.

Fox, W., Dessauer, H. C., and Maumus, L. T., 1961, Electrophoretic studies of blood proteins of two species of toads and their natural hybrid, *Comp. Biochem. Physiol.* **3:**52.

Frieden, E., et al., 1957, Changes in serum proteins in amphibian metamorphosis, *Science* **126:**559.

Fujikawa, K., Chung, D. W., Hendrickson, L. E., and Davie, E. W., 1986, Amino acid sequence of human factor XI, a blood coagulation factor with four tandem repeats that are highly homologous with plasma prekallikrein, *Biochemistry* **25:**2417.

Futamura, M., Terashi, Y., Okazaki, T., and Shukuya, R., 1982, Haemoglobin transition in early developmental stages of the tadpole, *Rana catesbeiana, Biochem. Biophys. Acta* **704:**37.

Grabowski, E. F., Didisheim, P., Lewis, J. C., Franta, J. T., and Stropp, J. Q., 1977, Platelet adhesion to foreign surfaces under controlled conditions of whole blood flow: Human vs. rabbit, dog, calf, sheep, pig, macaque, and baboon, *Trans. Am. Soc. Artificial Intern. Organs* **23:**141.

Gray, J. E., Snoeyenbos, G. H., and Reynolds, I. M., 1954, The hemorrhagic syndrome of chickens, *J. Am. Vet. Med. Assoc.* **125:**144.

Green, R. A., 1981, Hemostasis and disorders of coagulation. *Vet. Clin. North Am. (Small Anim. Pract.)* **11:**289.

Greene, C. E., Tsang, V. C. W., Prestwood, A. K., and Meriwether, E. A., 1981, Coagulation studies of plasmas from healthy domesticated animals and persons, *Am. J. Vet. Res.* **42:**2170.

Griminger, P., 1965, Blood coagulation, in: *Avian Physiology,* 2nd ed. (P. D. Sturkie, editor), p. 21, Bailliere, Tindall & Cassell, London.

Hann, C. S., 1966, Fibrinopeptides in one lizard and four bird species, *Biochim. Biophys. Acta* **124:**398.

Harris, J. A., 1972, Seasonal variation in some hematological characteristics of *Rana pipiens, Comp. Biochem. Physiol.* **43A:**975.

Hathaway, W. E., Hathaway, H. S., and Belhasen, L. P., 1964, Coagulation factors in newborn animals, *J. Lab. Clin. Med.* **63:**784.

Hawkey, C. M., 1974, The relationship between blood coagulation and thrombosis and atheroscelerosis in man, monkey, and carnivores, *Thromb. Diath. Haemorrh.* **31:**103.

Hawkey, C. M., 1975, *Comparative Mammalian Haematology: Cellular Components and Blood Coagulation of Captive Wild Animals,* Heinemann, London.

Herner, A. E., and Frieden, E., 1960, Biochemistry of anuran metamorphosis. VII. Changes in serum proteins during spontaneous and induced metamorphosis, *J. Biol. Chem.* **235:**2845.

Hesser, E. F., 1960, Methods for routine fish hematology, *Prog. Fish Cult.* **22:**164.

Hovig, T., Jorgensen, L., Packham, M. A., and Mustard, J. F., 1968, Platelet adherence to fibrin and collagen, *J. Lab. Clin. Med.* **71:**29.

Hwang, S. W., and Wosilait, W. D., 1970, Comparative and developmental studies on blood coagulation, *Comp. Biochem. Physiol.* **37:**595.

Irsigler, K., Lechner, K., and Deutsch, E., 1965, Studies on tissue thromboplatin. II. Species specificity, *Thromb. Diath. Haemorrh.* **9:**18.

Jain, N. C., 1975, A scanning electron microscopic study of platelets of certain animal species, *Thromb. Diath. Haemorrh.* **33:**501.

Janson, T. L., Stormorken, H., and Prydz, H., 1984, Species specificity of tissue thromboplastin, *Haemostasis* **14:**440.

Javaid, M. Y., and Akhtar, N., 1977, Haematology of fishes in Pakistan. II. Studies on fourteen species of teleosts, *Biologia* **23:**79.

Jordan, H. E., 1965, Comparative hematology, in: *Handbook of Hematology,* Vol. 2, (H. Downey, ed.), p. 700, Paul B. Hoeber, New York.

Jordan, R. E., 1983, Antithrombin in vertebrate species: conservation of the heparine-dependent anticoagulant mechanism, *Arch. Biochem. Biophys.* **227:**587.

Just, J. J., and Atkinson, B. G., 1972, Hemoglobin transitions in the bullfrog, Rana catesbeiana, during spontaneous and induced metamorphosis, *J. Exp. Zool.* **182:**271.

Kane, W. H., and Davie, E. W., 1986, Cloning of a cDNA coding for human factor V, a blood coagulation factor homologous to factor VIII and ceruloplasmin, *Proc. Natl. Acad. Sci. U.S.A.* **83:**6800.

Kaneko, J. J., 1980, *Clinical Biochemistry of Domestic Animals,* Academic Press, New York.

Kase, F., 1972, Comparison of some coagulation tests in man and four mammal species (rabbit, dog, cow and pig), *Vet. Med. (Prague)* **17:**495.

Kase, F., 1978, The effect of homo- and heterologous thromboplastins on plasmas of man, seven mammalian and two avian species: A comparative study, *Comp. Biochem. Physiol. A* **61:**65.

Katayama, K., Ericsson, L. H., Enfield, D. L., Walsh, K. H., Neurath, H., Davie, E. W., and Titani, K., 1979, Comparison of amino acid sequence of bovine coagulation factor IX (Christmas factor) with that of other vitamin K–dependent plasma proteins, *Proc. Natl. Acad. Sci. U.S.A.* **76:**4990.

Kelenyi, G., and Nemeth, A., 1969, Comparative histochemistry and electron microscopy of the eosinophil leucocytes of vertebrates: A study of avian, reptile, amphibian and fish leucocytes, *Acta Biol. Acad. Hung.* **20:**405.

Kim, W. M., Merskey, C., Deming, Q. B., Adel, H. N., Wolinsky, H., Clarkson, T. B., and Lofland, H. B., 1976, Hyperlipidemia, hypercoagulability, and accelerated thrombosis: Studies in congenitally hyperlipidemic rats and in rats and monkeys with induced hyperlipidemia, *Blood* **47**:275.

Kiortsis, V., and Kiortsis, I. M., 1963, Electrophorese des proteines du serum de *Triturus cristatus* laur. *Triturus marmoratus* latr. et de leur hybride naturel *Triturus blasii* de l'isle, *Comptes rendus des seances de l'Academie des Sciences* **256**:814.

Knoll, W., 1932, Das morphologische Blutbild der Säugetiere. I. Allgemeine und spezielle Morphologie der kernhaltigen Blutzellen der Säugetiere, *Z. Mikrosk.-Anat. Forsch.* **30**:116.

Knoll, W., 1932, Untersuchungen über die Morphologie des Säugetierblutes, *Folia Haematol.* **47**:201.

Kurachi, K., Schmer, G., Hermodson, M. A., Teller, D. C., and Davie, E. W., 1976, Characterization of human, bovine, and horse antithrombin III, *Biochemistry* **15**:368.

Kuruma, I., Okada, T., Kataoka, K., and Sorimachi, M., 1970, Ultrastructural observation of 5-hydroxytryptamine-storing granules in the domestic fowl thrombocytes, *Z. Zellforsch. Mikrosk. Anat.* **108**:268.

Laverack, M. S., 1985, Physiological adaptations of marine animals, *Symp. Soc. Exp. Biol.*, No. 39.

Lechler, E., and Penick, G. D., 1963, Blood clotting defect in hibernating ground squirrels (*Citellus tridecemlineatus*), *Am. J. Physiol.* **205**:985.

Lewis, J. H., 1974, Comparative hematology: Mammalian platelets, *Excerpta Med. Int. Congr. Ser.*, No. 357.

Lewis, J. H., 1975, Comparative hematology: Mammalian platelets, in: *Platelets: Recent Observations in Basic Research and Clinical Aspects* (O. N. Ulutin, ed.), pp. 18–23, American Elsevier, New York.

Lewis, J. H., 1976, Cross-immunoreactivities of mammalian fibrinogens, *Thromb. Res.* **9**:473.

Lewis, J. H., and Shirakawa, M., 1962, Studies on the thrombocytes of lower vertebrates, *Blood* **20**:795.

Lewis, M. S., Chung, S. I., and Gladner, J. A., 1976, Characterization of lamprey fibrinogen, *Fed. Proc. Fed. Am. Soc. Exp. Biol.* **35**:1486.

Leytus, S. P., Foster, D. C., Kurachi, K., and Davie, E. W., 1986, The gene for human factor IX: A blood coagulation factor whose gene organization is essentially identical with that of factor X and protein C, *Biochemistry* **25**:5098.

Loeb, W. F., and Quimby, F. W., 1989, *The Clinical Chemistry of Laboratory Animals*, Pergamon Press, Oxford.

Lopaciuk, S., McDonagh, R. P., and McDonagh, J., 1978, Comparative studies on blood coagulation factor XIII, *Proc. Soc. Exp. Biol. Med.* **158**:68.

Manikeri, S. R., Raghu, C. N., Padkar, V. N., and Sheth, U. K., 1980, Species variation in platelet function and blood coagulation, *Indian J. Med. Res.* **71**:44.

Mann, F. D., and Hurn, M., 1952, Species specificity of thromboplastin, *Proc. Soc. Exp. Biol. Med.* **79**:19.

Maupin, B., 1969, *Blood Platelets in Man and Animals*, Vols. 1 and 2, Pergamon Press, Oxford.

McCutcheon, F. H., 1936, Hemoglobin function during the life history of the bullfrog, *J. Cell. Comp. Physiol.* **8**:63.

Melby, E. C., Jr., and Altman, N. H., 1976, *Handbook of Laboratory Animal Science*, Vol. III, CRC Press, Cleveland.

Meyers, K. M., Katz, J. B., Clemmons, R. M., Smith, J. B., and Holmsen, H., 1980, An evaluation of the arachidonate pathway of platelets from companion and food-producing animals, mink and man, *Thromb. Res.* **20**:13.

Meyers, K. M., Holmsen, H., and Seachord, C. L., 1982, Comparative study of platelet dense granule constituents, *Am. J. Physiol.* **243**:R454.

Mitruka, B. M., Rawnsley, H. M., and Vadehra, B. V., 1978, *Clinical Biochemical and Hematological Reference Values in Normal Experimental Animals,* Masson Publishing, New York.

Moss, B., and Ingram, V. M., 1968, Hemoglobin synthesis during amphibian metamorphosis. I. Chemical studies on the hemoglobins from the larval and adult stages of *Rana catesbeiana, J. Mol. Biol.* **32:**481.

Murano, G., Walz, D., Williams, L., Pindyck, J., and Mosesson, M. W., 1977, Primary structure of the amino terminal regions of chicken fibrin chains, *Thromb. Res.* **11:**1.

Murphy, R. C., and Seegers, W. H., 1948, Concentration of prothrombin and Ac-globulin in various species, *Am. J. Physiol.* **154:**134.

Mustard, J. F., and Packham, M. A., 1970, Factors influencing platelet function: Adhesion, release and aggregation, *Pharmacol. Rev.* **22:**97.

Nachman, R. L., Levine, R., and Jaffe, E. A., 1977, Synthesis of factor VIII antigen by cultured guinea pig megakaryocytes, *J. Clin. Invest.* **60:**914.

Nelson, C. E., and Guttman, S. I., 1973, Serum protein electrophoresis of some amphibia (*Caecilidae, Rhinophrynidae, Microhylidae*), *Comp. Biochem. Physiol.* **44B:**423.

Niewiarowski, S., and Glinska-Gliniak, T., 1964, Studies of contact activation in plasma of various vertebrates and patients with congenital PTA and Hageman factor deficiencies, *Bull. Acad. Pol. Sci.* **12:**191.

Nilsson, D., and Waaler, B. A., 1965, Observations on factor VIII (antihaemophilic A factor) stability in rabbit and rat plasma: Estimation of factor VIII in rat blood by a new method, *Thromb. Diath. Haemorrh.* **14:**374.

Nolf, P., 1938, The coagulation of the blood, *Medicine (Baltimore)* **17:**381.

Osbaldiston, G. W., 1968, Cytology of the blood of domestic animals—the numbers of blood elements in normal animals, *Veterinarian* **4:**33.

Osbaldiston, G. W., Stowe, E. C., and Griffith, P. R., 1970, Blood coagulation: Comparative studies in dogs, cats, horses, and cattle, *Br. Vet. J.* **126:**512.

Paulus, J. M., 1967, Multiple differentiation in megakaryocytes and platelets (mammals), *Blood* **29:**407.

Plakke, R. K., and Pfeiffer, E. W., 1970, Urea, electrolyte and total solute excretion following water deprivation in the opossum (*Didelphis marsupialis virginiana*), *Comp. Biochem. Physiol.* **34:**325.

Pringle, H., and Tait, J., 1910, Natural arrest of hemorrhage in the tadpole, *J. Physiol.* **40:**lvi.

Pritchard, W. R., Malewitz, T. D., and Kitchen, H., 1963, Studies on the mechanism of sickling of deer erythrocytes, *Exp. Mol. Pathol.* **2:**173.

Que, B. G., and Davie, E. W., 1986, Characterization of a cDNA coding for human factor XII (Hageman factor), *Biochemistry* **25:**1525.

Quick, A. J., 1937, The coagulation defect in sweet clover disease and in the hemorrhagic chick disease of dietary origin: a consideration of the source of prothrombin, *Am. J. Physiol.* **118:**260.

Quick, A. J., 1938, Qualitative differences in the prothrombin, thromboplastin, and thrombin of different species, *J. Biol. Chem.* **123:**99.

Quick, A. J., 1941, The prothrombin concentration in the blood of various species, *Am. J. Physiol.* **132:**239.

Quick, A. J., and Stefanini, M., 1948, Concentration of labile factor of prothrombin complex in human, dog, and rabbit blood: Its significance in determination of prothrombin activity, *J. Lab. Clin. Med.* **33:**819.

Quick, A. J., Collins, R., Taketa, F., and Hussey, C. V., 1961, Variations in the prothrombin complex of different species, *Am. J. Physiol.* **200:**609.

Ratnoff, O. D., 1966, The biology and pathology of the initial stages of blood coagulation, *Prog. Hematol.* **5:**204.

Ratnoff, O. D., and Rosenblum, J. M., 1958, Role of Hageman factor in the initiation of clotting by glass: Evidence that glass frees Hageman factor from inhibition, *Am. J. Med.* **25:**160.

Raymond, S. L., and Dodds, W. J., 1979, Plasma antithrombin activity: A comparative study in normal and diseased animals, *Proc. Soc. Exp. Biol. Med.* **161:**464.

Rotblat, F., Hawkey, C., O'Brien, D. P., and Tuddenham, E. G. D., 1982, Immunologic studies of factor VIII coagulant activity (factor VIII:C). 2. Factor VIII in selected vertebrates, *Thromb. Res.* **25:**425.

Rothstein, R., and Hunsaker, D., II, 1972, Baseline hematology and blood chemistry of the South American woolly opossum *Caluromys derbianus, Lab. Anim. Sci.* **22:**227.

Rowsell, H. C., 1966, Hemostasis in domestic animals, *Can. J. Med. Technol.* **28:**2.

Rowsell, H. C., and Mustard, J. F., 1963, Blood coagulation in some common laboratory animals, *Lab. Anim. Care* **13:**752.

Saito, H., and Ratnoff, O. D., 1979, Comparative studies of Hageman factor (factor XII) in mammalian plasmas by immunological techniques, *Proc. Soc. Exp. Biol. Med.* **161:**412.

Saito, H., Goldsmith, G., and Ratnoff, O. D., 1974, Fletcher factor activity in plasma of various species, *Proc. Soc. Exp. Biol. Med.* **147:**519.

Saito, H., Goldsmith, G., and Waldmann, R., 1976, Fitzgerald factor (high molecular weight kininogen) clotting activity in human plasma in health and disease and in various animal plasmas, *Blood* **48:**941.

Sanderson, J. H., and Phillips, C. E., 1981, *An Atlas of Laboratory Animal Haematology,* Clarendon Press, Oxford.

Scarborough, R. A., 1930–1931, The blood picture of normal laboratory animals, *Yale J. Biol. Med.* **3:**63.

Schalm, O. W., 1967, *Veterinary Haematology,* Lea and Febiger, Philadelphia.

Schalm, O. W., Jain, N. C., and Carroll, E. J., 1975, *Veterinary Haematology,* 3rd ed., Lea and Febiger, Philadelphia.

Schermer, S., 1967, *The Blood Morphology of Laboratory Animals,* 3rd ed., p. 43, F. A. Davis, Philadelphia.

Schmaier, A. H., Gustafson, E., Idell, S., and Colman, R. W., 1984, Plasma prekallikrein assay: Reversible inhibition of CI inhibitor by chloroform and its use in measuring prekallikrein in different mammalian species, *J. Lab. Clin. Med.* **104:**882.

Schwartz, M. L., Pizzo, S. V., Sullivan, J. B., Hill, R. L., and McKee, P. A., 1973, A comparative study of crosslinked and noncrosslinked fibrin from the major classes of vertebrates, *Thromb. Diath. Haemorrh.* **29:**313.

Schwarz, F. P., 1974, Microphotometrical investigations on the nuclei of spindle cells (thrombocytes) in frog blood, *Zool. Anz.* **193:**189.

Schwarz, F. P., 1975, Further investigations on the nuclei of spindle cells (thrombocytes) in frog blood, *Zool. Anz.* **195:**225.

Seal, U. S., and Erickson, A. W., 1969, Hematology, blood chemistry and protein polymorphism in the white-tailed deer (*Odocoileus virginianus*), *Comp. Biochem. Physiol.* **30:**695.

Shum, Y., and Griminger, P., 1972, Prothrombin time stability of avian and mammalian plasma: Effect of anticoagulant, *Lab. Anim. Sci.* **22:**384.

Silver, M. D., 1966, Microtubules in the cytoplasm of mammalian platelets, *Nature (London)* **209:**1048.

Sinakos, Z., and Caen, J., 1967, Coagulation of platelets in various animal species: Photometric and microscope study, *Nouv. Rev. Fr. Hematol.* **7:**120.

Skjorten, F., and Evensen, S. A., 1973, Induction of disseminated intravascular coagulation in the factor XII–deficient fowl: Morphological effects of liquoid, bacterial endotoxin and tissue thromboplastin in the normal and anticoagulated fowl, *Thromb. Diath. Haemorrh.* **15:**25.

Smith, N., and Engelbert, V. E., 1969, Erythropoiesis in chicken peripheral blood, *Can. J. Zool.* **47:**1269.

Sorbye, O., 1962, Studies on the coagulation of chicken blood. III. Differentiation of x-, o_1- and o_2-factors by adsorption, *Acta Chem. Scand.* **16:**799.

Sorbye, O., 1962, Studies on the coagulation of chicken blood. IV. Adsorption of o- factor activity, *Acta Chem. Scand.* **16:**903.

Sorbye, O., 1962, Studies on the coagulation of chicken blood. IX. Labile factor activity of fresh oxalated plasma: the combined activity of sixteen discrete labile coagulation factors, *Acta Chem. Scand.* **16:**2411.

Sorg, D. A., and Buckner, B., 1964, A simple method for obtaining venous blood from small laboratory animals, *Proc. Soc. Exp. Biol. Med.* **115:**1131.

Spitzer, J. J., and Spitzer, J. A., 1952, The blood coagulation mechanism of frogs, with respect to the species specificity of thromboplastin, to intracardial thrombin injection, and to the effect of seasonal changes, *Canad. J. M. Sc.* **30:**420.

Spurling, N. W., 1981, Comparative physiology of blood clotting, *Comp. Biochem. Physiol. A* **68:**541.

Spurrier, W. A., and Dawe, A. R., 1973, Several blood and circulatory changes in the hibernation of the 13-lined ground squirrel, *Citellus tridecemlineatus, Comp. Biochem. Physiol.* **44:**267.

Stormorken, H., 1957, Species differences of clotting factors in ox, dog, horse, and man: Proaccelerin and accelerin, *Acta Physiol. Scand.* **39:**121.

Stormorken, H., 1957, Species differences of clotting factors in ox, dog, horse, and man: Prothrombin, *Acta Physiol. Scand.* **41:**101.

Stormorken, H., 1957, Species differences of clotting factors in ox, dog, horse, and man: Thromboplastin and proconvertin, *Acta Physiol. Scand.* **41:**301.

Strong, D. D., Moore, M., Cottrell, B. A., Bohonus, V. L., Pontes, M., and Evans, B., 1985, Lamprey fibrinogen chain: Clotting, cDNA sequencing and general characterization, *Biochemistry* **24:**92.

Sturkie, P. D., 1965, *Avian Physiology,* 2nd ed., Comstock, Ithaca, NY.

Svihla, A., Bowman, H. R., and Ritenour, R., 1951, Prolongation of clotting time in dormant estivating mammals, *Science* **114:**298.

Tait, J., and Burke, H. E., 1926, Platelets and blood coagulation, *Quart. J. Exper. Physiol.* **16:**129.

Teger-Nilsson, A., and Blomback, B., 1974, Rate of thrombin–fibrinogen reaction in some mammalian species, *Thromb. Res.* **5:**223.

Theil, E. C., 1970, Red blood cell replacement during the transition from embryonic (tadpole) hemoglobin to adult (frog) hemoglobin in *Rana catesbeiana, Comp. Biochem. Physiol.* **33:**717.

Tocantins, L. M., 1938, Mammalian blood platelets in vitro, *Medicine (Baltimore)* **17:**155.

Tooze, J., and Davies, H. G., 1967, Light- and electron-microscope studies on the spleen of the newt *Triturus cristatus:* The fine structure of erythropoietic cells, *J. Cell. Sci.* **2:**617.

Topp, R. C., and Carlson, H. C., 1972, Studies on avian heterophils. II. Histochemistry, *Avian Dis.* **16:**369.

Topp, R. C., and Carlson, H. C., 1972, Studies on avian heterophils. III. Phagocytic properties, *Avian Dis.* **16:**374.

Traill, K. N., Bock, G., Boyd, R., and Wick, G, 1983, Chicken thrombocytes: Isolation, serological and functional characterisation using the fluorescence activated cell sorter, *Dev. Comp. Immunol.* **7:**111.

Tremoli, E., Donati, M. B., and de Gaetano, G., 1978, Washed guinea pig and rat platelets possess factor-X activator activity, *Br. J. Haematol.* **37:**155.

Tschopp, T. B., Baumgartner, H. R., and Studer, A., 1971, Effect of Congo red on blood platelets and leucocytes of rabbits and cats, *Thromb. Diath. Haemorrh.* **26:**488.

Ur, A., 1974, The blood coagulation curve of some mammals, *Res. Vet. Sci.* **106:**204.

Veloso, D., Shilling, J., Shine, D., Fitch, W. M., and Colman, R. D., 1986, Recent evolutionary divergence of plasma prekallikrein and factor XI, *Thromb. Res.* **43:**153.

Wagner, E. D., Brinkhous, K. M., and Smith, H. P., 1939, Plasma prothrombin levels in various vertebrates, *Am. J. Physiol.* **125:**296.

Wagner, E. D., Brinkhous, K. M., and Smith, H. P., 1939, The prothrombin conversion rate in various species, *Proc. Soc. Exp. Biol. Med.* **40:**197.

Wagner, R. H., Richardson, B. A., and Brinkhous, K. M., 1957, A study of the separation of fibrinogen and antihemophilic factor (AHF) in canine, porcine, and human plasmas, *Thromb. Diath. Haemorrh.* **1:**3.

Warner, E. D., Brinkhous, K. M., and Smith, H. P., 1939, Plasma prothrombin levels in various vertebrates, *Am. J. Physiol.* **125:**296.

Warner, E. D., Brinkhous, K. M., and Smith, H. P., 1939, Prothrombin conversion rate in various species, *Proc. Soc. Exp. Biol. Med.* **40:**197.

Washburn, K. W., and Guill, R. A., 1972, Comparison of hematology between leghorn-type and heavy-type egg production strains, *Poult. Sci.* **51:**946.

Wells, J. J., and Sutton, J. E., Jr., 1916, Blood counts in the frog, the turtle, and twelve different species of mammals, *Am. J. Physiol.* **39:**31.

White, J. G., 1981, Morphological studies of platelets and platelet reactions, *Vox. Sang.* **40:**8.

White, J. G., 1983, Ultrastructural physiology of platelets with randomly dispersed rather than circumferential band microtubules, *Am. J. Pathol.* **110:**55.

White, J. G., and Gerrard, J. M., 1976, Ultrastructural features of abnormal blood platelets: A review, *Am. J. Pathol.* **83:**590.

Wintrobe, M. M., Lee, G. R., Boggs, D. R., Bithell, T. C., Athens, J. W., and Foerster, J., 1974, *Clinical Hematology,* 7th ed., Lea and Febiger, Philadelphia.

Wintrobe, M. M., Lee, G. R., Boggs, D. R., Bithell, T. C., Athens, J. W., and Foerster, J., 1981, *Clinical Hematology,* 8th ed., Lea and Febiger, Philadelphia.

Wuepper, K. D., and Cochrane, C. G., 1972, Plasma prekallikrein: Isolation, characterization, and method of activation, *J. Exp. Med.* **135:**1.

Youatt, W. G., Fay, L. D., Howe, D. L., and Harte, H. D., 1961, Hematologic data on some small mammals, *Blood* **18:**758.

INDEX